Water Management in Rice-based Ecosystems and Aquaculture

NIPA® GENX ELECTRONIC RESOURCES & SOLUTIONS P. LTD.
New Delhi-110 034

About the Authors

Dr. Krishna Gopal Mandal is currently Director of the ICAR-Mahatma Gandhi Integrated Farming Research Institute (MGIFRI), Motihari, Bihar; formerly Principal Scientist in Agronomy at ICAR-Indian Institute of Water Management (IIWM), Bhubaneswar. He was born on 1967 in village Kumardubi, Bankura district of West Bengal. He obtained his B.Sc. (Ag.) Hons with First Class First (Gold Medalist) in 1991 and M.Sc. (Ag) in Agronomy in 1993 and Faculty Topper from Palli Siksha Bhavana (Institute of Agriculture), Visva-Bharati, Sriniketan; Ph.D. in Agronomy (1998) from BCKV, Mohanpur, West Bengal; Post-Doctoral research (2005-06) at China Agricultural University, Beijing.

He joined the ICAR-Agricultural Research Service in 1998 and started his professional career as a Scientist and Scientist-SS in Agronomy (1998-2007) at ICAR-Indian Institute of Soil Science (IISS), Bhopal. Then he moved to ICAR-Indian Institute of Water Management (IIWM), Bhubaneswar in the year 2007 as Senior Scientist (Agronomy), then as Principal Scientist (Agronomy) in 2013, continued in the same Institute upto 2021. Since July 2021 he is holding the Research Management Position (RMP) as Director of ICAR-MGIFRI, Motihari, Bihar.

In his scientific career of about 26 years, he has contributed significantly to research on natural resource management: water, soil & nutrient management, biomass & bioenergy in crop production, conservation tillage, irrigation, agro-climatology and integrated farming systems (IFS). He has developed methodology for analysing crop production energetics; methodology of assessing soil environment, soil organic carbon and developed pedotransfer functions; analysing rainfall patterns and dry/ wet spells for a canal command in Odisha; developed impactful technologies viz. integrated nutrient management for oilseeds soybean and Indian mustard; water x nutrient interactions and increasing WUE of wheat and Indian mustard, crop diversification, conservation tillage and management of crop residues for soybean-wheat system in vertisols, improved planting method and irrigation water saving; participatory on-farm water management and development of IFS models in a canal command; on-farm water harvesting and IFS models for rainfed farmers in Odisha, rice water management & aerobic rice production technology- crop establishment, irrigation and fertilizer-N application, diversification in rice and drip irrigation in rice-based systems; currently working on development of IFS models for flood-prone and waterlogged ecosystems.

Dr. Mandal has contributed more than 220 publications, including research papers and review articles in high impact factor journals, research bulletins, training manuals, authored book, edited book and book chapters, popular articles and conference proceedings etc. with a large number of citations and high h-index.

Dr. Mandal has received several awards, honours and recognitions: ICRISAT Doreen Mashler Award (2006), BKJF-INCSW Sookshma Sinchan Puraskar (2018) by CWC-INCSW & BKJ Foundation, Dr. K.G. Tejwani Award (biennium 2016-2017) of IASWC, Dr. B.P. Pal Memorial Award (2019) of NABS, K.C. Das Memorial Award of IoE, Bhubaneswar; IIWM Scientist Proficiency Award (2011), ISA Certificate of Honour (2021), ISA Gold Medal Award (2021), Overseas Fellowship in Agronomy (2005-06) by MoHRD, Govt. of India and CSC, Govt. of P.R. China; Associate Editor (2016-2022) of Agronomy Journal, American Society of Agronomy, USA; Editor (2016-2021 & 2024-2025) of Indian Journal of Agronomy, ISA; Sectional Recorder (2015 & 2016), Agriculture & Forestry Science Section of ISCA; President of Plant, Agriculture & Forestry Science Section (2024-25) of ISCA.

Dr. Mandal has been conferred with the Fellowship of National Academy of Biological Sciences (NABS), Indian Society of Agronomy (ISA), West Bengal Academy of Science and Technology (WAST), Indian Water Resources Society (IWRS), Indian Society of Oilseeds Research (ISOR), Indian Society of Coastal Agricultural Research (ISCAR), and Member of the National Academy of Sciences, India; Life Member of 10 Professional Societies/ Academies in India.

Dr. Amod Kumar Thakur is a Principal Scientist in Plant Physiology at ICAR-Indian Institute of Water Management, Bhubaneswar, Odisha, India. He was born in 1969 in Darbhanga district of Bihar. He obtained his B.Sc. (Ag.) in 1992 from the Institute of Agricultural Sciences, Banaras Hindu University, Varanasi and his M.Sc. (Ag) in Plant Physiology in 1994 and Ph.D. in Plant Physiology (1999) from ICAR-Indian Agriculture Research Institute (IARI), New Delhi.

He joined the ICAR-Agricultural Research Service in 1999 and started his professional career as a Scientist at ICAR-Indian Institute of Water Management (IIWM), Bhubaneswar till today. In his scientific career of about 25 years, he has pioneered SRI research, understanding the mechanisms, plant morphology, physiology and agronomy in increasing rice productivity.

Dr. Thakur has contributed more than 100 publications, including research papers and review articles in high-impact factor journals, research bulletins, training manuals, authored books, edited book and book chapters, popular articles and conference proceedings etc. with a large number of citations and high h-index.

Dr. Thakur is a recipient of the USIEF Fulbright-Nehru Senior Research Fellowship 2012-2013 and USDA's Norman E. Borlaug International Agricultural Science and Technology Fellowship 2011 at Cornell University, Ithaca, NY, USA. He has received several awards, honours and recognitions: ICAR Award for Outstanding Multidisciplinary Team Research in Agriculture and Allied Sciences for the Biennium 2003-2004; Best Poster Presentation Award 2009, 2012; ICAR-DWM Proficiency Award 2011; J.J. Chinoy Gold Medal Award-2015; Outstanding Scientist Award 2023; etc.

Dr. Thakur has been conferred with the Fellowship of the Indian Society for Plant Physiology (ISPP), New Delhi and the National Academy of Biological Sciences (NABS), Chennai. He is a Life Member of several Professional Societies/ Academies in India. During his professional career, he has visited several countries, viz., USA, Vietnam, UK, Thailand, Singapore, Malaysia, The Netherlands, Philippines, etc.

Dr. Rajeeb K. Mohanty was born on 1969 in Balasore district of Odisha. He holds a M.Sc. in Fisheries Management (ICAR-Central Institute of Fisheries Education, Mumbai), Ph.D. in Aquaculture and D.Sc. in Environmental Science. He joined Agricultural Research Service of the ICAR and started his professional career as a Scientist since 1997 and presently he is Principal Scientist (Aquaculture) at ICAR-Indian Institute of Water Management, Bhubaneswar.

Dr. Mohanty has outstanding research contributions to aquaculture water use efficiency, water budgeting, nutritional water productivity, water footprint, environment impact assessment (EIA), bioresource technology, compensatory growth performance of Indian major carps (IMC), run-off recycling aquaculture, self-reliant integrated aquaculture, multiple use management of various aquaculture production systems, periphyton-based carp- Small Indigenous Species (SIS) polyculture, semi-intensive shrimp monoculture in improving aquaculture productivity and livelihood of small and marginal farmers.

Significant works done by Dr. Mohanty is reflected through implementation of 17 research projects of national importance as Project Leader, more than 140 publications in reputed high-impact journals, 8 commercialized technologies, recipient of SEVEN national awards such as ICAR-Team Research Award (2001-02 & 2003-04), Vasantrao Naik Award (2001 & 2007), Indira Gandhi Rajbhasha Puraskar, (2005-06), Dr. Hiralal Chaudhuri Young Scientist Award (2003), BKJF-INCSW Team Award- 2018, Fellowship National Environment Science Academy NESA -2019 and the National Academy of Biological Sciences (NABS), Chennai Fellowship-2020 and five other society awards. Besides, Dr. Mohanty has guided research work of 2 Ph.D. scholars and 8 M.Sc. students in the field of Aquaculture, water management and wetland ecology.

Dr. Souvik Ghosh is currently Professor (Agricultural Extension) at Department of Agricultural Extension, Visva-Bharati University, West Bengal; formerly Senior Scientist at ICAR-Indian Institute of Water Management (IIWM), Bhubaneswar. He was born on 1972 in Rampurhat, Birbhum district of West Bengal. He graduated as Gold Medallist in B.Sc. (Agril.) Hons. from Visva-Bharati University in the year 1995 and passed M.Sc. (Dairying) in Dairy

Extension Education with Director's Gold Medal at National Dairy Research Institute, Karnal in 1997. He obtained Ph.D. (Agricultural Extension) from Indian Agricultural Research Institute, New Delhi with distinction of Significant Post-Graduate Research in 2000.

He joined ICAR-Agricultural Research Service on 10th July 2000 and worked as Scientist and Senior Scientist at ICAR-Indian Institute of Water Management, Bhubaneswar during 2000-09 and 2009-13, respectively. Thereafter, he joined as Professor (Agricultural Extension) at Department of Agricultural Extension, Visva-Bharati University, West Bengal, on June 2013. He served as Head of the Department from November 2016 to October 2019. He was positioned as Visiting Scientist at College of Agriculture and Biological Sciences, South Dakota State University, USA during 2016.

In his research and academic career of about 24 years, he has contributed significantly to Agricultural Extension research and education. Notable contributions include development of methodologies for integrated assessment of recommended agricultural water management technologies from multi-layer multi-stakeholders' perspectives, formulation of impact assessment framework for participatory irrigation management (PIM) reforms, evaluation of irrigation systems from the farmers' perspectives following fuzzy logic, construction of water user group dynamics index, assessment of irrigation-agriculture-livelihood-poverty links in eastern Indian states, mapping of vulnerability on the basis of livelihood sensitivity, resilience and adaptation of farm households to climate change, various issues of agricultural extension management, communication, diversified farming and rural livelihoods. Dr. Ghosh has completed eight projects as PI and 22 projects as Co-PI. He has guided eight Ph.D. Thesis and 17 M.Sc. Thesis works. A total of 290 publications includes research papers in high impact journals, technical/ popular articles and books.

His outstanding contribution to research in Agricultural Extension has brought many laurels including Lal Bahadur Shastri Outstanding Scientist Award 2011 of the ICAR, ICAR Outstanding Team Research Award 2006, BKJF-INCSW Sookshma Sinchan Puraskar 2018 of INCSW, Central Water Commission, Govt. of India, Indira Gandhi Rajbhasha Puraskar (First Prize) 2012 of Ministry of Home Affairs, Govt. of India, Dr. O.P. Dahama Memorial Award 2022-23 of Indian Society of Extension Education (ISEE), Prof. Y.P. Singh Memorial Award 2018 of ISEE, Dr. K.N. Singh Memorial Award 2017 of ISEE, Best Scientist Award of ICAR-Indian Institute of Water Management 2008, Institution Award 2006 of Institution of Engineers (India), Odisha, ISEE Young Scientist Award 2005. He is Fellow of ISEE, IARI, New Delhi and Society for Community Mobilization for Sustainable Development, IARI, New Delhi.

He coordinated many national level trainers' training programmes by Ministry of Water Resources and Ministry of Agriculture, Government of India for states' line departments, irrigation departments, NGO officials of different states as well as farmers' trainings. He delivered more than a hundred invited lectures in training programmes, invited/guest lectures in different forums and events organized by ICAR institutes, Central and State Agricultural Universities, Central and State Government organizations, international organizations. He has also been involved in editorial assignments and reviewing of national and international journals. He has made significant contributions for advancing Agricultural Extension research and education.

Dr. Pothula Srinivasa Brahmanand is currently Project Director of the Water Technology Centre (WTC), ICAR-Indian Agricultural Research Institute (IARI), New Delhi, India; formerly Director Research at Dr. Rajendra Prasad Central Agricultural University, Pusa, Bihar and Principal Scientist in Agronomy at ICAR-Indian Institute of Water Management (IIWM), Bhubaneswar. He was born on 26th July 1973 in Eluru, West Godavari district of Andhra Pradesh. He obtained his B.Sc. (Ag.) and M.Sc (Ag) in Agronomy with First Class from Acharya N.G. Ranga Agricultural University, Bapatla, Andhra Pradesh and Ph.D. in Agricultural and Food Engineering (2009) from IIT, Kharagpur, West Bengal; bestowed with two Guinness World Records in 2012 and 2016.

Dr. Brahmanand joined ICAR-Agricultural Research Service in December 1997 and contributed professionally as Scientist, Scientist-SS, Scientist-SG, Senior Scientist and Principal Scientist (1997-2022) at ICAR-Indian Institute of Water Management (IIWM), Bhubaneswar. Thereafter, he joined in Research Management Position i.e. Director Research at Dr. Rajendra Prasad Central Agricultural University, Pusa, Samastipur, Bihar on 22nd April 2022 and continued upto 2nd February 2023. Since 3rd February 2023, he is holding the Research Management Position (RMP) as Project Director of WTC, ICAR-Indian Agricultural Research Institute (IARI), New Delhi, India.

During his professional career of 26 years, he has contributed immensely in water resource assessment and management, climate resilient interventions, waterlogged area and flood management, river health and rejuvenation, water quality, phytoremediation, ecosystem services, yield gap assessment of rice and ICAR flexi rubber checkdams. As national PI of ICAR collaborative project with International Water Management Institute (IWMI), Sri Lanka, developed an index based on depth and duration of flood water for assessment of crop insurance for flood-prone regions of Bihar. Provided leadership to water management research at Dr. Rajendra Prasad central Agricultural University, Bihar and assisted Government of Bihar in preparation of Bihar Krishi Road Map and technical guidance for getting GI tag to Marcha rice. Presently as Project Director at WTC, he is contributing to national water policy and research by technical support in precision irrigation, revival of water bodies, irrigation automation, groundwater recharge and wastewater treatment and its safe use in agriculture. He is a member of National Steering Council on Per Drop More Crop (PDMC), Ministry of Agriculture and farmers Welfare, Government of India.

Dr. Brahmanand has published about 92 research papers in refereed international and national journals (including Nature Communications (NAAS Score: 20.00)) and more than 75 other publications in agricultural water management including technical bulletins, training manuals, popular articles, extension folders/leaflets and book chapters.

Dr. Brahmanand was conferred with two international awards (Guinness World Record Holder in Longest Speech Marathon in 2012 and Guinness World Record Holder in Fastest Time to Identify All Elements of Periodic Table in 2016), five national awards (Indira Gandhi Rajbhasha Puraskar for best original book writing (Twice in 2007 and 2012); ICAR Outstanding team research award for multi-disciplinary research in Agriculture and Allied Sciences (Twice for the biennium 2001-02 and 2013-2014); Vasant Rao Naik National Award for best research in water management in drylands in India 2008) and several Professional Society awards including Eminent Water Resource Scientist Award of IWRS (IIT, Roorkee) for the year 2023.

Dr. Brahmanand has been bestowed with the Fellowship of Indian Water Resources Society (IWRS), IIT, Roorkee, Andhra Pradesh Academy of Sciences (APAS), Government of Andhra Pradesh, International Society for Development & Sustainability (ISDS) and National Environment Science Academy (NESA) and Life Member of 8 Professional Societies/ Academies in India.

Water Management in Rice-based Ecosystems and Aquaculture

A Text-*cum*-Reference Book

Krishna Gopal Mandal, Ph.D., Post-Doc
Director, ICAR-Mahatma Gandhi Integrated Farming Research Institute (MGIFRI)
Motihari, Bihar, India
Formerly Principal Scientist, ICAR-Indian Institute of Water Management (IIWM)
Bhubaneswar, Odisha, India

Amod Kumar Thakur, Ph.D.
Principal Scientist, ICAR-Indian Institute of Water Management (IIWM)
Bhubaneswar, Odisha, India

Rajeeb Kumar Mohanty, Ph.D., D.Sc.
Principal Scientist, ICAR-Indian Institute of Water Management (IIWM)
Bhubaneswar, Odisha, India

Souvik Ghosh, Ph.D.
Professor, Palli Siksha Bhavana (Institute of Agriculture)
Visva-Bharati, Sriniketan, West Bengal, India
Formerly Senior Scientist, ICAR-Indian Institute of Water Management (IIWM),
Bhubaneswar, Odisha, India

Pothula Srinivasa Brahmanand, Ph.D.
Project Director, Water Technology Centre (WTC)
ICAR-Indian Agricultural Research Institute (IARI), New Delhi, India
Formerly Principal Scientist, ICAR-Indian Institute of Water Management (IIWM)
Bhubaneswar, Odisha, India

NIPA® GENX ELECTRONIC RESOURCES & SOLUTIONS P. LTD.
New Delhi-110 034

NIPA® GENX ELECTRONIC RESOURCES & SOLUTIONS P. LTD.

101,103, Vikas Surya Plaza, CU Block
L.S.C.Market, Pitam Pura, New Delhi-110 034
Ph : +91 11 27341616, 27341717, 27341718
E-mail: newindiapublishingagency@gmail.com
www: www.nipabooks.com

For customer assistance, please contact
Phone: + 91-11-27 34 17 17
Fax: + 91-11-27 34 16 16

Print ISBN: 978-93-58874-83-9
ebook ISBN: 978-93-58879-16-2

Composed and Designed by NIPA®.

डॉ. हिमांशु पाठक
DR. HIMANSHU PATHAK
सचिव (डेयर) एवं महानिदेशक (आईसीएआर)
Secretary (DARE) &
Director General (ICAR)

भारत सरकार
कृषि अनुसंधान और शिक्षा विभाग एवं
भारतीय कृषि अनुसंधान परिषद
कृषि एवं किसान कल्याण मंत्रालय, कृषि भवन, नई दिल्ली–110 001

GOVERNMENT OF INDIA
DEPARTMENT OF AGRICULTURAL RESEARCH AND EDUCATION (DARE)
AND
INDIAN COUNCIL OF AGRICULTURAL RESEARCH (ICAR)
MINISTRY OF AGRICULTURE AND FARMERS WELFARE
Krishi Bhavan, New Delhi 110 001
Tel: 23382629 / 23386711 Fax: 91-11-23384773
E-mail: dg.icar@nic.in

FOREWORD

Water is the most important natural resource. Globally, out of total volume of water, only 2.5% is the fresh water. India accounts for only 4% of world's renewable water resources, but the country supports about 18% of the world's human population and 15% of livestock. There is increased demand of food and fresh water because of growing population and changing climate. One of the national priorities is reducing the usage of water by at least 25% in all agricultural production systems. This could be achieved through efficient water management and increasing water productivity. The majority of the food grains production happens in rice-based ecosystems in the country, and it consumes greater proportion of freshwater. In order to manage this precious resource, it is required to understand the challenges in rice ecosystems, issues of water resources and availability, technology options, rice ecosystem services, water governance, participatory irrigation management, and water policy.

It gives me immense pleasure that the book entitled '*Water Management in Rice-based Ecosystems and Aquaculture: A Text-cum-Reference Book*' has been written by a team of senior level Scientists of the ICAR, who have many years of research and development experience in the field of agricultural water management, especially in rice-based ecosystems and aquaculture. I'm sure this book will be useful to researchers, students, teachers, farmers and other stakeholders to understand the water management issues, strategies and technology options for rice-based agroecosystems and also aquaculture.

(Himanshu Pathak)

Dated the 7th May, 2024
New Delhi

भारतीय कृषि अनुसंधान परिषद
कक्ष क्र. 101, कृषि अनुसंधान भवन-II, नई दिल्ली-110 012, भारत
INDIAN COUNCIL OF AGRICULTURAL RESEARCH
Room No. 101, Krishi Anusandhan Bhavan-II, Pusa, New Delhi-110012, India

डॉ. सुरेश कुमार चौधरी
उप महानिदेशक (प्राकृतिक संसाधन प्रबंधन)
Dr. Suresh Kumar Chaudhari
Deputy Director General (Natural Resources Management)

06.05.2024

Foreword

Agricultural water management presents significant challenges. Spatial and temporal rainfall variability leads to both scarcity and excess of water, escalating the demand for irrigation water and exacerbating the supply-demand gap. This situation is further compounded by the increased rate of groundwater utilization, insufficient recharge of aquifers in hard-rock areas, and low irrigation efficiency in canal systems. Additionally, extreme weather conditions and climatic variability impact water resources and agricultural productivity, contributing to a continuous decline in per capita availability of fresh water. In our country, the net irrigated area comprises approximately 75 million hectares out of a total of 140 million hectares of net sown area. Sustainable water management is vital for rice-based agroecosystems, including rice-fallows, which are pivotal in the country's major grain-producing agriculture. Moreover, water management in aquaculture and rice-fish co-culture demands focused attention. Engaging in participatory irrigation management, improving water governance, and formulating effective policies are crucial for enhancing water use efficiency in the command areas.

The book titled *'Water Management in Rice-based Ecosystems and Aquaculture: A Text-cum-Reference Book'* consists of 10 chapters that offer an in-depth understanding of the subject. It covers topics such as the significance of rice for food and nutrition security, export earnings, challenges in rice ecosystems, issues concerning water resources and irrigation, technological options available, water management in aquaculture, ecosystem services in rice cultivation, and water governance and policy, specifically tailored for rice-based systems. The authors bring many years of experience in water management research, on-farm technology development, and dissemination through training and capacity-building of farmers and stakeholders. This book serves as an essential resource for researchers, educators, students, farmers, and other stakeholders engaged in agricultural water management.

I commend the authors for their meticulous efforts in synthesizing, composing, and publishing this comprehensive book, which aligns with the national priority of efficient water management and reduction of water use.

(S.K. Chaudhari)

Ph.: +91-11-25848364 Fax: +91-11-25848366 Email: ddg.nrm@icar.gov.in Website: www.icar.org.in

Preface

Rice farming is the livelihood of millions of people in the World. It is the staple food and primary source of nutrition for about half of the world's population. Rice provides 60-70% of the total calorie requirements for more than 2 billion people in Asia through different rice-based food products. The crop is very important considering its economic, social and cultural significance in India and other Asian countries. Rice-based ecosystems and aquaculture are the lifeline of millions of people in India.

Globally, rice is grown on about 165 million ha with an annual paddy production of about 787 million tonnes. There is wide variation in the production of paddy in different rice-growing regions of the world due to differential geographic location, genetic potential, weather elements, irrigation facilities and crop management systems. About 87% of the world's area and 90% of the world's paddy production occurred from Asia. The demand for food will increase by about 60% to feed the world population of 9.3 billion by 2050, and rice is a major contributor to total food grain production, it is very challenging with a farming-as-usual approach. In India, food demand is projected to rise to about 400 million tonnes by 2050. This calls for efficient management of water and natural resources in rice-based ecosystems and aquaculture-based systems.

The authors have conceptualized, outlined and started writing this book during the COVID-19 pandemic and finished recently with the latest available data and evidence-based technologies for increasing productivity and water productivity of rice-based ecosystems and aquaculture. This text-*cum*-reference book delves into the importance of rice as food and nutrition value, rich cultural heritage associated with rice in India and other Asian countries, trends in regional variation in area and production of rice in the world; area, production and yield (APY) of rice in India, rice economies through export earnings, water resources in India and its irrigation potential for rice-based ecosystems, diversified rice-systems in different eco-regions in India, water management and associated technology options in rice farming and post-rainy season crops for enhancing water productivity, rice-fish integrated farming systems, high-value aquaculture options for increasing income, business opportunities especially for rice and fish growing farmers and people especially in eastern and southern India, water management and advances in rice-farming in waterlogged ecosystems, water governance and participatory irrigation management (PIM) for irrigation commands, and ecosystems services associated with rice-based farming.

This book is a reference-*cum*-text book in the field of water management in rice and and rice-based eco-systems, and it consists of very well-defined ten chapters: Chapter-1: Rice: staple food, governs the economy and challenges in rice-based systems, Chapter-2: Water resources, issues and challenges in water management, Chapter-3: Rice agro-ecosystems in india, Chapter-4: Water management strategies and technologies in rice, Chapter-5: Water management strategies for dry-season crops under rice-based system, Chapter-6: Integrated rice-fish farming systems, Chapter-7: Pond-based integrated farming system, Chapter-8: High-value aquaculture under IFS, Chapter-9: People's awareness of ecosystem services, and Chapter-10: Water governance and participatory irrigation management (PIM) for rice-based agro-ecosystems. The authors have included a wide range of sub-topics very systematically and comprehensively. All chapters have been written with in-depth discussion, with updated data, figures and evidences available on the water management and associated technologies with rice-based ecosystems and aquaculture. Authors are well-recognized and have many years of experience in the field of agricultural water management, aquaculture and agricultural extension.

The authors wish to mention the encouragement received from Dr. Trilochan Mahapatra, Former Secretary, DARE & Director General, ICAR, and Dr. Himanshu Pathak, presently Secretary, DARE & Director General, ICAR, Dr. Suresh Kumar Chaudhari, Deputy Director General (Natural Resource Management), ICAR, Dr. Ashwani Kumar, Dr. S.K. Ambast, former Directors of ICAR-Indian Institute of Water Management (IIWM), and Dr Atmaram Mishra, colleagues and friends of ICAR-IIWM, Bhubaneswar. Authors were inspired and greatly encouraged by their parents, better-halves and children, and friends.

The authors wish to mention the contributions made by the publisher, The New India Publishing Agency (NIPA), New Delhi for processing of the manuscript, typesetting, production of this book, and marketing support. Authors have duly acknowledged the sources of data, information and evidence, and duly cited in the text and included in the reference at the end of each chapter; any omission therein is inadvertent. This text-*cum*-reference book will be useful to post-graduate students, researchers, line department officials, extension persons and policymakers.

Authors

Contents

1

Rice: Staple Food, Governs the Economy and Challenges in Rice-Based Systems

1.1. Introduction: India's present population, growth rate & projections

India is the second-most populous country in the world after China with its population of about 1.38 billion in the year 2020. The Population Division under the Department of Economic and Social Affairs, of the United Nations, has projected the country's population to be growing to about 1.50 billion by the year 2030, and 1.64 billion by 2050, of course with its decreasing annual growth rate from 1.04 to 0.25% or so during the period of next thirty years (Fig. 1.1; UNDESA, 2019). The projection says the Indian population will exceed China by the year 2030 even with its decreasing rate i.e. India will be the most populous country by then. As agriculture and allied sectors are the backbone of this country and the largest livelihood source, it support this growing human and livestock population with diversified natural resources and varied agro-climatic conditions. However, the sustainability of agricultural production should be maintained in the face of perturbation and stress on the agro-ecosystem especially the degraded natural resource base and changing climate.

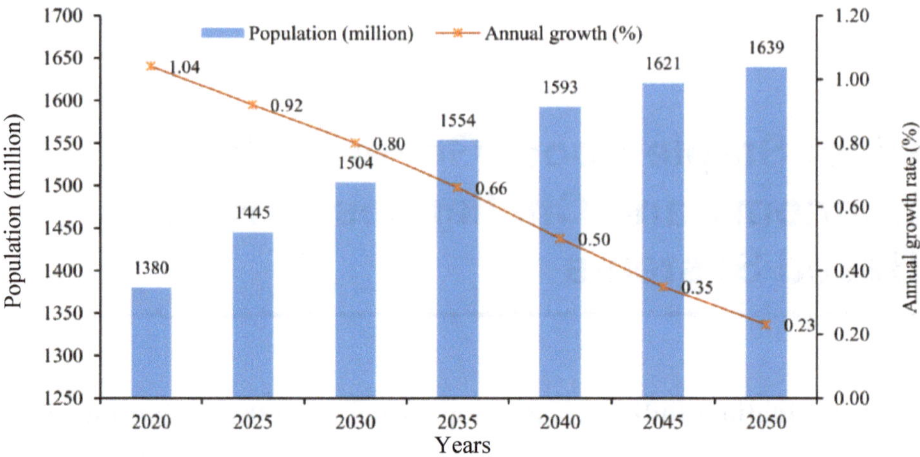

Fig. 1.1. India's population in the year 2020, projected population, and annual growth rate; *Data source:* UNDESA, 2019

1.2. Achievements in the production of food grains and agricultural allied commodities in India

India ranks second after the USA in the arable land area of about 156.46 million ha, which is 11.25% of the World. India's strength and potential are enormous in the agriculture and allied sectors. The country has shown its excellence in this sector through innovative research and development over the last 75 years i.e. after independence. There has been a quantum jump in the food grains production, cereals, and pulses combined, from a mere 50.83 million tonnes in the year 1950-51 to 315.72 million tonnes in 2021-22; rice production from 20.58 to 130.29 million tonnes, wheat from 6.46 to 106.84 million tonnes and pulses from 8.41 to 27.69 million tonnes during the period (Fig. 1.2). India ranks first in the world in its production of pulses, onion, jute and mesta (10.32 million bales of 180 kg each in 2021-22), buffalo population (109.85 million heads in 2019), and milk production (209.96 million tonnes in 2020-21); second in the world in the production of paddy, wheat, groundnut, sugarcane, tea, cotton, fruits and vegetables, potato, and cattle and goat population (DES, 2023). The main source of livelihood is the agriculture and allied sectors; about 70% of the rural population still depends on these sectors. About 68.45% of total land holdings belong to marginal (1-2 ha land) and 17.62% to small category (up to 1 ha land) of landholders, i.e. small and marginal combined (landholding up to 2 ha) occupies about 86% of all landholdings (DES, 2023).

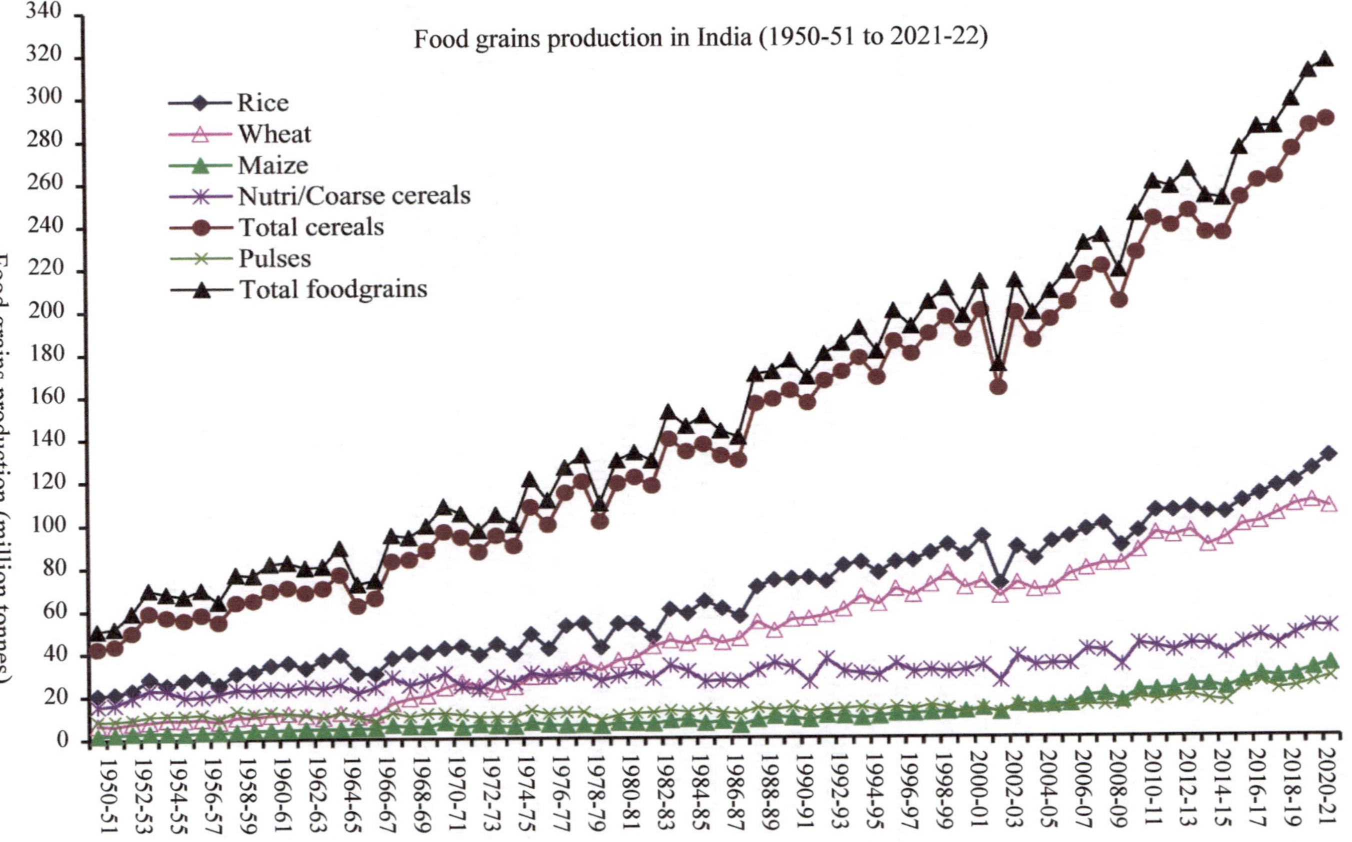

Fig. 1.2. Foodgrains production in India (1950-51 to 2021-22); *Data source:* DES (2023)

Rice, wheat and maize are the major cereals in the world. Globally, these three crops together contribute about 2.58 billion tonnes (average of 10 years, 2012 to 2021) of production to total cereal production of 2.88 billion tonnes (average of 10 years, 2012 to 2021), i.e. about 89.5% of global cereal production comes from these three crops (FAOSTAT_data_en_8-29-2023). Thus these three crops are global stable food crops. Among these, rice is the staple food for more than half of the world's population. Most importantly rice is the main food crop for low-income and low-to-medium income countries in the world. In India, rice is closely associated with the daily major food consumption, a primary source of nutrition. The majority of people follow rice farming as the only livelihood option. Enumerable festivals and religious ceremonies linked to rice are embedded in the cultural heritage of the Indian people over thousands of years.

Nutricereals is a group of crops comprising sorghum (Jowar), pearl millet (Bajra), finger millet (Ragi/Mandua) and small millets *viz*; little millet (Kutki), kodo millet (Kodo), barnyard millet (Sawa/Jhangora), foxtail millet (Kangni/ Kakun), proso millet (Cheena) all together comes under Millets which is now called as 'Nutri-Cereals' due to their high nutritive value.

1.3. Rice and cultural heritage in India

The Rigveda, the ancient Indian Vedic literature in Sanskrit, has a reference of rice in its hymns. The Sanskrit words 'dhana', 'dhanaa', 'dhanya', which mean rice, are found in Rigveda (Nene, 2012). Susruta who was a sage of ancient times recognized 'dhanya' and it was corroborated by Monier-Williams (1872). The Latin name, *Oryza sativa* is the dominant and cultivated species for rice. It belongs to the family, Poaceae. This species was domesticated from wild grass viz. *Oryza rufipogon,* about 10,000 to 14,000 years ago. The two Asian subspecies are: '*indica*' and '*japonica*' which are prevalent in tropical regions, subtropical and temperate regions of East Asia, respectively. *Oryza glaberimma* was domesticated later in West Africa. The Asian subspecies occurred about 8,200 to 13,500 years before in China. Archeological evidence suggests that *O. sativa* was cultivated in the middle basins of the Yangtze river and the upper basin of the Huai river in China. The implements for rice farming dates at least 8,000 years back. Later on, rice cultivation spreads in the river basins of all Asian countries.

Historically, it is believed that rice was widely cultivated in river valleys of South Asia and Southeast Asia about 10,000 years ago (Gnanamanickam, 2009), and it is believed to have originated probably in India. Domestication of rice in India is mainly attributed to the Indus Valley civilization about 3000–1500 BC (Possehl, 2002; Nene, 2005); however, the evidence of rice cultivation in India has been pushed to 4,000 years ahead with the discovery of rice grains and early pottery found in the site of Lahuradewa, Uttar Pradesh, situated in the middle Ganges plains dating to c. 6409 BC (Fuller *et al.*, 2007; Tewari *et al.*, 2008).

Both in ancient times and today, the intense labour needed for rice cultivation- to reclaim land, build and maintain terrace systems, to synchronize cropping patterns against soil erosion, landslides and flooding - has required villages to work together. The unifying effect of rice on people is particularly evident in the vast Mekong river delta, where villages with distinct cultures joined to tame the land and cultivate their staple crops. Furthermore, the need for standing water in rice-based systems has widely shaped the landscapes of rural areas and villages. In these ways, the struggle to cultivate and harvest rice has moulded community structures and a corresponding culture of rice.

The relationship between rice and people has inspired songs, paintings, stories and other modes of communication. Festivals have been dedicated to rice and rice cultivation- for example, the land opening festival in China, which marks the beginning of the rice season. Rice was considered divine by many Asian emperors and kings in ancient times. The Japanese, even today, refer to rice as their "mother" and regard rice farmers as the guardians of their culture and the countryside.

A movement to western India and south to Sri Lanka was also accomplished very early. Rice was a major crop in Sri Lanka as early as 1000 B.C. The crop may well have been introduced to Greece and the neighbouring areas of the Mediterranean by returning members of Alexander the Great's expedition to India around 344-324 B.C. From a centre in Greece and Sicily, rice spread gradually throughout southern Europe and to a few locations in northern Africa.

Rice is mixed with religion and tradition. Ahuja and Ahuja (2006) expressed that the relationship of rice and religion was similar to that of Calcium and bones in humans and animals.

Rice is associated with social and cultural events i.e. rituals, ceremonies and festivals in India, especially in the rural areas. Harvest of paddy is celebrated in the form of festivals, which are known by local names in the states (Table 1.1). These are Onam, Bihu, Sankranthi, Thai Pongal, Makara Sankranthi, Nabanna etc.

Rice is significant in the culture of people in eating rice in different forms, and it is the cultural heritage even in the world. Special guests are honoured with a meal of rice in village areas in Senegal. In China, Thailand people used to say 'Have you taken rice today' to greet others in the old era; New Year greetings by saying 'May your rice never burn'. Rice is sacred to people of all religions. Rice is considered the goddess of sun in Japan, Java. The names of people and places were derived from rice *viz.* 'Java' means the island of rice; 'Denjong' in Sikkim means rice valley, 'Nellore' means the town of rice. There are different names in the rituals while marriage gifts and the feasts *viz.* Dhanpan, Bhat, Bhat Bharna, Tichauli etc. In Japan, there are several such names that were derived based on rice: Toyota (bountiful rice field), Mizomonokumme (land of luxurious rice), Sawada (rice field), Honda (main rice field), Nakasoni (middle rice field), Narita (developing rice field), Tanaka (middle rice field), etc. (Ahuja and Ahuja, 2006). There are several stories, folklore, myths, poetry, and songs associated with rice in rice-growing countries *viz.* China, Japan, Indonesia, Malaysia, Korea, Thailand, Indian states like Tamil Nadu, Assam, Manipur, Himachal Pradesh, Punjab, West Bengal, Odisha

The Hindu people in West Bengal and Odisha, India consider the rice plant as the Lakshmi, the Goddess of wealth; a container is filled to its brim with rice to worship Lakshmi in Odisha. This ritual is popularly known by the name 'Manabasa' in Odisha. Rice whole grains are one of the essential ingredients for all kinds of puja, Durga Puja, Dashera, Kali puja, Deepawali, Makkar Sankranti, and religious ceremonies in India. Annapurna is the Hindu Goddess for people all over the country. Rice is offered to The God, and deities in religious festivals in different forms *viz.* unhusked rice, unbroken grains, yellow rice made with turmeric or vermilion, cooked rice, rice flour, popped rice, parched rice and even rice beverages depending upon the religious belief of local people. The tribal people use rice beverages to offer to God. Several rice varieties, named with 'bhog' as suffixed i.e. the rice 'prasad' offered to the Gods and Goddesses during puja rituals, are still growing by farmers of eastern India. Some names of such varieties are Govindabhog, Krishnabhog, Sitabhog, Rambhog, Laksmibhog, Gopalbhog, Laksminarayanbhog, Narsinghbhog, etc. Daily harvested rice was used to offer Lord Jagannath and as mahaprasad at Puri. In many places, rice flour is made as a paste and used for ritual diagrams, wall paintings, floor decoration *viz.* Alpana in West Bengal, rangoli in Maharashtra, 'Mandana' in Madhya Pradesh and Rajasthan, 'Sathia' in Gujarat, 'Jinuti' in Odisha, 'Kolam' in south India, 'Mugli' in Andhra Pradesh, 'Aipan' in Uttar Pradesh, 'Ariapan' in Bihar, 'Chowk Puran in Punjab, Haryana and Uttar Pradesh, 'Likhnu' in Himachal Pradesh, etc. (Ahuja and Ahuja, 2006). In Odisha and West Bengal, there are several festivals within a year, which are related to rice cultivation. Rice grains and hulled rice is the indispensable items during rites of human life- birth, marriage ceremony, bhumi puja, tilak of turmeric rice, death and final rites, and post-death rituals.

Table 1.1. Socio-cultural festivals in rural India/ religious festivals linked to rice

Practices	Indian states	Local name of festivals
Rice sowing	Odisha West Bengal Andhra Pradesh Assam Arunachal Pradesh West Bengal, Jharkhand	Akshaya Tritiya Ambubachi Eruvaku Purnima Bohag Bihu Mopin Damurai
Rice transplanting	Odisha West Bengal, Assam	Rajaparb Ambubachi
Rice growing	Assam Odisha	Kati Bihu Koda yatra, Asadi Paara, Chitangi amavasya, Karam Puja, Garbha sankranti, satpuri Purnima rituals
Weeding in rice	Odisha Himachal Pradesh Mizoram West Bengal	Chitau amavasya Sanskranti of sawan and bhadon Chitla ritual Kojagari and Kartik puja
Harvest of paddy	Assam Andhra Pradesh Tamil Nadu Karnataka Kerala West Bengal, Odisha, Chhattisgarh, Tripura	Bihu Sankranthi Thai Pongal Huttari, Makara Sankranthi Onam Pous, Nabanna
Rice threshing	Manipur Chhattisgarh	Phoukouba rituals Kharihari Puja
Rice storage	Mizoram Manipur	Sikisa sacrifice ritual Khurumba ritual
Religious festivals and preparations from rice	West Bengal	'Khichudi bhog' in Pousa Sankranti
	Uttarakhand, West Bengal, Kerala, Bihar, Odisha, and most of the states	Annaprashan which means "feeding rice", also called 'Bhaat khulai', 'Mukhe Bhaat', 'Choroonu'
	Jharkhand	'Karam Parva' in the month of Ashwin or Kartik; 'Karam Bough' is offered with non-boiled rice and gur or molasses
	Andhra Pradesh	Payasam, ondrallu, laddulu, paravannam, arshalu, etc. are some of the preparations from rice, and are offered to the God; thalambralu and akshanthalu while mixing in turmeric powder and also used as vadibiyyam
	West Bengal	Various types of pitha- pati sapta, gar gare, anska pitha, pitha payas, pur pitha
	South India	Dosa, Idli
	Assam	Na-Khuwa Bhooj festival
	Rajasthan	Navagrahapooja.

Source: Ahuja and Ahuja (2006)

1.4. Rice bowls in India

There are many places or regions, which are historically called rice bowls in India. The agro-ecological conditions and management practices are very suitable in those places or regions for rice cultivation over several generations of farmers. The maximum production is obtained from those regions. These exclusive regions or ecological niches should be given importance to sustainable rice production in the country.

Among states, Andhra Pradesh is known as the rice bowl of India. Within the state, East Godavari district lies in the main delta region of Krishna and Godavari rivers where production of rice is the maximum because of its agro-ecological suitability to rice cultivation with its best crop management practices by the farmers. Thanjavur area of Tamil Nadu is called the granary of South India and also the rice bowl of the state. Similarly, Nanchinadu in Kerala is the harbinger of rice, and is called a rice bowl of the former Travancore kingdom; a place in Kerala, called Kuttanadu, is also famous for rice farming and many designate this place as the rice bowl of Kerala. Nanchinadu is famous for plough culture by oxen or bullocks and rice fields were considered royal. Some regions under Bellari, Koppal and Raichur are popularly called the rice bowls of Karnataka, where 'Sona Mashuri' rice is cultivated in the irrigation command areas of Tungabhadra. Its good quality grains have high demand among people even across the state.

In eastern India, the east Burdwan district is known as the rice bowl of West Bengal due to its highest yield and also grain quality. Like Andhra Pradesh, Chhattisgarh is also called the rice bowl of India because this state is also one of the productive states in the country, and currently export of rice from the state is quite high. In Odisha, Bargarh is said to be the rice bowl as it has the highest production sites due to its rich and fertile soil, irrigation sources from Hirakund irrigation command, and very good crop management by progressive farmers. The district Rohtas is the rice bowl of Bihar.

For Basmati rice, the state of Punjab can be considered as the Basmati bowl of India. In the state, Basmati districts are Gurdaspur, Hoshiarpur, Amritsar, Jalandhar, Kapurthala, and Nawanshahar. These districts have been identified by the Government of India for further promotion of Basmati rice in the country.

1.5. Importance of rice as food and nutrition

Rice is the major source of calorie and protein intake per day. It is a primary source of dietary energy for the people of seventeen countries in Asia including the Pacific, nine countries in the USA including Latin America, and also the people of eight African countries (Rathna Priya *et al.*, 2019). Rice and its different rice-based food products provide 60-70% of the total calories required for more than 2000 million people in Asia alone. In Africa, rice has gained popularity as a food item among low-income as well as food-insufficient countries. Rice has become a significant crop for food and nutrition security in many African countries.

World rice statistics of the International Rice Research Institute (IRRI), Philippines show that the decadal average rice calorie intake was about 646 to 733 kcal per day per capita, which is about 28 to 33% of the total calorie intake per day (Fig. 1.3, Table 1.2). Rice fat contributes to an individual about 2.95 to 4.83% of total fat intake per day. Similarly, a major share of protein comes from rice in daily food consumption. The decadal average value shows that rice protein intake is about 21.88 to 25% of total protein consumed per day per capita.

Rice grain is a source of plenty of carbohydrates. It contains more than 70 g of carbohydrates per 100 g of the rice grain, be it raw milled rice, parboiled, polished white rice, or brown rice (Table 1.3). It contains a considerable amount of protein and a small fraction of fats. Among different parts of the grain, bran contains the highest amount of fat and protein. It is a good source of minerals and vitamins *viz.* Ca, P, Zn, Fe, and Ca-Mg salt of phytic acid i.e. phytin. It is also the source of vitamins viz. niacin, thiamine, riboflavin, and α-tocopherols (Table 1.4). The nutrient contents are comparable with other cereals viz. wheat and maize, and tuber crops *viz.* potatoes. The bran layer is retained in brown rice during the process of polishing, which is why brown rice and rough rice are more nutritious. Rice grain has amino acids and it is a source of protein. The lysine content of protein is 3.5 to 4.0% which is the highest among cereals and is superior to wheat, maize, millets, and sorghum (Eggum, 1979). Linoleic acid is a major component of fatty acids present in rice bran. Cholesterol is not present. Of course, the nutrient contents and proximate composition vary among different rice varieties and degrees of polishing during milling. Even the variation has been reported for different rice fractions in particular rice varieties. Therefore, sustaining and improving the production of rice is essential for global as well as Indian food and nutrition security.

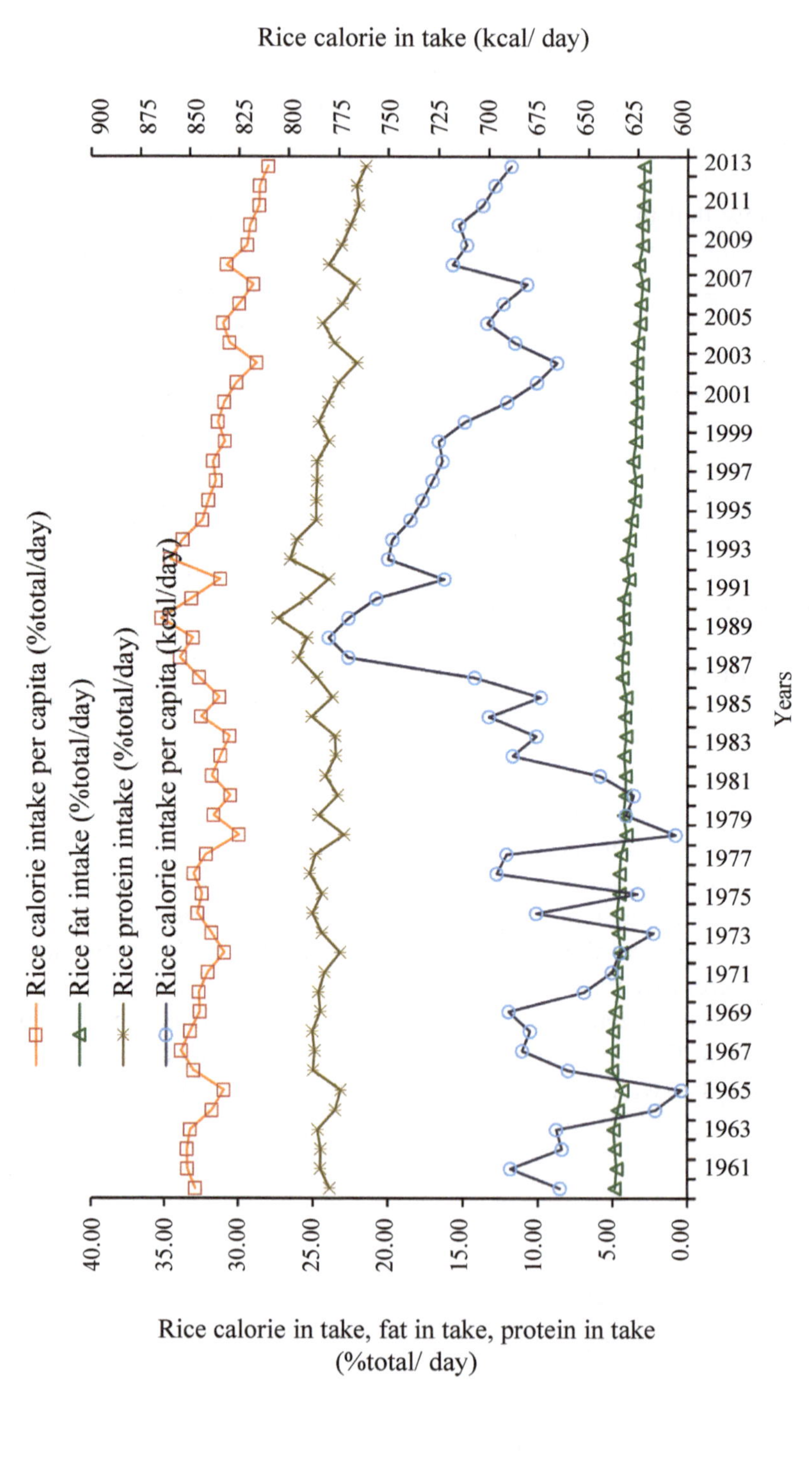

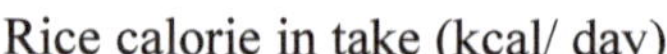

Fig. 1.3. Trend of rice nutrition- calorie, fat and protein intake

Source: World Rice Statistics Online Query Page http://ricestat.irri.org:8080/wrs accessed on 10.03.2021

Table 1.2. Average rice calorie, rice fat, and protein intake in India during 1961-2013 in different decades

Period	**Rice calorie intake/capita**		**Rice fat intake (%total/day)**	**Rice protein intake (%total/day)**
	kcal/day	%total/day		
1961-70	661.30	32.88	4.83	24.38
1971-80	646.60	31.98	4.51	24.37
1981-90	703.60	32.32	4.23	24.71
1991-00	733.60	32.33	3.77	25.00
2001-10	693.90	30.05	3.24	23.26
2011-13	696.33	28.43	2.95	21.88

Source: Computed from data source of World Rice Statistics Online Query Page http://ricestat.irri.org:8080/wrs accessed on 10.03.2021

Table 1.3. Carbohydrate, protein and fat contents in rice (in 100 g)

Rice	**Carbohydrate (g)**	**Protein (g)**	**Fat (g)**	**Reference**
Raw milled rice	74.80±0.85	9.16±0.75	1.24±0.08	Longvah *et al.* (2017)
Parboiled rice	77.16±0.76	7.81±0.63	0.55±0.08	Longvah *et al.* (2017)
Polished white rice	76.80-86.56	5.8-7.7	0.5-2.3	Devi *et al.* (2015)
Rice bran	34.0-62.0	11.3-14.9	15.0-19.7	Juliano (1983), Pedersen and Eggum (1983), Eggum *et al.* (1982)
Rice hull	22.0-34.0	2.0-2.8	0.3-0.8	Eggum *et al.* (1982)
Brown rice	71.31-84.75	5.9-11.0	1.6-2.8	Devi *et al.* (2015)

Table 1.4. Components of minerals and vitamins of different rice grains (in 100 g with 14% moisture)

Rice	Ca (mg)	P (g)	Zn (mg)	Fe (mg)	Phytin (g)	Niacin (mg)	Thiamine (mg)	Riboflabin (mg)	Tocopherol (mg)
Milled rice	10.0-30.0	0.08-0.15	0.6-2.3	0.20-2.8	0.02-0.07	1.3-2.4	0.02-0.11	0.02-0.06	0.075-0.30
Rice bran	30.0-120.0	1.1-2.5	4.3-25.8	8.6-43.0	0.9-2.2	26.7-49.9	1.20-2.40	0.18-0.43	2.60-13.3
Rice hull	60.0-130.0	0.03-0.07	0.9-4.0	3.9-9.5	-	1.6-4.2	0.09-0.21	0.05-0.07	-
Brown rice	10.0-50.0	0.17-0.43	0.6-2.8	0.2-5.2	0.13-0.27	3.5-5.3	0.29-0.61	0.04-0.14	0.90-2.50

Source: Rathna Priya *et al.* (2019)

1.6. Trends in area, production and yield of rice in major growing regions in the World

Worldwide, rice is grown on 165.25 million ha currently, with an annual production of about 787.29 million tonnes of paddy. About 87% of the world's area and 90% of the world's paddy production occurred from Asia in the year 2021 (Fig. 1.4 and 1.5). Regionally and leading country-wise, rice area is 143.06 million ha in Asia, 64.81 in South Asia, 45.08 in Southeast Asia, 32.75 in East Asia, 46.38 in India and 30.15 million ha in China with paddy production of 708.15 million t in Asia, 279.22 in South Asia, 231.99 in East Asia, 194.47 in South-East Asia, 195.42 in India and 214.40 million t in China in the year 2021 (FAOSTAT as on 29.08.2023). The average paddy yield is 4.76 t ha^{-1} in the world, 4.95 t ha^{-1} in Asia, 4.31 t ha^{-1} in South Asia, 4.31 t ha^{-1} in Southeast Asia, 7.08 t ha^{-1} in East Asia, 4.21 t ha^{-1} in India and 7.11 t ha^{-1} in China in 2021 (Fig. 1.6, Table 1.5). This shows that there exists a wide variation in the production of paddy per unit area in different regions of the world. This variation is primarily due to differential genetic potential, soils, weather elements especially temperature, rainfall and sunshine hours, irrigation facilities, and crop management systems.

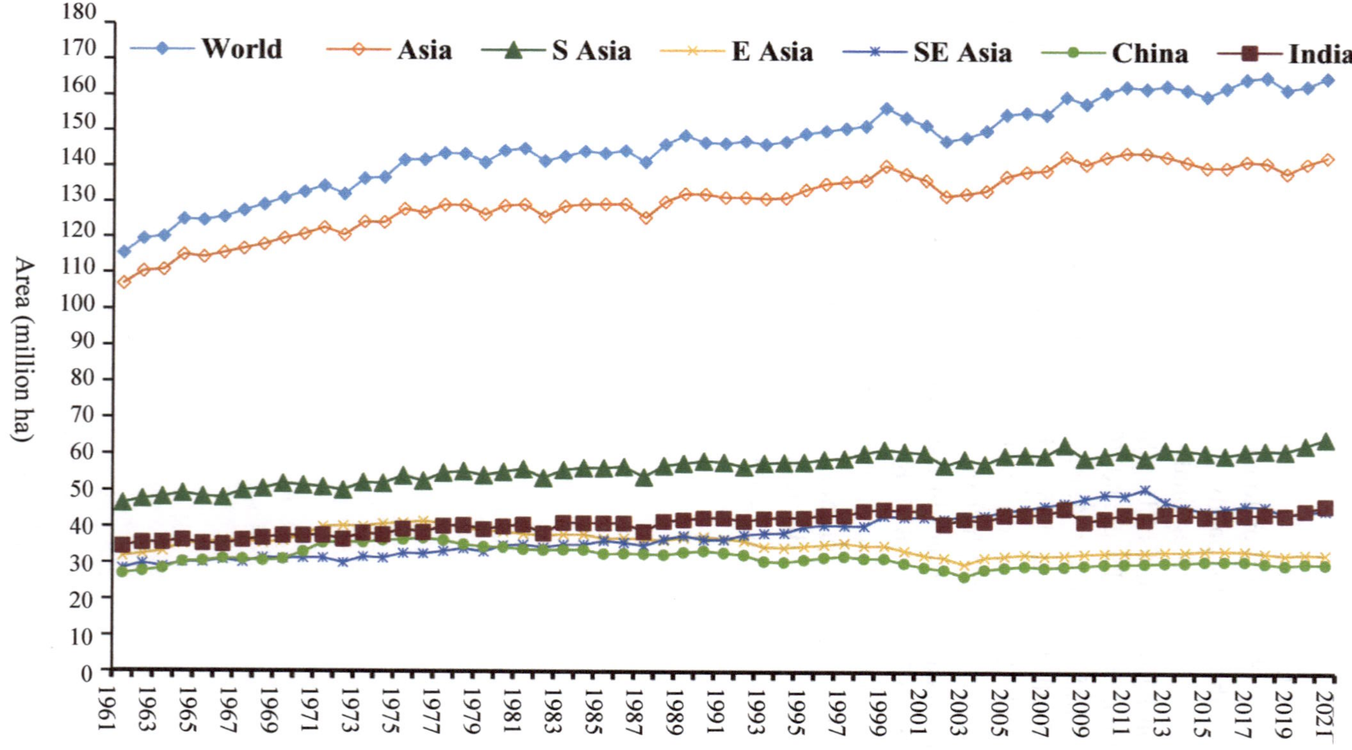

Fig. 1.4. Trends of the rice area in the world
Data source: FAOSTAT_data_en_4-30-2021, FAOSTAT_data_en_8-29-2023

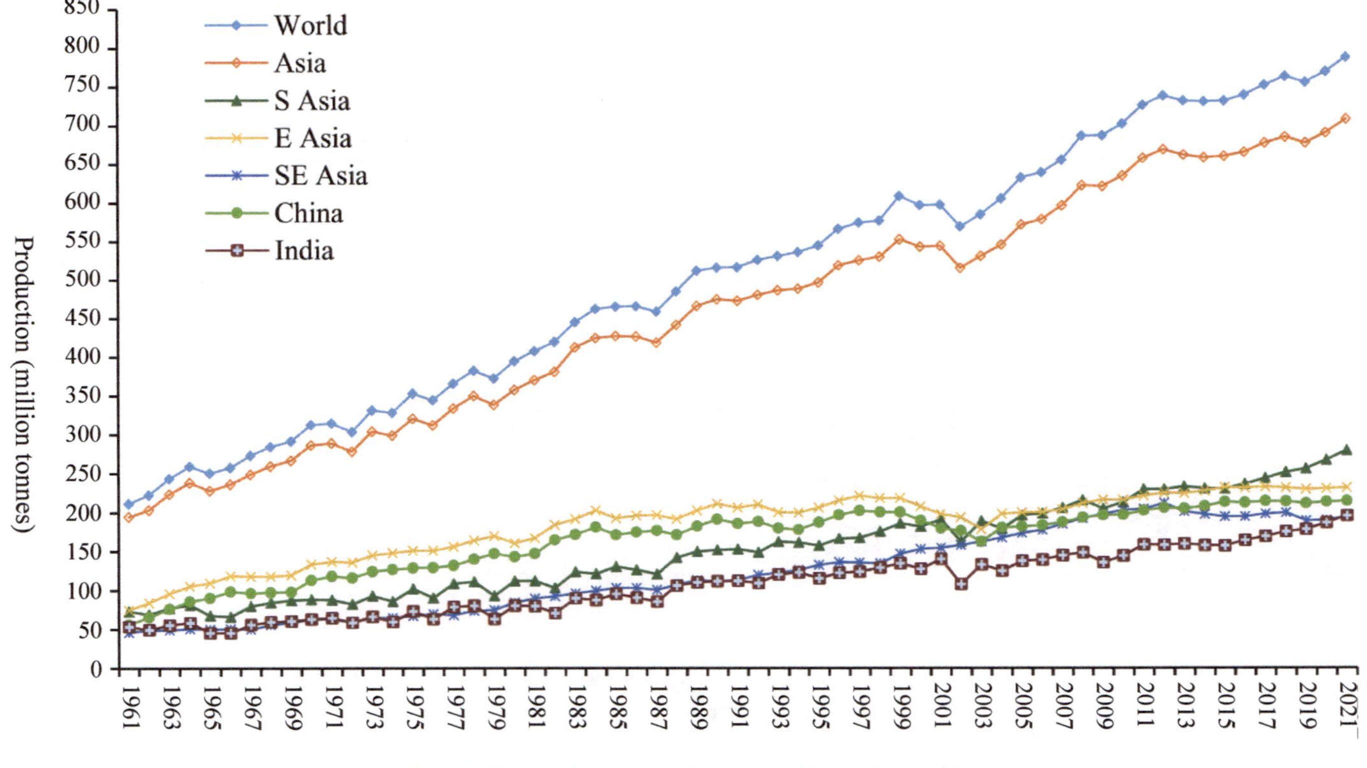

Fig. 1.5. The trend of rice production (paddy) in the world

Data source: FAOSTAT_data_en_4-30-2021, FAOSTAT_data_en_8-29-2023

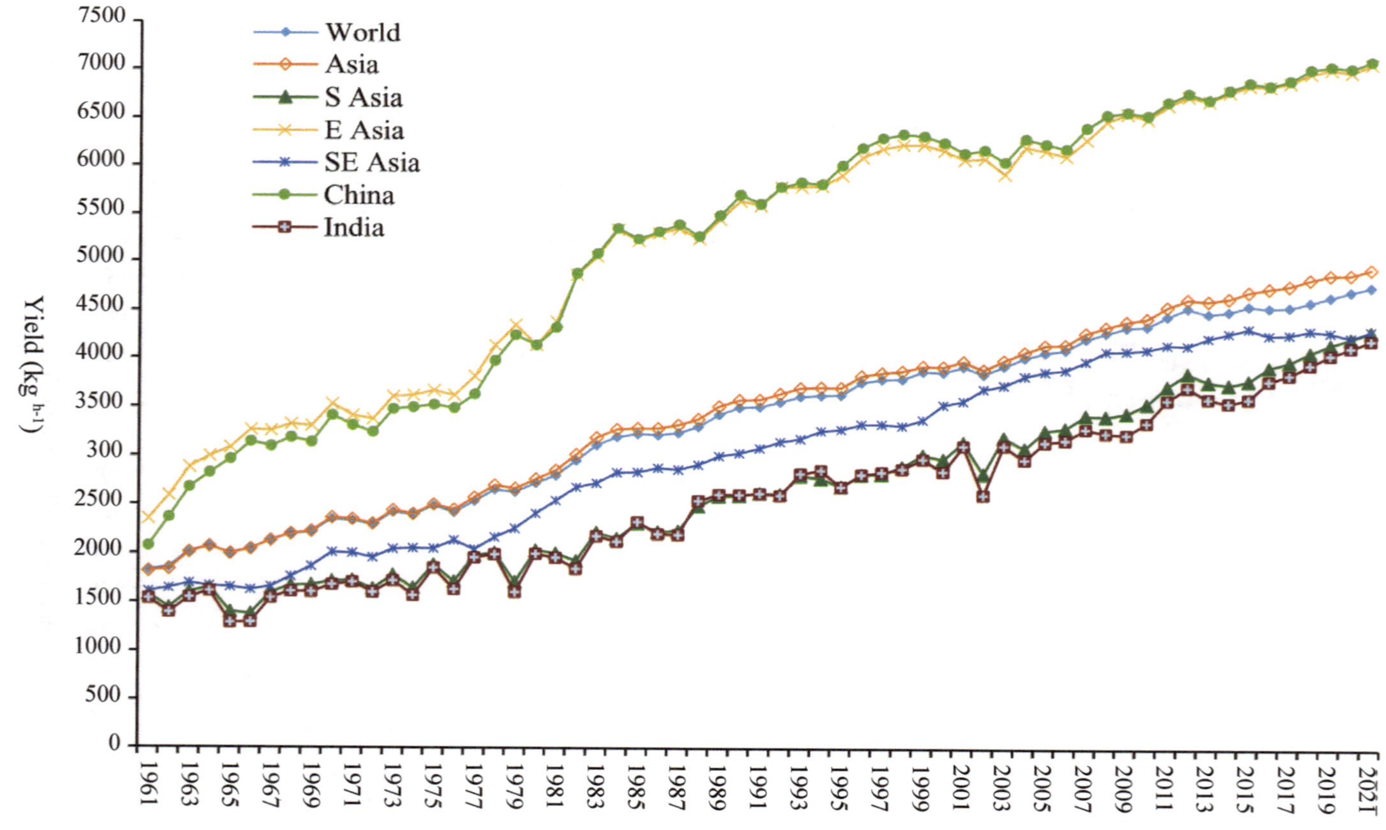

Fig. 1.6. Trend of rice (paddy) yield in the world

Data source: FAOSTAT_data_en_4-30-2021, FAOSTAT_data_en_8-29-2023

Table 1.5. Average paddy yield (last 5 years, 2017-2021)

	Paddy yield (kg ha^{-1}) in 2021	**Paddy yield (kg ha^{-1}) 5-Year average (2017-2021)**
China	7112	7030
India	4213	4043
S Asia	4308	4158
E Asia	7085	6996
SE Asia	4313	4290
Asia	4949	4869
World	4764	4662

S Asia: Afghanistan, Bangladesh, Bhutan, India, Iran, Sri Lanka, Nepal, Pakistan
E Asia: China-main land, China-Hong Kong, China-Taiwan, Japan, DPR Korea, Rep Korea
SE Asia: Brunei, Myanmar, Indonesia, Cambodia, Laos PDR, Malaysia, Philippines, Timor-Leste, Thailand, Vietnam
Data source: FAOSTAT_data_en_4-30-2021, FAOSTAT_data_en_8-29-2023

1.7. Trends in area, production and yield of rice in India

The area under rice in India is the largest in the world. According to the data source of the Department of Agriculture, Cooperation and Farmer's Welfare of the Government of India, the current rice area is 46.38 million ha with its rice production of 130.29 million tonnes and the average yield is 2.81 t ha^{-1} in the year 2021-22 (Fig. 1.7). Although rice area is the largest in the world, its production is only next to China, and yield level is much lower than China. The trend of area, production, and yield shows that the rice area has stagnated at around 42.7 million ha for the last 45 years. Its expansion will not be possible in the future. However, ample potential exists in increasing the yield from its current level to a greater with the evolving of high-yielding genotypes, better crop management, and most importantly irrigation and better water management practices.

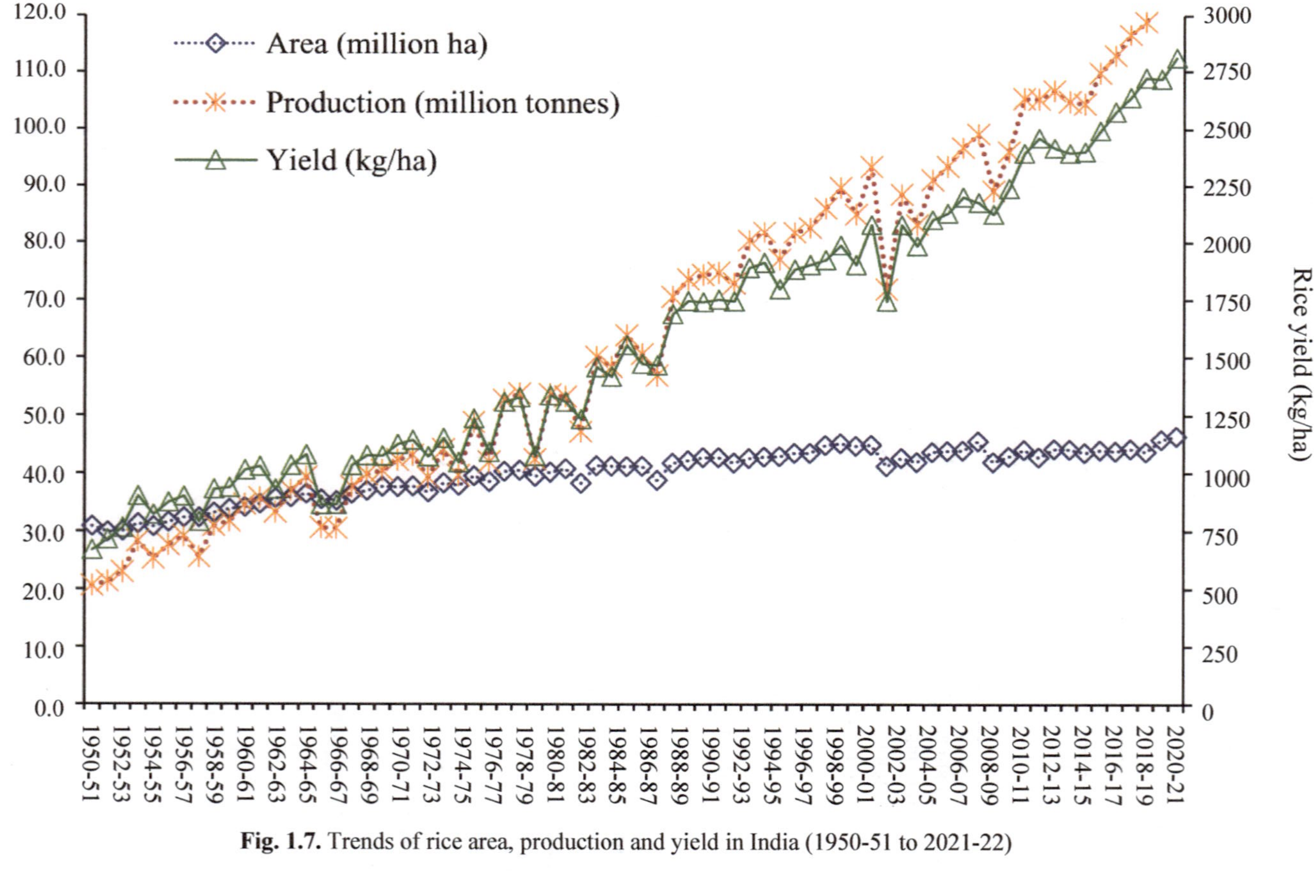

Fig. 1.7. Trends of rice area, production and yield in India (1950-51 to 2021-22)

Data source: DES (2023)

The contribution to rice is about 40% to the total foodgrains production in the country, and wheat contributes 35% (Fig. 1.8).

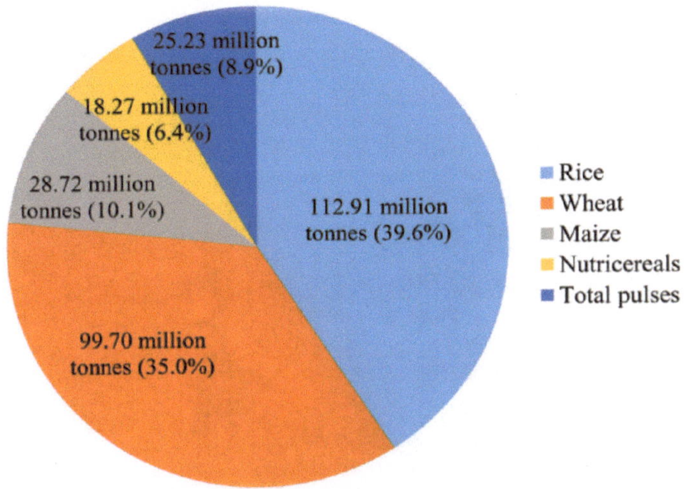

Fig. 1.8. Contribution of rice to total foodgrains production in the country (year 2021-22); *Data source:* DES (2023)

1.8. Area, production and yield of rice in major rice-growing states in India

The average rice area (5-year average, 2016-17 to 2020-21) is 5.79 million ha in Uttar Pradesh, which is the maximum among rice-growing states in India, and the second-highest is in West Bengal (5.44 million ha), whereas rice production is the highest (15.74 million tonnes) in West Bengal. Major rice-growing states (>2 million ha) are Uttar Pradesh, West Bengal, Odisha, Telangana, Chhattisgarh, Bihar, Punjab, Assam, Andhra Pradesh and Madhya Pradesh (Fig. 1.9). Major production (>7 million tonnes) comes from West Bengal, Uttar Pradesh, Punjab, Andhra Pradesh, Telangana, Odisha and Bihar (Fig. 1.10). The highest rice yield (average of 5 years: 2016-17 to 2020-21) was reported from Punjab (4179 kg ha^{-1}) followed by Andhra Pradesh, Telangana, Tamil Nadu, Haryana, yield values of those are above 3000 kg ha^{-1} (Fig. 1.11).

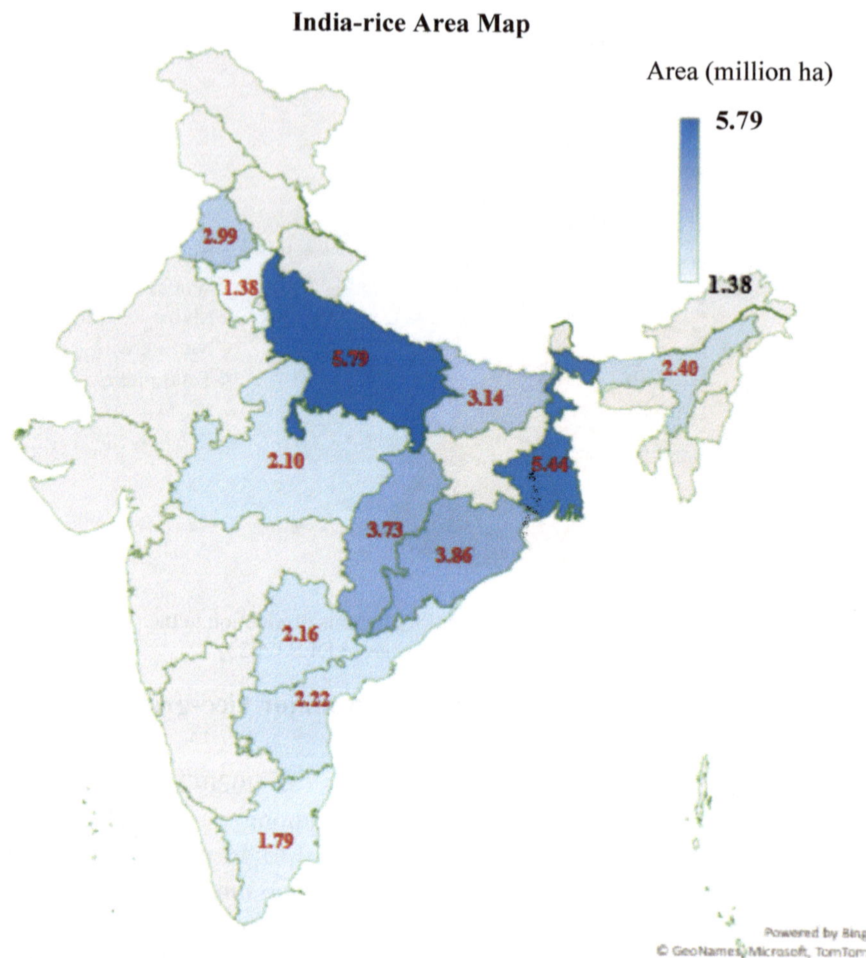

Fig. 1.9. Mapping of the rice area in major growing states in India (5-year average: 2016- 17 to 2020-21); *Data source*: DES (2019, 2020, 2021, 2022, 2023)

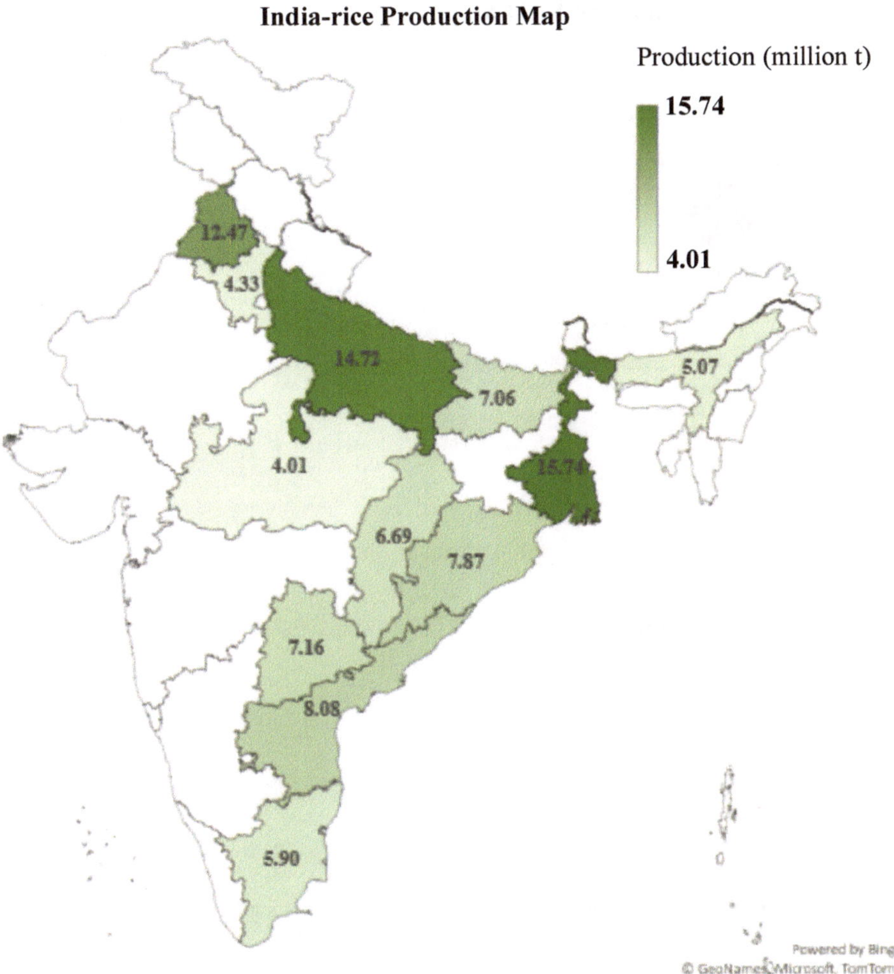

Fig. 1.10. Mapping of rice production in major growing states in India (5-year average: 2016-17 to 2020-21); *Data source:* DES (2019, 2020, 2021, 2022, 2023)

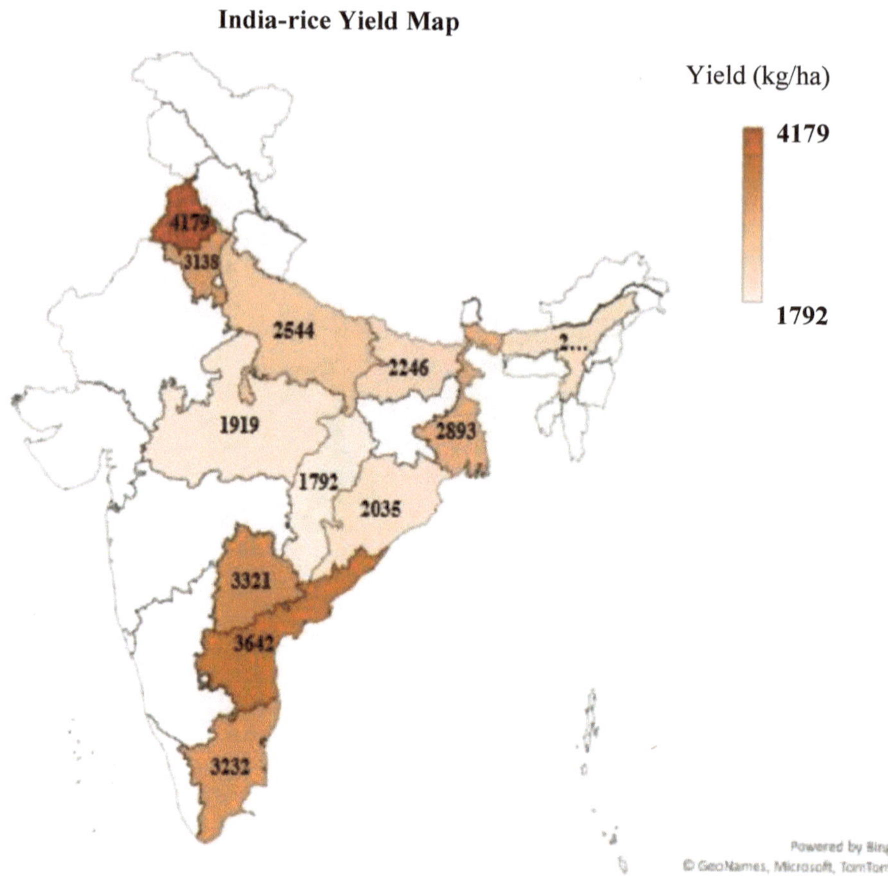

Fig. 1.11. Mapping of rice yield in major growing states in India (5-year average: 2016-17 to 2020-21); *Data source:* DES (2019, 2020, 2021, 2022, 2023)

1.9. Trends in India's export-import of rice

India is a leading exporter of rice in the world. India's share is around 40% of the global exports of rice to about 140 countries. The item is milled equivalent (ITCHS 100630) (Indian Trade Classification Harmonized System of coding, ITCHS 100630 is the code); other countries are Thailand (22%), Vietnam (12%), Pakistan (8%), the USA (6%) and others. The trend of export and import of rice shows differential and contrasting lines over the years in India from 1961 to 2021 (Fig. 1.12). It has four distinct phases. The import quantity and value were much greater than the export during the years 1961-1977 (Table 1.6, 1.7 & 1.8). Afterwards, there has been a phenomenal increase in the export of milled rice, and the average export was about 12.25 million tonnes with earning monetary value of 7276.20 million USD annually for the last ten years. On the contrary, rice import decreased drastically to about 136 thousand tonnes with an expenditure of 34.99 million USD per year during 1978-1994, and then to 2.48 million tonnes at the expense of 2.03 million USD annually only during 2012-2021.

The share of India basmati rice to the total quantity of milled rice exported to other countries every year (5-year average: 2018-19 to 2022-23) was less than the non-basmati rice, whereas the monetary value earned through the export of basmati rice was marginally less because of the greater price of basmati rice than non-basmati (Fig. 1.13).

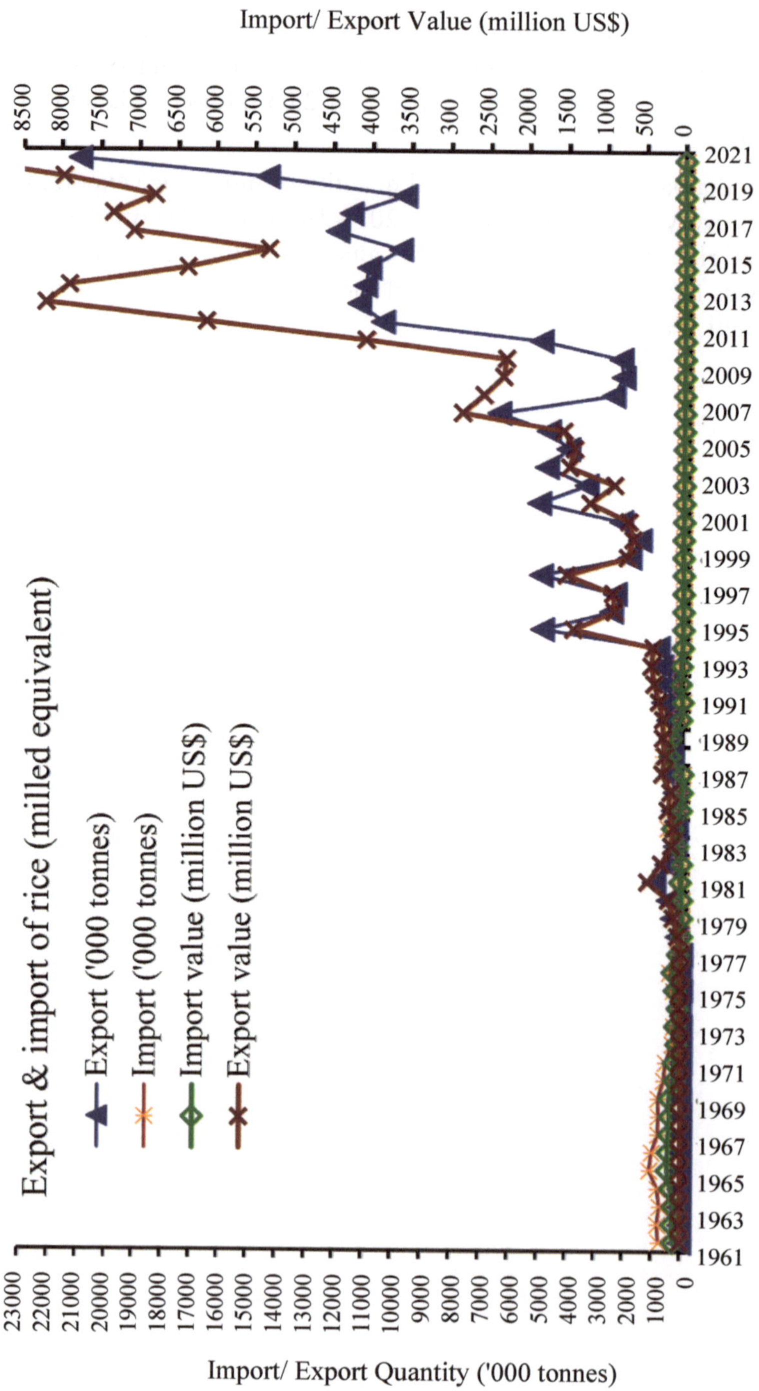

Fig. 1.12. Trends of export and import of rice (milled equivalent) by India during 1961 - 2021

Data source: FAOSTAT_data_en_4-30-2021, FAOSTAT_data_en_8-29-2023 https://www.fao.org/faostat/en/#data/QCL

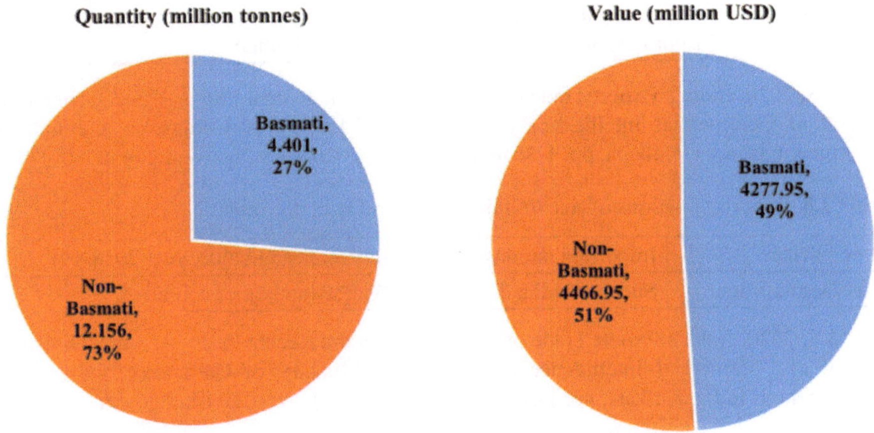

Fig. 1.13. Quantity and value of export of basmati and non-basmati rice by India (last 5-year average, 2018-19 to 2022-23)

Data source: DGCI&S, 2023. Directorate General of Commercial Intelligence and Statistics, Ministry of Commerce and Industry, Government of India, dated 29.08.2023 www.dgciskol.gov.in

Table 1.6. Decadal average of export and import of rice (milled equivalent) by India

Period	Export		Import	
	Average quantity per year ('000 tonnes)	Average value per year (million US$)	Average quantity per year ('000 tonnes)	Average value per year (million US$)
1961-1977	13.274	4.832	589.073	100.641
1978-1994	472.557	242.179	136.094	34.991
1995-2011	3490.430	1629.879	3.597	0.893
2012-2021	12255.541	7276.201	2.484	2.027

FAOSTAT, 2023. Export quantity and Export value of Rice, paddy (rice milled equivalent). http://www.fao.org/faostat/en/#rankings/countries_by_commodity_exports [accessed on 9.03.2021]

FAOSTAT_data_en_4-30-2021, FAOSTAT_data_en_8-29-2023 https://www.fao.org/faostat/en/#data/QCL

Table 1.7. Major exporting nations of rice from India, in general over the years

Type of rice	Export from India to nations (in order of decreasing value, year 2018-19)
Basmati rice	Iran, Saudi Arabia, Iraq, United Arab Emirates, Yemen Republic, Kuwait, USA, UK, Oman, Qatar, and others

Non-Basmati rice	Nepal, Benin, Bangladesh, Senegal, Guinea, Cote d'Ivoire, United Arab Emirates, Somalia, Indonesia, Liberia and others

DGCI&S, 2020. India's Export Trade of Milled Rice in the Recent Past – Aspects, Directorate General of Commercial Intelligence and Statistics, Ministry of Commerce and Industry, Government of India, Kolkata, pp. 1-30. dated 17-Nov-2020 www.dgciskol.gov.in

Table 1.8. Importing nations of rice to India, in general over the years

Type of rice	Importing nations (in order of decreasing value, year 2018-19)
Non-Basmati rice	Sudan, Thailand, Spain, USA, Italy and other countries

DGCI&S, 2020. India's Export Trade of Milled Rice in the Recent Past – Aspects, Directorate General of Commercial Intelligence and Statistics, Ministry of Commerce and Industry, Government of India, Kolkata, pp. 1-30. dated 17-Nov-2020 www.dgciskol.gov.in

The Government of India has given importance to the production of Basmati rice and its export to other nations. Punjab has the exclusive zone for Basmati rice. The varietal improvement is essential along with its improved agronomic technology. The states *viz.* Punjab, Haryana, Jammu and Kashmir, Himachal Pradesh, Uttarakhand, and Western Uttar Pradesh are the major areas in India where basmati is mainly grown. The exclusive zone covers the Gurdaspur, Amritsar, Kapurthala, Jalandhar, Hoshiarpur, and Nawanshahar districts of Punjab. Agro-climatic condition of the specific geographical area. Specific geographical area, Himalayan foothills of the Indian sub-continent. Basmati rice is an extra-long slender grain, aromatic, soft and fluffy texture upon cooking, and delicious. The geographical area ecological niche of Punjab for Basmati rice. Govt is trying to tap the potential of the zone. The main varieties of Basmati rice as notified under the seed act 1966 are Basmati-386, Basmati-217, Ranbir Basmati, Karnal- local/Taraori Basmati, Basmati-370, Type-3 (Dehradoon Basmati), Pusa Basmati-1, Pusa Basmati-1121, Punjab Basmati-1, Haryana Basmati-1, Mahi Sugandha, etc.

1.10. Conclusion- challenges ahead for rice-based systems in India

Despite spectacular achievements in the previous decades, India has many other concerns. The contribution of agriculture and allied sectors has declined steadily to the national gross domestic product (GDP) because of comparatively faster growth in the other sectors *viz.* industry and manufacturing, and energy. Although India has overcome the begging bowl image during the 1960s and early 1970s, and achieved self-sufficiency in food production since the 1980s, India accounts for many hungry and poor and undernourished people. Poverty has not been eradicated from the country. The production systems in the country have become soil resource-based, dominated by cereal crops like rice and wheat. There has been pressure on soil resources and depletion of water resources in high productivity zones in the country. Overall sustainability is under threat, which calls for a paradigm shift in major policies.

To meet the global rice demand, it is estimated that about 114 million tonnes of additional milled rice will be needed to be produced by 2035, which is equivalent to an overall increase of about 26% in the future. India also needs to improve its management of agricultural practices on multiple fronts. Improvements in agriculture performance have a weak linkage in improving nutrition, the agriculture sector can still improve nutrition through multiple ways: increasing incomes of farming households, diversifying the production of crops, empowering women, strengthening agricultural diversity and productivity, and designing careful price and subsidy policies that should encourage the production and consumption of nutrient-rich crops. Diversification of agricultural livelihoods through agri-allied sectors such as animal husbandry, forestry and fisheries has enhanced livelihood opportunities, strengthened resilience and led to a considerable increase in labour force participation in the sector.

Water management is the key to creating sustainable rice-based production systems, particularly because rice is the only major cereal that can withstand water submergence. Improving the productivity of rice systems would contribute to hunger eradication, poverty alleviation, national food security and economic development. Irrigation and water management challenges and production technologies have been elaborated for rice-based systems in the following chapters.

References

Ahuja, S.C. and Ahuja, U. 2006. Rice in religion and tradition. In: Souvenir, 2nd International Rice Congress-2006, Oct 9-13, New Delhi, pp. 45-52.

DES, 2019. Agricultural Statistics at a Glance 2018, MoA&FW, DAC&FW, Directorate of Economics and Statistics, Govt of India, New Delhi, 502p.

DES, 2020. State-wise Percentage Coverage of Irrigated Area under Principal Crops during 2015-16 (Provisional). In: Agricultural Statistics at a Glance 2019, MoA&FW, DAC&FW, Directorate of Economics and Statistics, Govt of India, New Delhi, 335p.

DES, 2021. Agricultural Statistics at a Glance 2020, MoA&FW, DAC&FW, Directorate of Economics and Statistics, Govt of India, New Delhi, 280p.

DES, 2022. Agricultural Statistics at a Glance 2021, MoA&FW, DAC&FW, Directorate of Economics and Statistics, Govt of India, New Delhi, 457p.

DES, 2023. Agricultural Statistics at a Glance 2022, MoA&FW, DAC&FW, Directorate of Economics and Statistics, Govt of India, Govt of India, New Delhi, 280p.

Devi, G.N., Padmavathi, G., Babu, V.R. and Waghray, K. 2015. Proximate nutritional evaluation of rice (*Oryza sativa* L.). *Journal of Rice Research,* 8(1):23-32.

DGCI&S, 2020. India's Export Trade of Milled Rice in the Recent Past – Aspects, Directorate General of Commercial Intelligence and Statistics, Ministry of Commerce and Industry, Government of India, Kolkata, pp. 1-30. dated 17-Nov-2020 www.dgciskol.gov.in

DGCI&S, 2023. India Export Statistics: APEDA Products, https://agriexchange.apeda.gov.in/indexp/genReport_combined.aspx#content

Eggum, B.O. 1979. The nutritional value of rice in comparison with other cereals. In: Proceedings, Workshop on Chemical Aspects of Rice Grain Quality, IRRI. Los Banos, Laguna, The Philippines; p. 91–111.

Eggum, B.O., Juliano, B.O. and Maniñgat, C.C. 1982. Protein and energy utilization of rice milling fractions by rats. *Qual. Plant. Plant Foods Hum. Nutr*. 31: 371-376.

FAOSTAT, 2020. Export quantity and Export value of Rice, paddy (rice milled equivalent). http://www.fao.org/faostat/en/#rankings/countries_by_commodity_exports [accessed on 9.03.2021]

FAOSTAT, 2023 as on 29.08.2023 FAOSTAT_data_en_4-30-2021, FAOSTAT_data_en_8-29-2023 https://www.fao.org/faostat/en/#data/QCL

Fuller, D.Q., Qin, L. and Harvey, E. 2007. Evidence for a late onset of agriculture in the lower Yangtze region and challenges for an archaeobotany rice. In: Sanchez-Mazas A, Blench R, Ross MD, Peiros I, Lin M, editors. Past Human Migrations in East Asia, Matching Archaeology, Linguistics and Genetics. London: Routledge, p. 40–83.

Gnanamanickam, S.S. 2009. Rice and its importance to human life. In: Biological Control of Rice Diseases. *Progress in Biological Control*, 8:1-11.

Juliano, B.O. 1983. Rice: Chemistry and Technology. 2nd ed. St. Paul, MN: Am. Assoc. Cereal Chem; 1985b. p. 774.

Longvah, T., Ananthan, R., Bhaskarachary, K. and Venkaiah, K. 2017. Proximate principles and dietary fibre. Hyderabad, India: Indian Food Composition Tables, National Institute of Nutrition, Department of Health Research, Ministry of Health and Family Welfare, Government of India, p. 3

Monier-Williams, M. 1872. A Dictionary in English and Sanskrit. Motilal Banarasidass, New Delhi, India. 859 pp. (Reprint 1999.)

Nene, Y.L. 2005. Rice research in South Asia through ages. *Asian Agri-History*, 9 (2): 85-106.

Nene, Y.L. 2012. Rigveda has References to Rice? *Asian Agri-History*, 16 (4): 403-409.

Pedersen, B., Eggum, B.O. 1983. The influence of milling on the nutritive value of flour from cereal grains. *Plant Foods and Human Nutrition,* 33:267-278.

Possehl, G.L. 2002. The Indus Civilization: A Contemporary Perspective. Alta Mira Press. CA; Oxford: Walnut Creek; 7.

Rathna Priya, T.S., Nelson, Ann Raeboline Lincy Eliazer, Ravichandran Kavitha and Antony Usha 2019. Nutritional and functional properties of coloured rice varieties of South India: a review. *Journal of Ethnic Foods*, 6:11 https://doi.org/10.1186/s42779-019-0017-3

Tewari, R., Srivastava, R.K., Saraswat, K.S., Singh, I.B. and Singh, K.K. 2008. Early farming at Lahuradewa. *Pragdhara*, 18: 347-373.

UNDESA, 2019. United Nations, Department of Economic and Social Affairs, Population Division. World Population Prospects: The 2019 Revision, www.unpopulation.org.

2

Water Resources, Issues and Challenges in Water Management

2.1. Introduction

Water plays a pivotal role in this earth system. The withdrawal of water for irrigation is about 70% of the global freshwater. Water use efficiency is less than 50% in many countries. It has been forecasted by the Food and Agricultural Organization (FAO, 2015) that global water requirements will be increased by 50% by the year 2050 to meet the demands of agricultural water use. This is so because food demand will be higher for the growing human population. We have entered into an earth system where everything living has evolved and developed (Cosgrove and Loucks, 2015). There has been a new geological scenario, some say Anthropocene (Crutzen, 2002; Williams *et al.*, 2011). Consequently, freshwater will increasingly be a scarce and the costliest resource in the future, which needs to be managed properly.

There has been a mention in Atharva Veda (Atharva Veda 19.2.1) that 'One should take proper managerial actions to use and conserve water from mountains, wells, rivers and also rainwater for various purposes'.

Agriculture is an important sector of the Indian economy, accounting for about 16% of the nation's GDP in the current prices, 11.94% of its national exports (DES, 2023). About half of the population still relies on agriculture as its principal source of income; agriculture is a source of raw material for a large number of industries. The net sown area has remained about ~140 million ha since 40 years but the number of farmers has increased from 70 to 140 million even more. About 10 million farmers are being added in every five years. Hence, the country has the twin challenges of meeting its water needs along with sustaining pace of development.

Among cereals, rice has the lowest productivity per unit of water. During rainy season, irrigation is provided only during deficit period to make up the water requirement. Rice crop grown in dry season or in low rainfall areas, consumes large amount of water with very low water use efficiency. The amount of water added to rice crop under most irrigation system is very high, 2500 mm as against 480 mm of evapotranspiration (Bouman, 2009).

Due to climate change, Himalayan snow and ice, which provide vast amounts of water for agriculture, are expected to decline by 20% by 2030. Increased climate variability has made rainfall patterns more inconsistent and unpredictable. Year-to-year variation is continued to be a matter of serious concern: increased drought like situations due to overall decrease in the number of rainy days, increased flood events due to overall increase in rainfall intensity, effect on ground water quality in alluvial aquifers, affecting ground water recharge, increased saline water intrusion of coastal and island aquifers due to rising sea levels. Agriculture already faces uncertainty due to increasing regional variations in rainfall and temperature.

In addition to basic water requirements for the growth of rice, water is required to prepare the land. The amount varies depending upon soil texture, soil moisture content prior to land preparation and method of land preparation. Where puddling is practiced, water requirements between 125 and 200 mm have been reported to be necessary. After land preparation, there is an additional quantity of water required to carry out specific water management practices.

Total water requirement includes water needed to raise seedlings, prepare land and to grow the crop from transplanting to harvest. The amount will vary due to many factors *viz.* soil type, topography, depth of water table, duration of the crop, evaporative demand of the growing season etc. It is estimated that 150-200 mm of water is needed for nursery preparation, 250-400 mm to raise seedlings for 30-40 days, about 300 mm for land preparation which includes soaking, seepage, percolation losses and surface evaporation (Bouman *et al.*, 2007).

2.2. Water resources in general, and water demands in India

Water is the most important natural resource in this planet. In the world, total volume of water resource is 1.4 x 10^9 billion cubic metre (BCM) approximately. Out of which 1.365 x 10^9 BCM i.e. 97.5% of the total resources lies in oceans, and only 0.035 x 10^9 BCM i.e. 2.5% is the fresh water, approximate sources of which are glaciers and polar ice (24 x 10^6 BCM), groundwater (10.78 x 10^6 BCM), and lakes and rivers (0.105 x 10^6 BCM) (CWC, 2021). In India, total annual average water resource is 1999.2 BCM, out of which utilizable water is 1122 BCM (estimated 690 BCM from surface and 432 BCM from total groundwater resources) due to hydrological characteristics and topographical constraints (CWC, 2021). Although India has the plenty of surface water annually, there is mounting pressure on its land and water resources because of growing population, increased food demand and increased demand for fresh water over years and in the future. India accounts for only about 2.4% of the world's geographical area and 4% of world's renewable water resources, but the country has to support about 18% of the world's human population and 15% of livestock.

The irrigation for agriculture uses the largest volume of water for producing food and other agricultural commodities in India. Other sectors are domestic, industry, and energy for water use. Irrigation is the prime necessity for producing food, feed and fibre for the growing population in the country. The share of water for irrigation is currently being under threat due to the increasing competition with other sectors for the existing water resources due to rapid urbanization. As India plays a pivotal role in global food and water supply, water use for irrigation cannot be unjustified. Moreover, the changing pattern in food consumption has been playing a considerable impact on demand for food and water. Consequently, water demands were projected at the global (IWMI, 2000; Rijsberman, 2000; Rosegrant *et al.*, 2002; Seckler *et al.*, 1998) as well as national level (Bhalla *et al.,* 1999; Dyson and Hanchate, 2000; GoI, 1999). The projection shows a higher rate of increase in the demand for industry and energy sectors, which pushes down the rate of increase in the share for irrigation (Table 2.1). It clearly indicates that there will be the scarcity of water for irrigation purposes in the future. More food will required to be produced from less water.

2.3. Rainfall: the primary source of water

Rainfall is the primary source of all surface water in the country, and it is the key source to water flow in the river basins. India is fortunate to receive quite a high amount of rainfall in many areas because of favourable monsoons and physiographic features.

2.3.1 Spatial variability of rainfall over Indian states

The normal rainfall (average of 50-years, January to December, 1950-2000) in India is 1661 mm, which is received over all states and union territories (IMD, 2020). The highest rainfall occurs in Meghalaya, and the lowest in Rajasthan. The states and union territories receiving annual rainfall above 2000 mm are Meghalaya, Goa, Sikkim, Arunachal Pradesh, Andaman & Nicobar Islands, Kerala, Mizoram, Tripura, Dadra and Nagar Haveli, Assam and Manipur; less than 1000 mm are Rajasthan, Haryana, Punjab, Delhi, Gujarat, Tamil Nadu, Andhra Pradesh, Telangana; the remaining other states and union territories receive rainfall between 1000 and 2000 mm (Fig. 2.1).

India accounts for 33% of the net area sown in rainfall areas receiving less than 750 mm annual rainfall; 35% of the net area sown is under rainfall zone of 1125-2000 mm and the high rainfall area covers only 8% of the net sown area. Annual rainfall, including snowfall amounts to 4000 BCM. On the account of area and agricultural significance, 68% of the sown area is subjected to drought in varying degrees (35% of the area receives rainfall of 750-1125 mm and is drought prone, 33% of the area receives rainfall of <750 mm and is chronically drought prone); 21% of the sown area receives rainfall of <750 mm which is located in peninsular India and Rajasthan. Drought prone area lies in arid region (19.6%), semi-arid (37%) and sub-humid (21%) areas that occupy 77.6% of its total land area of 329 million ha. There is considerable spatial as well as temporal variability. Some area are drought prone, on the other hand some area are frequently flood-affected due to excess rainfall, flash floods and/ or riverine floods.

Table 2.1. Projected demand of water for different sectors/ purposes in India

Sectors				**Water demand (BCM)**					
	Projection by MoWR Standing Sub-Committee			**Projection by National Commission on Integrated Water Resources Development (NCIWRD)**					
	Year 2010	**Year 2025**	**Year 2050**	**Year 2010**		**Year 2025**		**Year 2050**	
				Low	High	Low	High	Low	High
Irrigation	688	910	1072	543	557	561	611	628	807
Domestic	56	73	102	42	43	55	62	90	111
Industry	12	23	63	37	37	67	67	81	81
Energy	5	15	130	18	19	31	33	63	70
Others	52	72	80	54	54	70	70	111	111
Total	813	1093	1447	694	710	784	843	973	1180

BCM, billion cubic metre, MoWR Ministry of Water Resources, Govt. of India; Data source: GoI (1999)

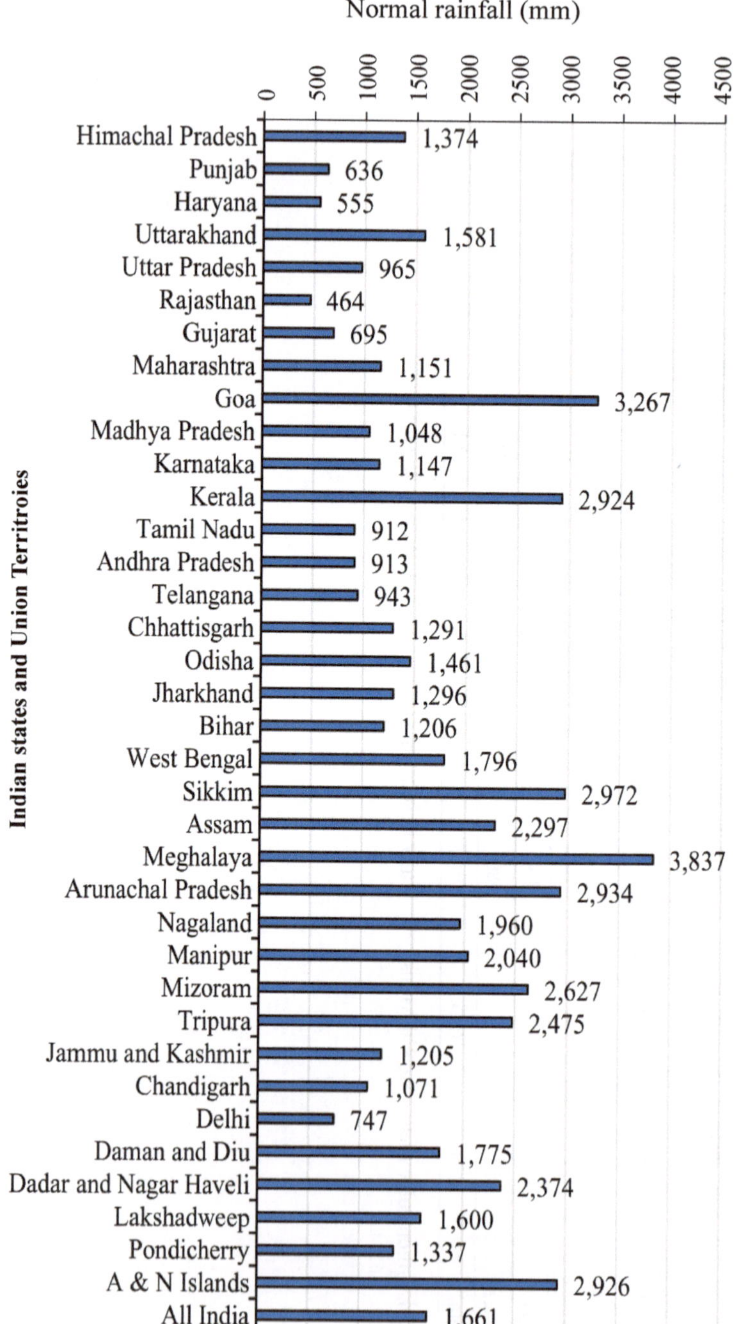

India: 50-years normal rainfall (Jan-Dec 1951-2000)

Fig. 2.1. Normal rainfall in different states and union territories of India (50-years: 1951-2000)

Data source: IMD (2020)

2.3.2 Temporal variability of rainfall over Indian states

There is temporal variability in rainfall distribution. Out of the total annual rainfall, about 75-80% of it occurs in about 100-120 days. Rainfall is monsoon dominated in India. South-west monsoon, north-east monsoon, cyclonic depressions and disturbances, and occasional cyclones contribute to rainfall in different degrees in various states or regions in the country. The rainfall occurs mostly during monsoon months i.e. June to September in the year. Of course there is also a considerable rainfall during pre-monsoon season (March to May), post-monsoon season (October to December), and winter season (January to February).

The Indian Meteorological Department (IMD), Pune have classified rainfall based on percent departure from long-period average (LPA). The IMD classification is mentioned in Table 2.2. The LPA data on rainfall (31 years: 1992-93 to 2022-23 up to December 2022) shows that there was more than 10% negative departure of monsoon rainfall from the normal in 5 years, and up to 10% negative departure occurred in 12 years in 31 years (Fig. 2.2a). The excess in monsoon rainfall i.e. positive departure also occurred in 12 years, of course within 10% positive departure. The negative or positive departure of rainfall from its LPA is more erratic during post-monsoon, winter season, and pre-monsoon season as evident from 31-years of data presented (Fig. 2.2b,c,d). There was more of the negative deviations than excess over normal rainfall. Overall, rainfall in India is erratic in about 4 out of 10 years.

Table 2.2. Rainfall classification based on percent departure from normal or long-period average

Rainfall terminology	**Percent departure of received rainfall from normal or long-period average (LPA)**
a) Weekly or seasonal rainfall distribution	
Excess	+20% or more departure from normal
Normal	-19% to +19% departure from normal
Deficient	-20% to -59% departure from normal
Scanty	-60% to -99% departure from normal
No rain	-100% departure from normal
b) Distribution of rainfall on all India scale	
Normal	within ±10% departure from LPA
Below normal	<10% departure from LPA
Above normal	>10% departure from LPA
Drought year- all India	>10% rainfall deficiency, and 20 to 40% spatial drought coverage
Severe drought year- all India	>10% rainfall deficiency, and >40% spatial drought coverage

Source: IMD (2020)

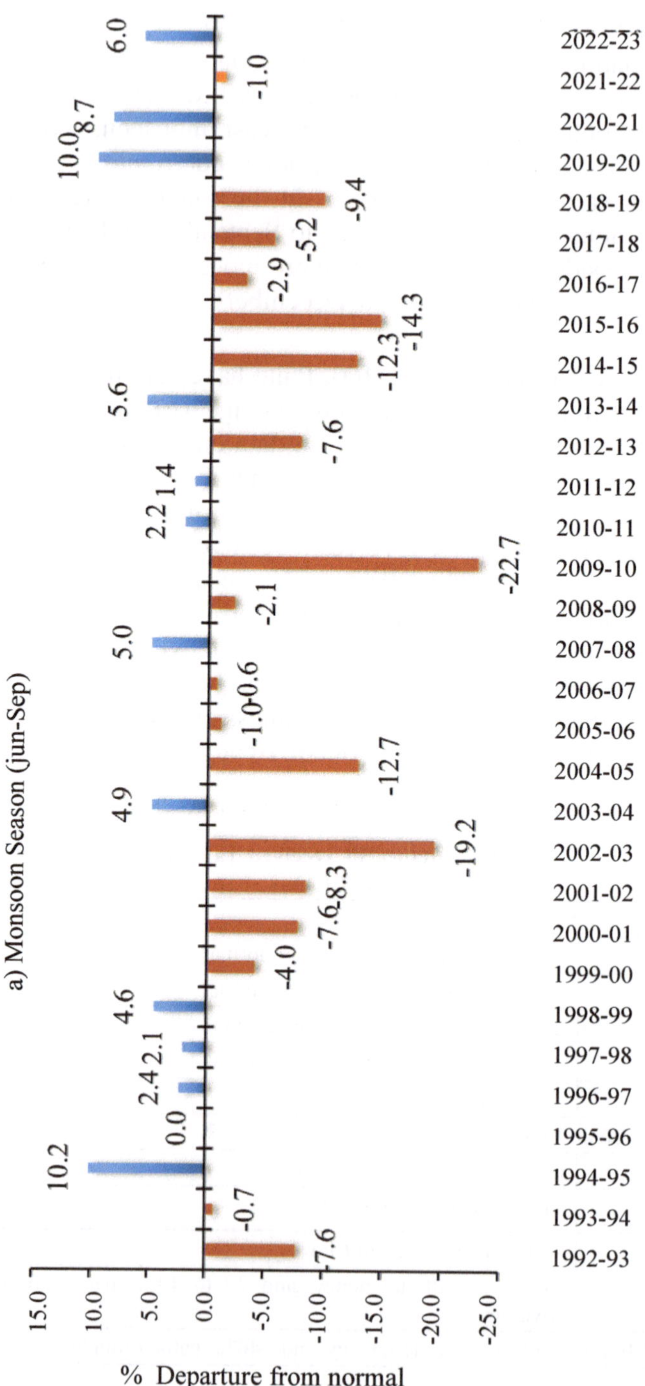

Fig. 2.2a. Departure of monsoon season rainfall from normal

Data source: DES (2023)

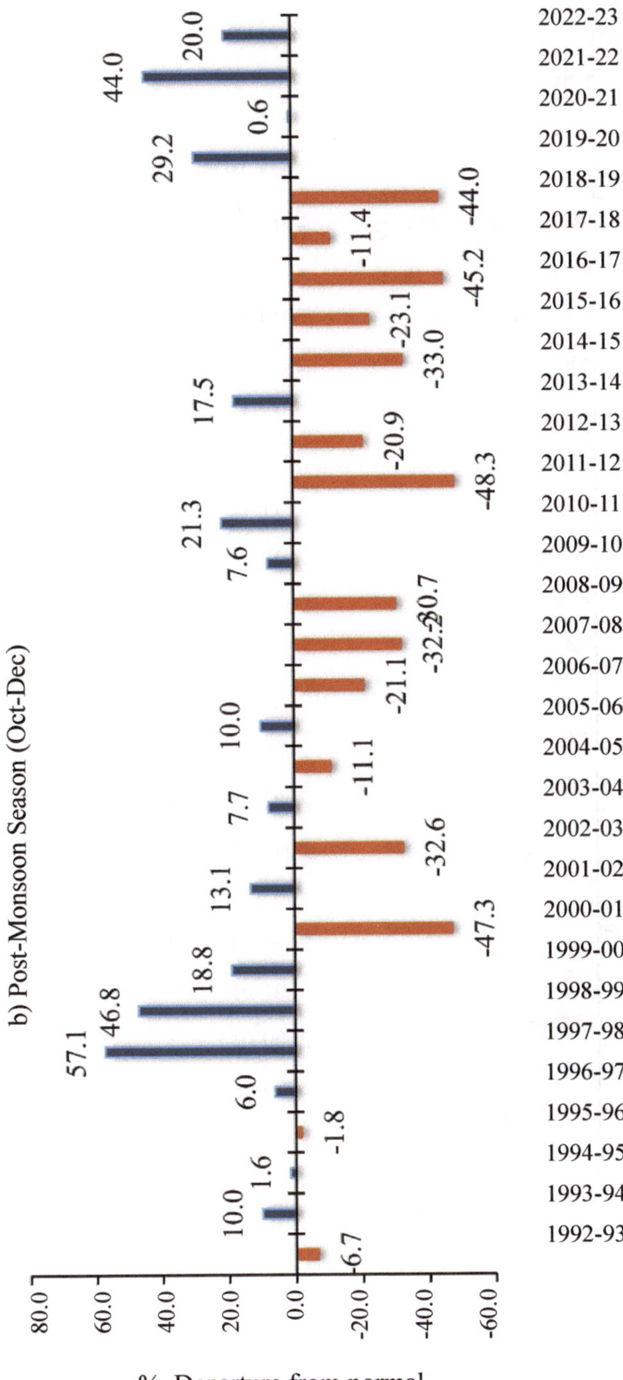

Fig. 2.2b. Departure of post-monsoon season rainfall from normal

Data source: DES (2023)

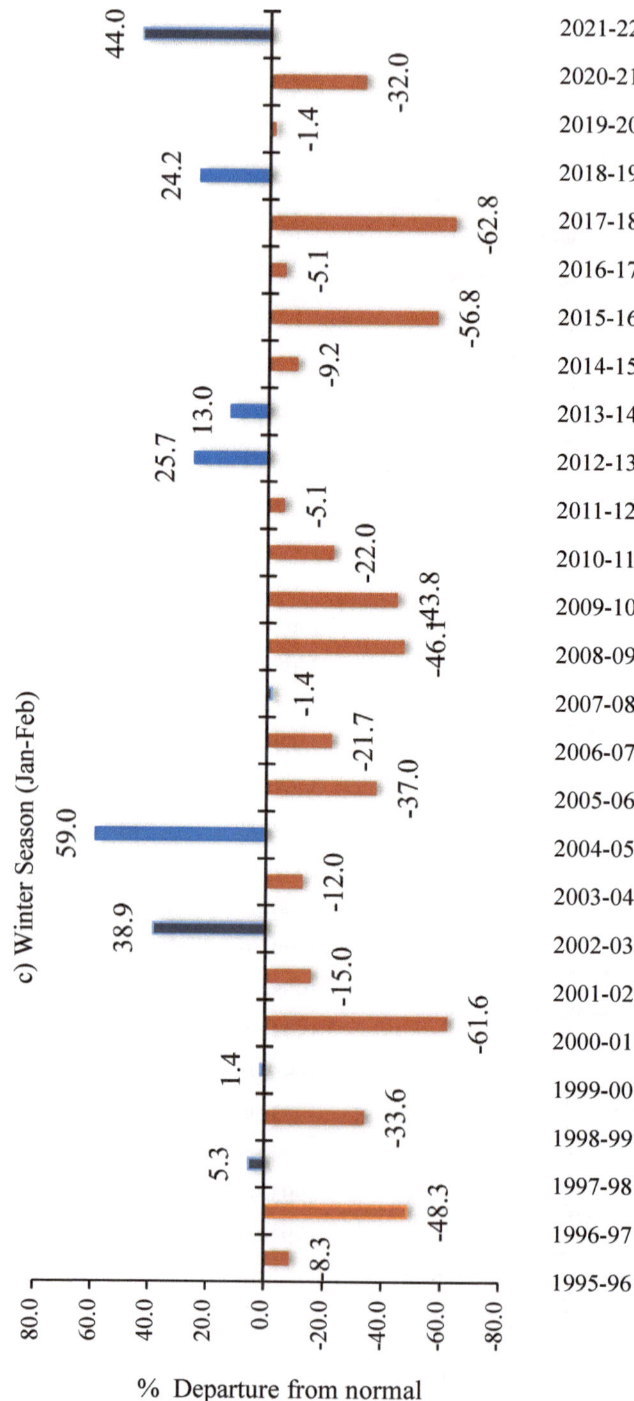

Fig. 2.2c. Departure of winter season rainfall from normal

Data source: DES (2023)

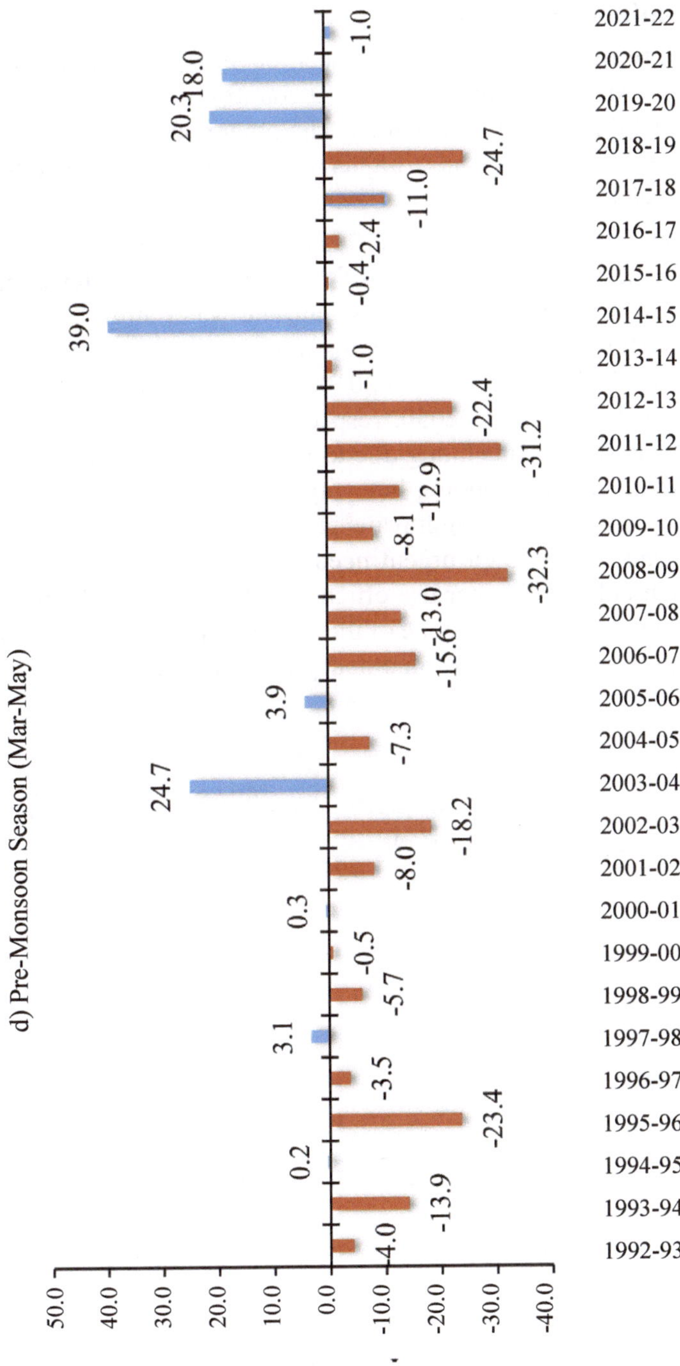

Fig. 2.2d. Departure of post-monsoon season rainfall from normal

Data source: DES (2023)

2.4. Per capita availability of water

Per capita availability of water per year has been declining in almost every country in the world. Among major rice-growing countries in Asia, Indonesia has the greater availability of water per capita. The other countries *viz.* Nepal, Philippines, Japan, Pakistan, China have higher water available per head than India and South Korea (Fig. 2.3).

According to Falkenmark and Widstrand (1992) a country is categorised as 'water stress' if the per capita availability of renewable fresh water comes down to less than 1700 m^3 in a year, and as 'chronic water stress', when it further drops below 1000 m^3. In India the availability is declining from 5177 m^3 in 1951 to 1820 m^3 in 2001, 1588 m^3 in 2010, and it is expected to decline to 1341 and 1140 m^3 by 2025 and 2050, respectively (Fig. 2.4). Per capita availability of water is steadily declining due to increase in population, rapid industrialization, urbanization, cropping intensity and declining groundwater table. The problems are likely to aggravate in future if suitable measures are not adopted. Hence, there is an urgent need for efficient water resource management through enhanced water use efficiency (Singh *et al.*, 2010) and waste water recycling. As irrigation is one of the critical inputs for agriculture, its use should be judicious to reduce blue and green water footprints (Rost *et al.*, 2008).

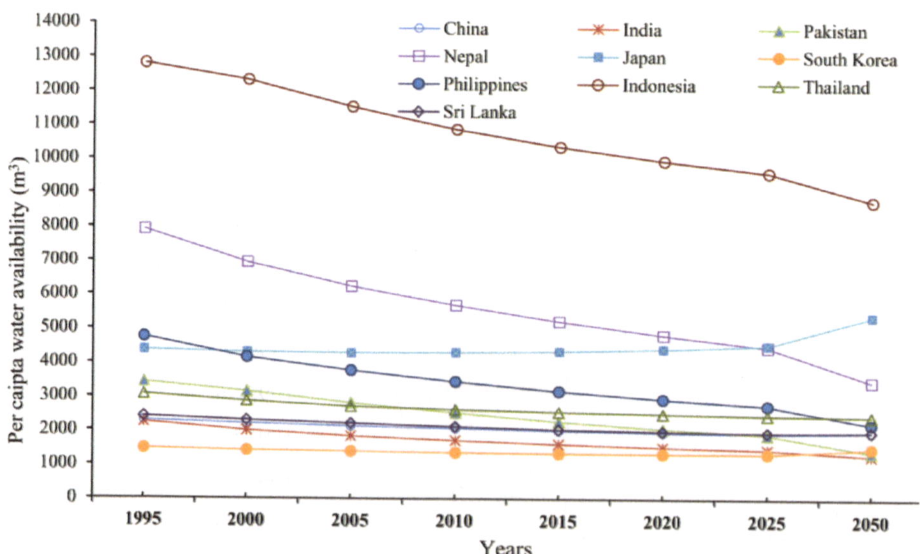

Fig. 2.3. Per capita availability of water in Asian major rice-growing countries; *Data source:* Bhatt *et al.* (2016), modified by Bhatt *et al.* (2016) from Gardner-Outlaw and Engelman (1997)

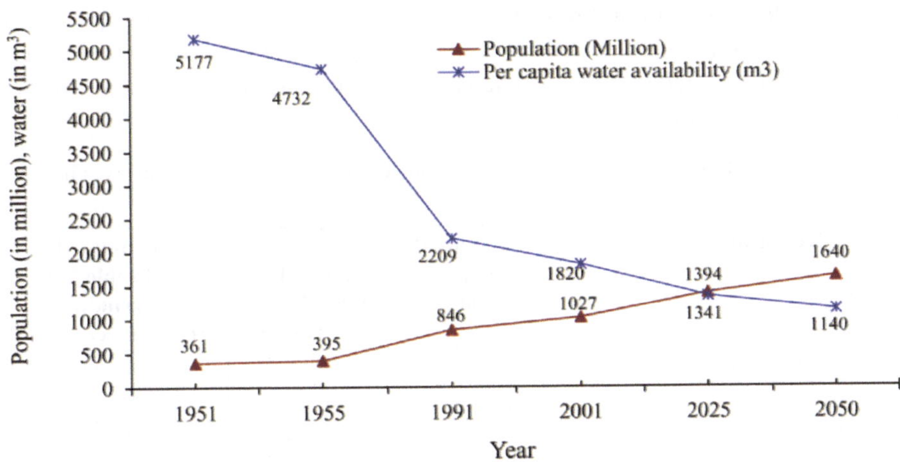

Fig. 2.4. Population increase and decrease in per capita availability of water in India
Source: CWC (2013), MoWR (2015)

2.5. Surface water resources in river basins of India

India is endowed with a large number of major, medium and small size rivers. River water constitutes a large volume of primary surface water resources. There are broadly four groups of river systems:

i) Himalayan rivers,
ii) Deccan rivers,
iii) Coastal rivers, and
iv) Rivers of the inland drainage basins.

River basin is land area where all rainwater, snow or ice-melted water converge to a single drainage point at a lower elevation i.e. the exit of a basin, and the water joins into a river.

There are twelve major river basins and eight combined or composite basins. The Himalayan Rivers have a continuous flow almost throughout the year, which derives water from rainfall-runoff, and melting of glaciers, snow or ice. These Himalayan Rivers are perennial and swell during the monsoon season with the receipt of heavy rainfall and frequent floods. The Deccan rivers are rain-dependent, and thus the volume of water flow varies. Many of them are not perennial. The coastal rivers, especially the west coast are short in length and have small catchment areas. Most of them are also not perennial. The rivers draining the west coast have large velocities of flow due to the steep slopes. They carry large volumes of water due to the high monsoon rains. The east coast flowing rivers are longer and have formed deltas which have extensive area and are suitable for intensive cultivation with appropriate soil and water management practices. These deltas constitute the principal rice bowls of India (Michael, 2011). The streams of inland drainage basins are few and located in western Rajasthan.

Table 2.3. Annual potential of surface water resources and extractable flow in river basins of India

Sl. No.	River basins	Catchment area (km^2)	Average annual potential in the river (BCM)	Estimated utilizable flow (excluding groundwater) (BCM)
1	Indus (upto border)	31,7708	45.53	46.00
2	a) Ganga	8,38,803	509.52	250.00
	b) Brahmaputra	1,93,252	527.28	24.00
	c) Barak and others	86,335	86.67	
3	Godavari	3,12,150	117.74	76.30
4	Krishna	2,59,439	89.04	58.00
5	Cauvery	85,167	27.67	19.00
6	Pennar	54,905	11.02	6.86
7	East flowing rivers between Mahanadi & Pennar	82,073	26.41	13.11
8	East flowing rivers between Pennar and Kanyakumari	1,01,657	26.74	16.50
9	Mahanadi	1,44,905	73.00	49.99
10	Brahmani & Baitarani	53,902	35.65	18.30
11	Subarnarekha	26,804	15.05	6.81
12	Sabarmati	31,901	12.96	1.93
13	Mahi	39,566	14.96	3.10
14	West flowing rivers of Kutch, Saurashtra including Luni	1,92,112	26.93	14.98
15	Narmada	96,660	58.21	34.50
16	Tapi	65,806	26.24	14.50
17	West flowing rivers from Tapi to Tadri	58,360	118.35	11.94
18	West flowing rivers from Tadri to Kanyakumari	54,231	119.06	24.27
19	Area of Island drainage in Rajasthan Desert	1,44,836	-	-
20	Minor river basins drainage to Bangladesh & Myanmar	31,382	31.17	-
	Total	32,71,953	1999.20	690.09

Source: CWC (2021)

The Central Water Commission (CWC), Ministry of Jal Shakti, previously Ministry of Water Resources, Government of India has listed 20 river major basins (Table 2.3). Of the total annual river water resources potential of 1999.2 BCM, estimated utilizable flow (excluding groundwater) is 690.1 BCM. The Ganga-Brahmaputra-Barak composite basin is the largest one with its total annual potential of about 1123.5 BCM, and estimated utilizable flow is 274 BCM (CWC, 2021). The second largest basin is that of Godavari which carries a potential of 117.74 BCM. The other major river basins are Indus, Krishna, Mahanadi and west flowing rivers from Tadri to Kanyakumari, west flowing rivers from Tapi to Tadri. On the basis of catchments, the river basins could be divided into three categories:

i) Large/ major river basins with 20,000 km^2 and above catchment areas,
ii) Medium river basins having 2,000 to 20,000 km^2 catchment areas, and
iii) Minor river basins having less than 2,000 km^2 catchment areas.

In addition to the major river basins, there are several medium river basins which are of east-flowing rivers, west flowing rivers, and river basins flowing into neighbouring countries like Bangladesh and Myanmar in the north eastern region. There is considerable scope for increasing the utilization of water in the Ganga-Brahmaputra basins by construction of storages at suitable locations in neighbouring countries. There is a vast difference between potential water resource and utilizable flow due to topographical and geological limitations. Other than rivers and canals, water bodies' *viz.* reservoirs, tanks, ponds, lakes etc. cover an area of about 7.3 million ha, and out of those, maximum area of 2.93 million ha is under reservoirs, 2.43 million ha is under tanks, ponds, lakes etc. (MoWR, 2015). However, total inland water resources is not evenly distributed in the states.

2.6. Groundwater resources in India

Groundwater is an important and worthy resource in India. It is used for drinking, other domestic purposes and irrigation in agricultural sector. It is a preferred source of water as its extraction is mostly under the direct control of the users. Groundwater resource can be categorised in to two components- static and dynamic. The aquifer below the zone of fluctuated water level holds the static groundwater, and its amount has been estimated as 10,812 BCM in India (Michael, 2011). Generally, this component of groundwater is not replenished annually. The dynamic component, which is recharged annually is considered the groundwater resource, its amount is about 436.15 BCM in India (CGWB, 2022). After deduction of the natural discharge from total recharged amount, extractable groundwater is 397.62 BCM annually (Table 2.4). Currently, total groundwater extraction is ~244.92 BCM for irrigation, industry and domestic purposes combined; for irrigation only total groundwater extraction is 217.61 BCM annually. The extraction percentage (as the percent of its extractable volume) i.e. annual groundwater withdrawal is 61.60% for all purposes, and for irrigation only the extraction is 54.73% in India (CGWB, 2022). The use of groundwater for irrigation has increased at a rapid rate, and it has become the backbone of agricultural security as well as drinking purposes since last 25 years or so (Vijay Shankar *et al.*, 2011). The country is the biggest groundwater users (Shah, 2009). About 85% in rural water and ~45% in urban water supply come from groundwater contribution. Overall, it is required to manage this precious groundwater resource efficiently for sustainable development (Saha and Ray, 2019).

Table 2.4. State-wise and all India groundwater resources, year 2020

States/ UTs	Total Annual ground water recharge (BCM)	Total natural discharge (BCM)	Annual extractable ground water resources (BCM)	Current total ground water extraction[$] (BCM)	Current ground water extraction for irrigation (BCM)	Net ground water availability for future use (BCM)	Ground water extraction (%)	Ground water extraction for irrigation (%)
Andhra Pradesh	24.15	1.21	22.94	7.63	6.60	15.91	33.26	28.77
Arunachal Pradesh	3.19	0.27	2.92	0.01	0.003	2.90	0.34	0.10
Assam	27.05	5.09	21.96	2.58	1.97	19.33	11.75	8.97
Bihar	28.05	2.60	25.45	13.02	10.33	12.23	51.16	40.59
Chhattisgarh	12.65	1.11	11.54	5.35	4.53	6.25	46.36	39.25
Delhi	0.32	0.03	0.29	0.29	0.07	0.02	100.00	24.14
Goa	0.40	0.08	0.32	0.08	0.02	0.24	25.00	6.25
Gujarat	26.81	1.90	24.91	13.30	12.65	12.52	53.39	50.78
Haryana	9.53	0.90	8.63	11.61	10.47	0.97	134.53	121.32
Himachal Pradesh	1.07	0.10	0.97	0.36	0.20	0.62	37.11	20.62
Jammu & Kashmir	4.68	0.46	4.22	0.89	0.20	3.32	21.09	4.74
Jharkhand	6.15	0.51	5.64	1.64	0.93	4.02	29.08	16.49
Karnataka	18.16	1.76	16.40	10.63	9.60	7.08	64.82	58.54
Kerala	5.65	0.53	5.12	2.65	1.16	2.13	51.76	22.66
Madhya Pradesh	36.16	2.78	33.38	18.97	17.33	15.25	56.83	51.92
Maharashtra	32.01	1.76	30.25	16.63	15.29	14.20	54.98	50.55
Manipur	0.51	0.05	0.46	0.02	0.003	0.44	4.35	0.65
Meghalaya	2.04	0.22	1.82	0.08	0.03	1.73	4.40	1.65
Mizoram	0.22	0.02	0.20	0.01	0.00	0.19	5.00	0.00

Nagaland	2.17	0.22	1.95	0.02	0.002	1.93	1.03	0.10
Odisha	17.08	1.37	15.71	6.86	5.50	8.74	43.67	35.01
Punjab	22.80	2.20	20.60	33.85	32.80	1.61	164.32	159.22
Rajasthan	12.24	1.17	11.07	16.63	14.37	0.99	150.23	129.81
Sikkim	0.96	0.10	0.86	0.01	0.00	0.85	1.16	0.00
Tamil Nadu	19.59	1.90	17.69	14.67	13.52	5.65	82.93	76.43
Telangana	16.63	1.60	15.03	8.01	7.13	7.14	53.29	47.44
Tripura	1.47	0.22	1.25	0.10	0.02	1.14	8.00	1.60
Uttar Pradesh	72.20	5.32	66.88	46.03	41.29	21.53	68.82	61.74
Uttarakhand	2.02	0.17	1.85	0.87	0.63	0.98	47.03	34.05
West Bengal	29.33	2.77	26.56	11.84	10.84	14.19	44.58	40.81
All States	436.15	38.51	397.62	244.92	217.61	184.56	61.60	54.73

[$]Current total groundwater extraction means groundwater use for irrigation, industry and domestic purposes

Source: CGWB (2022)

The hydrology in the country is a bit complicated because of differential geological formations, climatology, lithological and tectonic framework. The Central Ground Water Board broadly classified groundwater regions into two categories based on rock formations and geology (CGWB, 2012): i) porous, and ii) fissured formations. There are two sub-categories under porous formations- unconsolidated and semi-consolidated formations. The areas with quaternary rocks comprising old and new alluviums, and the coastal alluvium constitute the unconsolidated formations. Those are found in the Indo-Gangetic plains, Brahmaputra basins, inter-montane valleys of Kashmir, Himachal Pradesh, intra-crotonic basins *viz.* Narmada, Tapti, aolian sands in Rajasthan and Gujarat, and the coastal plains and deltaic tracks. The water yield is substantial in the areas of unconsolidated formations. These areas are in Punjab, Haryana, Uttar Pradesh, Bihar, West Bengal and Assam. These are the most important and potential groundwater resources in a large scale and for extensive development because aquifers have ample scope to get recharged. The semi-consolidated formations are composed primarily of sandstone, limestone, shale which may be found in the north-western regions. The sandstones of Gondwanas, Tipams, Lathis and Cuddalore are the most productive aquifers under this category. The water yield is moderate in those areas. Groundwater is under active exploitation through shallow aquifers from some important pockets along the eastern coast, in the deltas of Godavari, Krishna and Cauvery rivers.

The consolidated or the fissured formations are based on lithological units of sedimentary, crystalline, meta-sedimentary, igneous and metamorphic and carbonate rocks. This hard rock occupies almost two-thirds of the area in India. Hard rock formations lie south of the Indo-Gangetic plains and spread in the states of Rajasthan, Madhya Pradesh, Bihar, Gujarat, Maharashtra, Andhra Pradesh, Odisha, Karnataka, Tamil Nadu, Kerala. These formations have negligible porosity, except in some vesicular volcanic rocks. Groundwater is pumped through open wells. The water yield is low in those areas, and the supply of water from these sources is limited.

Groundwater resources vary widely in different regions or states in the country with respect to its total annual recharge, extractable volume, and current extraction for all purposes and for irrigation (Table 2.4). In some states, extraction exceeds the annual extractable volume, and in other states, there is ample scope to extract further than the current level. Accordingly there have been regional imbalances in groundwater depletion in the country. Excessive extraction i.e. over-exploitation (which is called *mining*) of groundwater has occurred in northern Indian states *viz.* Rajasthan, Punjab and Haryana (Delhi included). The water table has gone down over several years of such groundwater mining. During the first decade of this century, CGWB (2006) reported faster withdrawal of groundwater exceeding the rate of replenishment in north-west India.

Using observations of changes in terrestrial water storage (TWS) from the GRACE satellites of NASA (Tapley *et al.*, 2004) and hydrological modelling techniques, Rodell *et al.* (2009) in their legendary publication in 'Nature', have shown that the groundwater in Punjab, Rajasthan and Haryana (including Delhi) had depleted at an average rate of 4.0±1.0 cm per year, which was equivalent to a volume of 17.7±4.5 BCM per year, and the total depletion was ~109 BCM during the period August 2002 to October 2008. Using similar method, another estimate showed that the loss of groundwater was at the rate of 54 ± 9 BCM per year between April 2002 and June 2008 from the same region of India (Tiwari *et al.*, 2009). This over-exploitation caused a significant lowering of depth of water table (Hoque *et al.*, 2007, CGWB, 2006), this serious situation for the north India was well-recognised (Barnett *et al.*, 2005; Kumar *et al.*, 2005; Amarasinghe *et al.*, 2007).

With subsidized energy, farmers have few incentives to limit pumping. With rising energy prices groundwater pumping may become unaffordable, leaving smallholders dry and hungry, especially if subsidies are discontinued. Even the depletion of groundwater aquifer and reduced stream flows affect drinking water supplies, health and rural livelihoods. Regional variability exits in the degree of groundwater extraction: Punjab, Haryana, Rajasthan, parts of Gujarat, Tamil Nadu and Karnataka are over exploited; Himachal Pradesh, Maharashtra, Uttar Pradesh, Uttarakhand, Madhya Pradesh, Telangana have groundwater extraction more than 50%; eastern Indian states *viz.* Assam, Bihar, Chhattisgarh, Odisha and West Bengal have extraction between 25-50%, north-eastern states have very low extraction than the country average of 61.60% (CGWB, 2022).

2.7. Irrigation potential and utilization

Several irrigation projects have been developed and in operation in India since many years. According to the latest data sources from Central Water Commission, the ultimate irrigation potential (UIP) stands at 139.89 million ha (CWC, 2019; 2021); out of this 58.47 million ha under major & medium, 81.428 million ha under minor irrigation projects (Table 2.5). In the minor irrigation projects, 78% of the total ultimate potential is contributed from groundwater i.e. minor-lift irrigation. Up to XI Plan period (2007-2012) of India, the irrigation potential created (IPC) was 113.529 million ha (47.972 million ha under major & medium, 65.557 million ha under minor projects). The total irrigation potential utilized (IPU) was 87.92 million ha up to XI Plan period, whereas gross irrigated area was 96.62 million ha up to the year 2015-16. Therefore there was a gap of about 26.36 million ha between UIP and IPC, and about 25.61 million ha between IPC and IPU up to XI Plan period. There exists a wide variation of UIP, IPC and IPU among different states in India. Uttar Pradesh and Bihar have the most potential in major and medium projects. The states *viz.* Uttar Pradesh, Bihar, Madhya Pradesh, Maharashtra, Punjab, Haryana, Gujarat, Andhra Pradesh and Odisha account for ~75% of the total UIP in major and medium projects. The most UIP in minor-lift irrigation exists in Uttar Pradesh. Madhya Pradesh and Andhra Pradesh have much higher potential in UIP minor-flow irrigation than the other states. Hence, it is a big challenge among development personnel and scientists who are engaged in water management in the country to bridge the gap through development of irrigation infrastructure and use of different existing agricultural technologies and as well as evolving of new technologies.

Table 2.5. Irrigation potential in India

Irrigation projects	Ultimate irrigation potential (UIP) (million ha)	Irrigation potential created (IPC)* (million ha)	Irrigation potential utilized (IPU)* (million ha)
Major and medium irrigation projects	58.465	47.972	35.01
Minor irrigation projects$	81.428	65.557	52.91
Total	139.893	113.529	87.92

*up to XI Plan (2007-2012); for latest values for the period 2012-13 to 2019-20

$includes surface water and groundwater i.e. minor-flow (17.34 million ha) and minor-lift (64.09 million ha)

Data source: CWC (2021)

2.7.1. Variability of source-wise irrigated area

In the field level, irrigation water is supplied through canals, tube wells, open-wells, tanks and other sources. The trends of source-wise irrigation show that canal irrigated area increased from 8.295 million ha in 1950-51 to 16.183 million ha by the year 2014-15, tube well-irrigated area from zero to 31.6066 million ha, other wells from 5.978 to 11.354 million ha, tank irrigated area decreased from 3.613 to 1.723 million ha and other-source irrigated area from 2.967 to 7.519 million ha during the same period (Fig. 2.5).

Canal irrigated area increased to its maximum of 17.791 million ha (1991-92), thereafter it has been either constant with an average being 16.36 million during 1992-93 to 2014-15, or slightly decreasing. There was a decreasing trend in tank irrigated area, from 4.26 million ha in 1965-66 to 1.723 million ha in 2014-15. Tube well irrigated area steadily and phenomenally increased from a mere 0.135 million ha in 1960-61 to 31.61 million ha by the year 2014-15. The area under other wells increased to the maximum of 12.61 million ha (1998-99), then decreased and remained almost constant with an average of 10.46 million ha (2003-04 to 2014-15).

During the decade of 1960s, the share of canals and other wells combined was 69%, and groundwater extraction through tube-well (6%) and other wells (27%) combined was 33% (Fig. 2.6 a,b). But, the scenario has changed through the years with respect to the use of groundwater through tube wells. During the period of 2010-11 to 2014-15, the percentage share of tube-wells has been increased to 46% (Fig. 2.6 b). Now the principal irrigation source in India is through tube wells and other wells (62%) followed by canal irrigation (26%). Tank irrigation is confined to a few pockets accounting less than 5%.

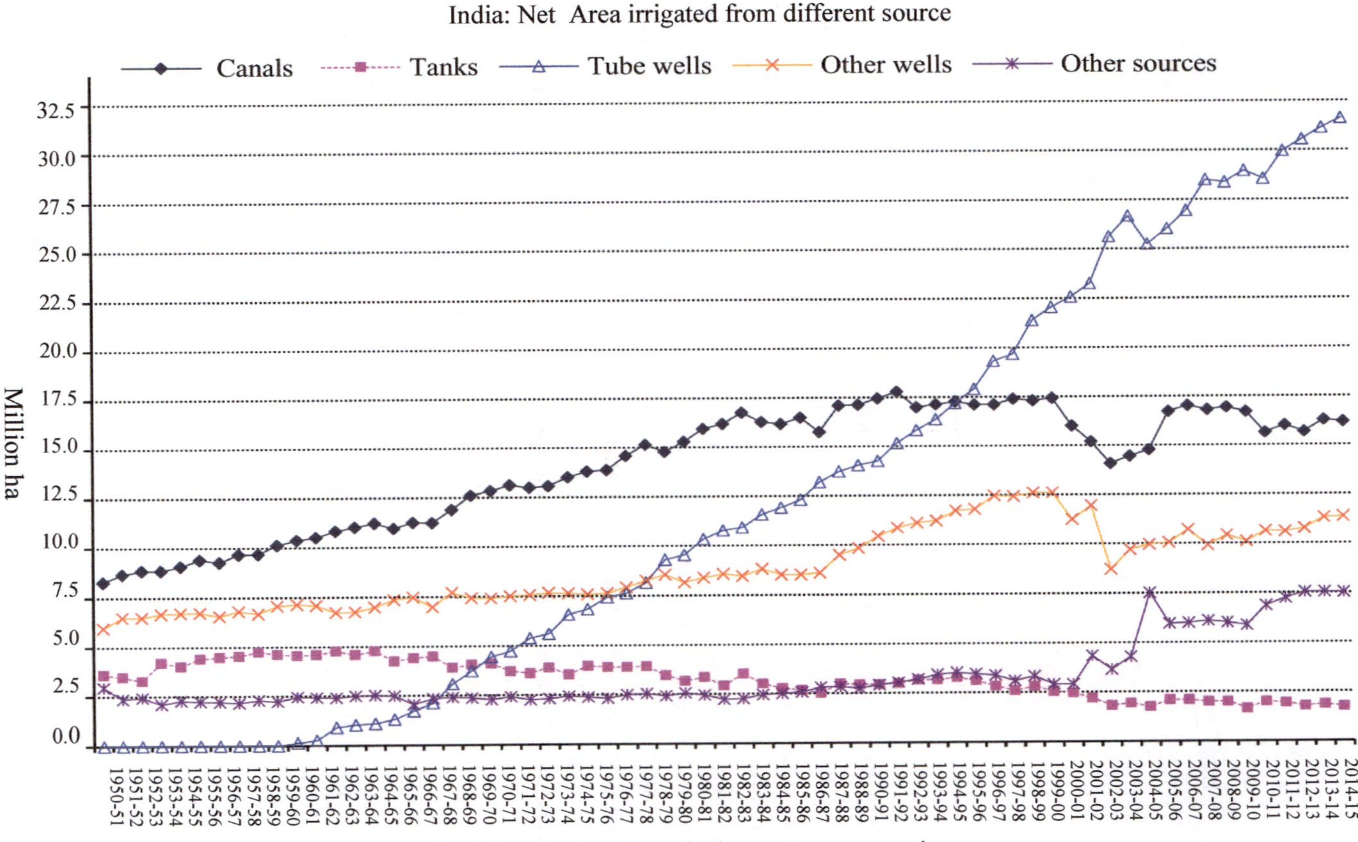

Fig. 2.5. Net irrigated area in the country- source wise

Data source: MSOPI (2018)

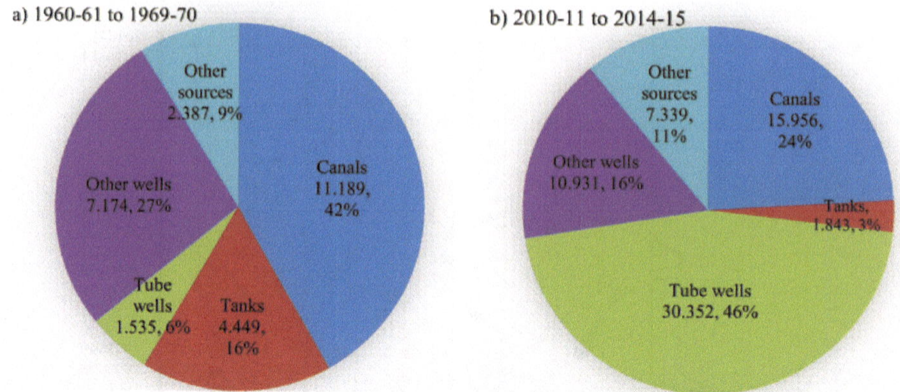

Fig. 2.6 a & b. Percentage share of different sources; *Data source:* MSOPI (2018)

2.7.2. Status of micro-irrigation in India

Currently micro-irrigation (drip and sprinkler) is adopted in about 14.49 million ha. The variation exists in different states (Table 2.6). It is very essential to adopt micro-irrigation due to increased scarcity of irrigation water in many states, and to increase the water use efficiency, and greater area coverage under irrigation. During the last about 10 years, there has been increased interest i.e. on 'More crop per drop'. These methods can bring degraded, undulated and uncultivable lands under cultivation through the use of micro-irrigation. For rice, micro-irrigation has been tried in several parts of the rice growing countries, but there are several concerns due to use of micro-irrigation *viz.* problem of weeds, diseases and soil moisture stress (Mandal *et al.*, 2019).

Table 2.6. State-wise area under micro-irrigation (as on 31.03.2022)

Sl. No	States of India	Drip irrigated area (ha)	Sprinkler irrigated area (ha)	Total (ha)
1	Andhra Pradesh	1400780	521165	1921945
2	Arunachal Pradesh	6245	4275	10520
3	Assam	7603	25671	33274
4	Bihar	15834	106995	122829
5	Chhattisgarh	35600	345634	381234
6	Goa	1386	1346	2732
7	Gujarat	921010	788672	1709682
8	Haryana	43868	639656	683524
9	Himachal Pradesh	7934	6403	14337
10	Jammu & Kashmir	1780	293	2073

11	Jharkhand	30056	18211	48267
12	Karnataka	824916	1604399	2429315
13	Kerala	24360	9289	33649
14	Madhya Pradesh	357328	301497	658825
15	Maharashtra	1414185	625598	2039783
16	Manipur	358	9933	10291
17	Meghalaya	308	307	615
18	Mizoram	5914	1808	7722
19	Nagaland	8040	7277	15317
20	Odisha	32354	142798	175152
21	Punjab	37682	16508	54190
22	Rajasthan	318742	1777985	2096727
23	Sikkim	7006	10518	17524
24	Tamil Nadu	867898	387723	1255621
25	Telangana	222603	91085	313688
26	Tripura	2304	3204	5508
27	Uttar Pradesh	57963	245138	303101
28	Uttarakhand	14941	11796	26737
29	West Bengal	10437	107259	117696
	Total	6679434	7812443	14491877

Data source: DES (2023)

2.7.3. Net and gross irrigated area in the country

The gross irrigated area in the country increased from 22.56 million ha in 1950-51 to 112.23 million ha in 2019-20 and the gross water demand by irrigation sector alone contributes >70% of the gross water demand by all users. It is also reported that net irrigated area in India is 75.46 million ha (2019-20) which accounts for about 53.9% of total net sown area (139.9 million ha) in 2019-20 (Fig. 2.7).

In spite of the fact that net sown area being constant at about 140 million ha for the last 6 decades, gross cropped area and gross irrigated area has been significantly increased during the same period. It was so because of extraction of groundwater to a greater extent through tubewells, consequently gross irrigated area has been increased to 112.23 million ha. Whereas, depletion of groundwater and limitation of surface water imply that not all of the net sown area is amenable to irrigation. Hence, we have to make critical analysis of challenges faced by us in agricultural water management in India and there is a strong need to focus on improved/ advanced water resource management techniques and supporting policies.

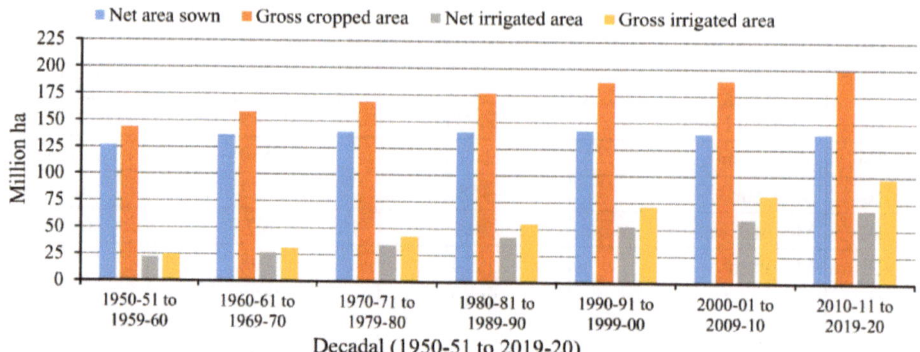

Fig. 2.7. Net irrigated area, gross irrigated area & net and gross cropped area

Data source: DES (2023)

2.7.4. State-wise irrigated area

The average irrigation coverage is 53.1% in India (2019-20). The highest percentage is in Punjab (98.6) and the lowest in Sikkim (8.8%). In north Indian states, irrigation is comparatively higher in Punjab, Haryana and Uttar Pradesh; Bihar and West Bengal in eastern India; 8 states are above the national average (53.1%), irrigation coverage is below the national average in remaining other states (Fig. 2.8).

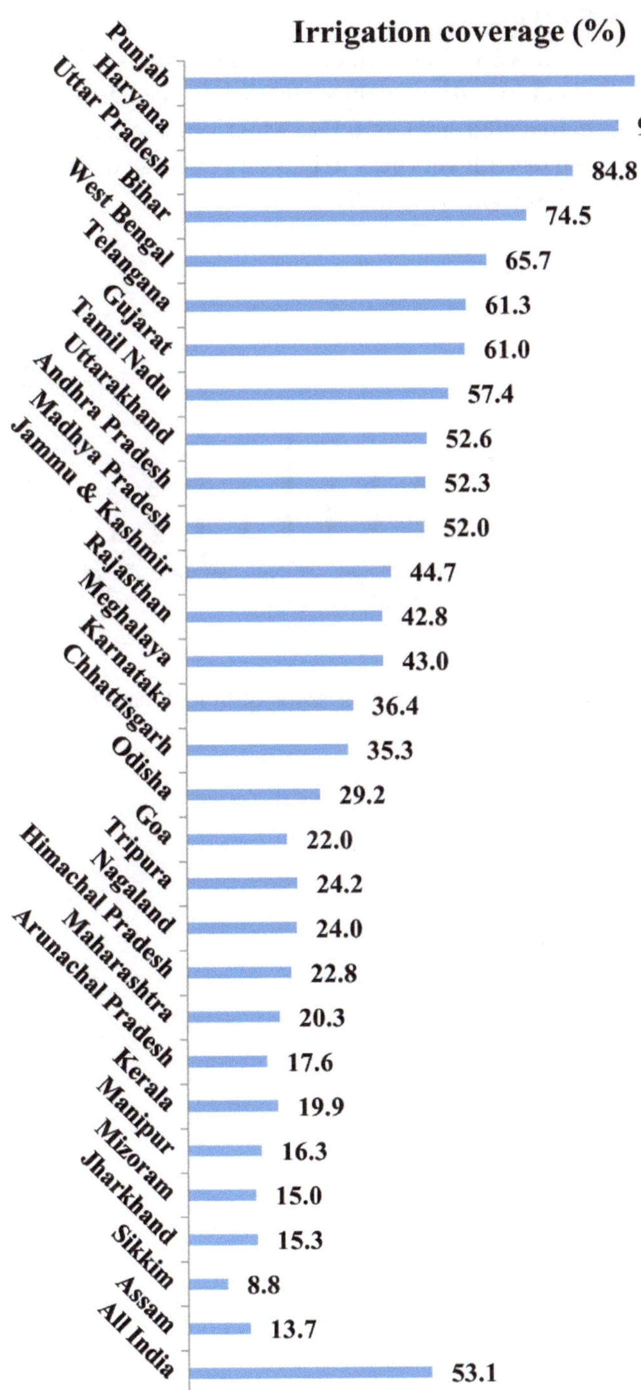

Fig. 2.8. State-wise irrigation coverage (%) in major crops of India (year 2019-20)

Data Source: DES (2023)

2.7.5. Crop-wise irrigated area

The average percentage covered is 53.1% for all crops under irrigation in India (2019-20). Sugarcane and wheat irrigation is above 90%; rapeseed & mustard and barley is 80-90%. Rice irrigated area is 65%. Most other oilseeds and pulses are grown either rainfed or with very less irrigation coverage in the rice-based systems or during *kharif* seasons (Fig. 2.9).

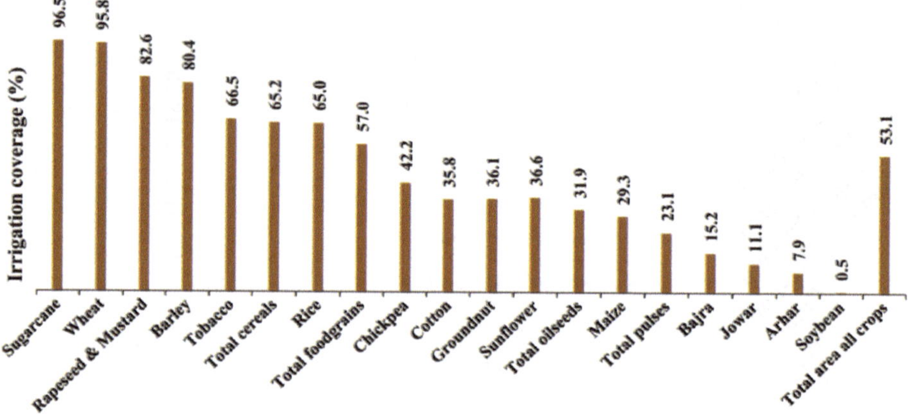

Fig. 2.9. Crop-wise irrigation coverage (%) in major crops of India (year 2019-20)
Data source: DES (2023)

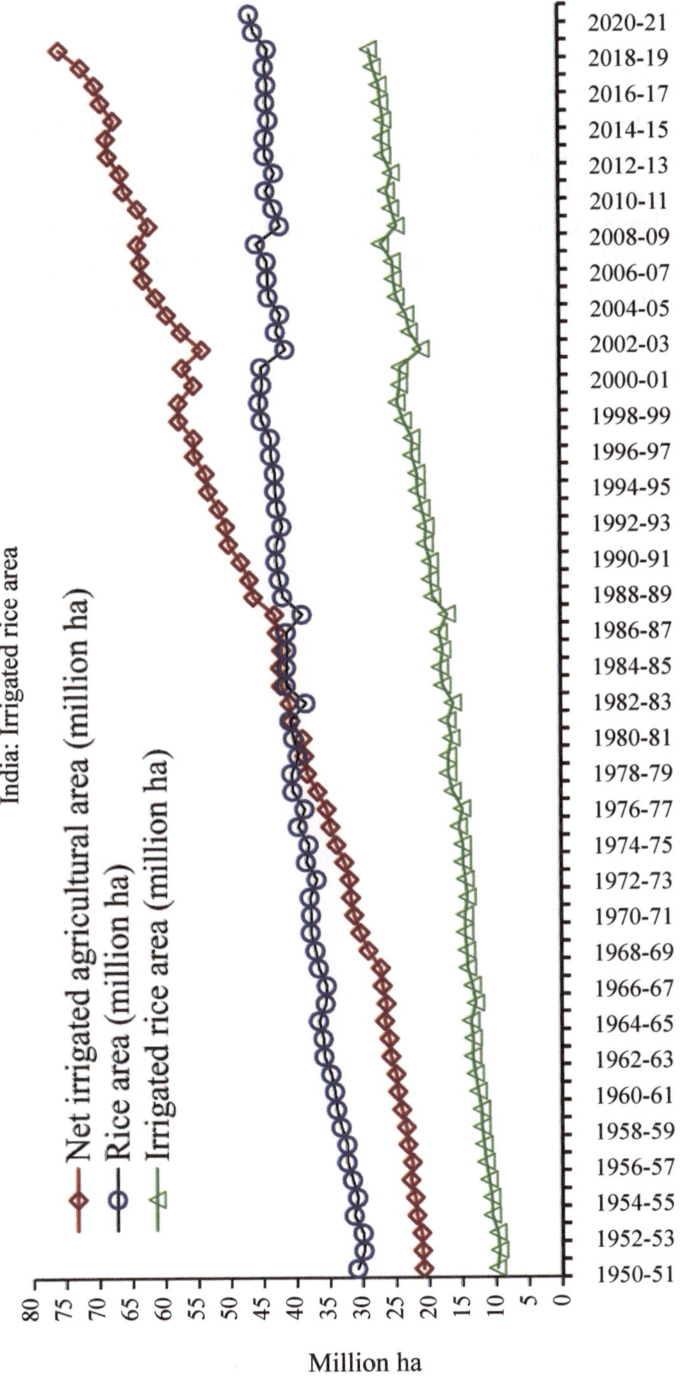

Fig. 2.10. Trends of net irrigated area, rice area and rice irrigated area in India (1950-51 to 2021-22)
Data source: DES (2023)

2.7.6. Rice irrigated area

The average rice area is about 42.9 million with a range of 40-46 million for the last 45 years (1977-78 to 2021-22). Out of this about 65% rice area is irrigated currently (2021-22), whereas net irrigated area is 75.46 million ha for all crops (2019-20). There is an increasing trend of net irrigated area and rice irrigated area and irrigated foodgrains area in the country (Fig. 2.10, 2.11). As most of the area is already irrigated (>95%), the trend is not increasing further in wheat. Five-year (2015-16 to 2019-20) average rice irrigation percentage varied widely in different states (Fig. 2.12). Rice producing states *viz.* West Bengal and Odisha is below the national average.

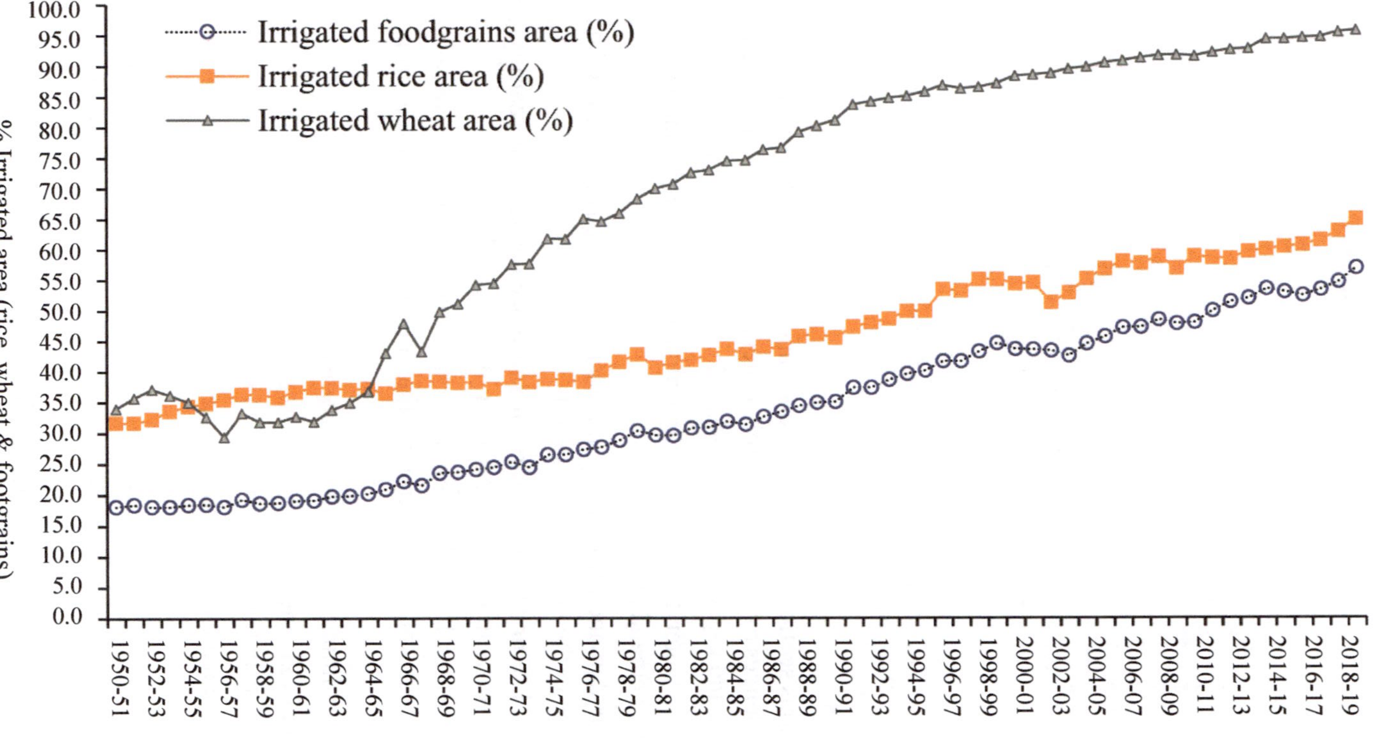

Fig. 2.11. Trends of irrigated total food grains area, rice and wheat irrigated area (%) in India

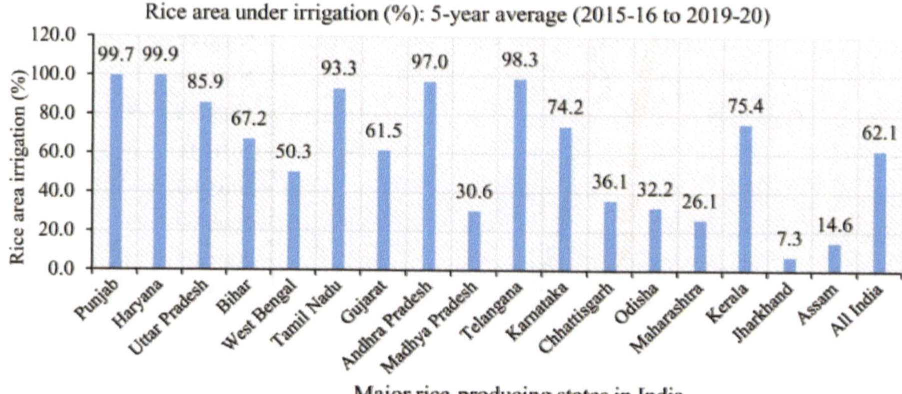

Fig. 2.12. State-wise rice area under irrigation (%): 5-year average (2015-16 to 2019-20)
Data source: DES (2019, 2020, 20221, 2022, 2023)

2.8. Challenges

From the previous sections having the trends and current water scenario, there are several challenges towards sustainable water management in the agriculture irrigation sector and especially for rice-based cropping systems.

i) Scarcity of water for irrigation and increasing water demand, and the gap in the demand and supply
ii) Increase in demand for industry and energy sector pushing down the rate of increase in the share for irrigation
iii) Spatial and temporal variability of rainfall causing both scarcity and excess of water in Indian states and union territories
iv) Increased water demand during post-rainy season crops to increase gross irrigated area for agricultural production in the rice-based cropping systems
v) Steadily declining per capita availability of water
vi) Siltation of existing dams and reservoirs causing decreased capacity of storage of surface water resources in river basins
vii) Vast difference exists between potential water resource and utilizable flow
viii) Increased rate of groundwater utilization; regional disparity in groundwater extraction for agricultural purposes, exceeding the annual replenishment
ix) In-sufficient recharge of aquifers in hard-rock areas
x) There is a gap between the irrigation potential created and its utilization

xi) Variability in source-wise irrigated area; with the surface source canal irrigated area remained static and increased extraction of groundwater

xii) Low irrigation efficiency and crop water use efficiency

xiv) Micro-irrigation (using drips and sprinklers) is still low compared to its potential

xv) Net irrigated area is only about 75 million ha, and gross cropped area 112 million ha whereas net cropped area is around 140 million ha

xv) Large area is under water scarcity and rainfed.

xvi) Wide variability of irrigation coverage, crop-wise irrigated area, rice irrigated area in Indian states

xvii) Pressing problems of weather extremes and climatic variability on water resources and agricultural performance

xix) Wastewater utilization is still low for agricultural purposes

2.9. Conclusion

There is a need to develop strategies for sustainable water management in agriculture. Any common strategy or plan may not work for the entire country because of the multiple nature of variabilities. Efforts should be continued to evolve water-saving agricultural practices/ irrigation and dissemination of existing technologies to be practiced by the farmers. Efficiency of surface water (i.e., the *blue water*) projects has been assessed to be about 30-40%, which can be increased up to 60% by adopting efficient management practices, proper maintenance and modernization of existing infrastructures, command area development, participatory irrigation management and efficient irrigation and agricultural practices (Jain *et al.*, 2019). Similarly, the efficiency of ground water facilities can be increased from the present level of about 65% to about 75%. This measure alone can save considerable water to meet the major portion of demand-supply gap. The efficiency of water use in agriculture can be improved with choice of most appropriate cropping pattern and the specific crop variety; soil moisture (*green water*) conservation, water saving irrigation like SRI method of rice cultivation as it is fast catching up in southern, it reduces water consumption by about 29-30% and yield increases by about 30-40%.

Formulation of strategies would be required to prevent and arrest the implications due to climate change like decline in the glaciers and the snowfields in the Himalayas, increased drought-like situations due to overall decrease in the number of rainy days, increased flood events due to overall increase in the rainy day intensity, effect on groundwater quality in alluvial aquifers due to increased flood and drought events, influence on groundwater recharge due to changes in precipitation and evapotranspiration, and increased saline water intrusion of coastal and island aquifers due to rising sea levels.

There is definite need for focused research and studies directed towards improving efficiency in water use. As to development of technology, the key principles for improving water productivity at the field, farm and basin levels are: (i) increase in the marketable yield of the crop for each unit of water used, (ii) reduction of all outflows (drainage, seepage and percolation), and (iii) increase in the effective use of rainfall, stored water and water of marginal quality. New irrigation technologies that will improve field level water application efficiencies will be critical components of the demand-side management. Developing improved water stress tolerant varieties- new plant types are required to be in place to raise the yield potential under water stress conditions. Biotechnology could amend many abiotic and biotic constraints to sustainable crop production including drought stress and tolerance to high temperature.

Improved agronomic management is required crop diversification and integrated farming systems, changing the crop planting date and making more effective use of rainfall, reducing water use for land preparation, land levelling, reducing water use during crop growth, reducing weeds, improving fertilization application efficiencies etc. The efficiency of water use in agriculture can also be improved through better system of water application like micro irrigation which aims at increasing the area under efficient methods of irrigation viz. drip and sprinkler irrigation; supplementary irrigation of rainfed crops. The potential technological options have been described in the following chapters for rice and rice-based crop production systems.

References

Amarasinghe, Upali; Shah, Tushaar; Turral, Hugh; Anand, B. K. 2007. India's water future to 2025-2050: business-as-usual scenario and deviations. Colombo, Sri Lanka: International Water Management Institute (IWMI). 41p. (IWMI Research Report 123) doi: http://dx.doi.org/10.3910/2009.124

Barnett, T.P., Adam, J.C. and Lettenmaier, D.P. 2005, Potential impacts of a warming climate on water availability in snow-dominated regions, *Nature*, 438, 303– 309, doi:10.1038/nature04141

Bhalla, G.S., Hazell, P. and Kerr, J. 1999. Prospects of India's cereal supply and demand to 2020. Food, Agriculture and the Environment Discussion Paper 29, International Food Policy Research Institute (IFPRI), Washington, D.C., USA.

Bhatt, R., Kukal. S.S., Busari, M.A., Arora, S. and Yadav, M. 2016. Sustainability issues on rice–wheat cropping system. *International Soil and Water Conservation Research*, 4: 64-74

Bouman, B.A.M. 2009. How much water does rice use? *Rice Today*, 2009, 8, 28–29.

Bouman, B.A.M., Lampayan, R.M. and Toung, T.P. 2007. Water management in irrigated rice: coping with water scarcity. Los Baños (Philippines): International Rice Research Institute. 54 p.

CGWB, 2006. Dynamic Ground Water Resources of India (as on March 2004), Central Ground Water Board, Ministry of Water Resources, Government of India, New Delhi

CGWB, 2012. Manual on Aquifer Mapping, Central Ground Water Board, Ministry of Water Resources, Government of India, CHQ, Faridabad, P. 35.

CGWB, 2022: CGWB Year Book India 2021-22, Govt of India, Ministry of Jal Shakti, Dept of Water Resources, River development & Ganga Rejuvenation, Faridabad, page 99-100.

Cosgrove, W.J., and Loucks, D.P. 2015, Water management: Current and future challenges and research directions, *Water Resources Research* 51, 4823–4839, doi:10.1002/2014WR016869.

Crutzen, P.J. 2002, Geology of mankind, *Nature*, 415, 23.

CWC, 2013. Annual Report 2012-13, Central Water Commission (CWC), Ministry of Water Resources, Government of India, New Delhi, 2013, pp. 1-14

CWC, 2019. Water and Related Statistics. Directorate Information System Organisation, Water Planning & Project Wing, Central Water Commission (CWC), Department of Water Resources, River Development & Ganga Rejuvenation, Ministry of Jal Shakti, Govt. of India, New Delhi. Pp. 1-169.

CWC, 2021. Water Resources at a Glance 2021. Central Water Commission (CWC), Department of Water Resources, River Development & Ganga Rejuvenation, Ministry of Jal Shakti, Govt of India, Sewa Bhawan, R.K. Puram, New Delhi. 20-22.

DES, 2019. Agricultural Statistics at a Glance 2018, MoA&FW, DAC&FW, Directorate of Economics and Statistics, Govt of India, New Delhi, 502p.

DES, 2020. State-wise Percentage Coverage of Irrigated Area under Principal Crops during 2015-16 (Provisional). In: Agricultural Statistics at a Glance 2019, MoA&FW, DAC&FW, Directorate of Economics and Statistics, Govt of India, New Delhi, 335p.

DES, 2021. Agricultural Statistics at a Glance 2020, MoA&FW, DAC&FW, Directorate of Economics and Statistics, Govt of India, New Delhi, 280p.

DES, 2022. Agricultural Statistics at a Glance 2021, MoA&FW, DAC&FW, Directorate of Economics and Statistics, Govt of India, New Delhi, 457p.

DES, 2023. Agricultural Statistics at a Glance 2022, MoA&FW, DAC&FW, Directorate of Economics and Statistics, Govt of India, Govt of India, New Delhi, 280p.

Dyson, T. and Hanchate, A. 2000. India's demographic and food prospects: State level analysis. Economic and Political Weekly 11: 4021-4036 (November).

Falkenmark, M. and Widstrand, C. 1992. Population and water resources: A delicate balance. Population Bulletin No. 3. Population Reference Bureau, Washington DC. http://www.ircwash.org/sites/default/files/276-92PO-10997.pdf.

FAO, 2015. Water at a Glance- The relationship between water, agriculture, food security and poverty, Food and Agriculture Organization (FAO) of the United Nations, Rome, 15p.

Gardner-Outflow, T. and Engelman, R. 1997. Sustaining water, easing scarcity: a second update, Population Action International- Population and Environment Programme, Washington, DC 20036, USA, pp. 1-20. www.populationaction.org

GoI, 1999. Report of the Commission for Integrated Water Resource Development, Volume I. Ministry of Water Resources, New Delhi, India, In: Chapter 6 Water, The Ministry of Statistics and Programme Implementation, Govt of India, www.mospi.nic.in, p.199.

Hoque, M.A., Hoque, M.M. and Matin Ahmed, K. 2007. Declining groundwater level and aquifer dewatering in Dhaka metropolitan area, Bangladesh: Causes and quantification, *Hydrogeol. J.*, 15, 1523– 1534, doi:10.1007/s10040-007-0226-5.

IMD, 2020. India Meteorological Department, State rainfall normals (mm)- Annuals (Jan-Dec: 1951-2000), Hydromet Division, India Meteorological Department, New Delhi, https://mausam.imd.gov.in.

IWMI, 2000. World water supply and demand 1995 to 2025. International Water Management Institute 2000, www.cgiar.org/iwmi/pubs/WWVison/WWSDOpen.htm

Jain, R., Kishore, P. and Singh, D.K. 2019. Irrigation in India: Status, challenges and options. *Journal of Soil and Water Conservation*, 18(4): 354-363.

Kumar, R., Singh, R.D. and Sharma, K.D. 2005. Water resources of India, *Current Science*, 89, 794– 811.

Mandal, K.G., Thakur, A.K. and Ambast, S.K. 2019. Current rice farming, water resources and micro-irrigation. *Current Science*, 116 (4): 568-576.

Michael, A.M. 2011. Irrigation- Theory and Practice. Second Edition. Vikas Publishing House Pvt. Ltd., Noida, UP. pp. 1-123

MoWR, 2015. A year of Inclusive Development in Water Resources Sector. National Water Mission, Ministry of WR, RD&GR, Govt of India. http://www.nationalwatermission.gov.in/sites/ default/files/COP%2021%20MoWR%20Booklet. pdf pages 1-18 (accessed 7 June 2021).

MSOPI, 2018. Net area under irrigation by source, In: Irrigation- Statistical Year Book India 2018, Ministry of Statistics and Programme Implementation (MSOPI), Govt of India, New Delhi. GoI, (http://mospi.nic.in/statistical-year-book-india/2018/181)

Rijsberman, F.R. 2000. World Water Scenarios: Analysis. London, UK: Earthscan Publications, 350p.

Rodell, M., Velicogna, I., and Famiglietti, J.S. 2009. Satellite-based estimates of groundwater depletion in India. *Nature* 460, 999-1002. https://doi.org/10.1038/nature08238

Rosegrant, M.W., Cai, X. and Cline, S.A. 2002. World Water and Food to 2025: Dealing with scarcity, International Food Policy Research Institute (IFPRI), Washington, D.C., USA. 338p.

Rost, S., Gerten, D., Bondeau, A., Lucht, W., Rohwer, J., and Schaphoff, S. 2008. Agricultural green and blue water consumption and its influence on the global water system. *Water Resource Research*, 44, W09405. http://dx.doi.org/10.1029/ 2007WR006331.

Saha, D. and Ray R.K. 2019. Groundwater Resources of India: Potential, Challenges and Management. In: Sikdar P. (eds) Groundwater Development and Management. Springer, Cham. https://doi.org/10.1007/978-3-319-75115-3_2

Seckler, D., Amarasinghe, U.A., Molden, D., de Silva, R. and Barker, R. 1998. World water demand and supply 1990 to 2025: Scenarios and Issues. Research Report No. 19. Colombo, Sri Lanka: International Water Management Institute, 50p.

Shah, Tushaar. 2009. Taming the anarchy: groundwater governance in South Asia. Washington, DC, USA: Resources for the Future; Colombo, Sri Lanka: International Water Management Institute (IWMI). 310p.

Singh, R., Kundu D.K. and Bandyopadhyay, K.K. 2010. Enhancing agricultural water productivity through enhanced water use efficiency. *Journal of Agricultural Physics*, 10: 1-15.

Tapley, B.D., Bettadpur, S., Ries, J.C., Thompson, P.F. and Watkins, M.M. 2004. GRACE measurements of mass variability in the Earth system. *Science* 305, 503–505.

Tiwari, V.M., Wahr, J. and Swenson, S. 2009. Dwindling groundwater resources in northern India from satellite gravity observations, *Geophys. Res. Lett.*, 36, L18401: 1-5,doi:10.1029/2009GL039401.

Vijay Shankar, P.S., Kulkarni, H. and Krishnan, S. (2011). India's Groundwater Challenge and the Way Forward. *Economic and Political Weekly*, XLVI(2), January 8. Pp. 37-45

Williams, M., J. Zalasiewicz, A. Haywood, and M. Ellis (2011), The Anthropocene: A new epoch of geological time, *Philos. Trans. R. Soc. A*, 369, 835–1111.

3

Rice Agro-Ecosystems in India

3.1. Introduction

Rice is grown in varying geographical locations and climatic situations. Most of the rice area lies between the equator and 40^0 N latitude and between 70^0 and 140^0 E longitude. It grows well from tropical to sub-tropical warm and temperate climatic regions up to 40^0 S and 50^0 N of the equator. However, the highest yields are obtained in the region between 30^0 and 45^0 N of the equator. The average yield generally increases as the countries are situated further away from the equator (Chatterjee and Maiti, 1985). In India, rice grows in regions from 8^0 N to 35^0 N latitudes under widely varying conditions of the land, rainfall, altitude and climatic conditions. It is grown at the sea level in river deltas as in parts of Kerala, as deepwater in states of West Bengal, Assam and Bihar; at high altitudes of 914 to 1524 m or even more as in Kashmir, and in the slopes of the Himalayas. A temperature range of 20^0 C to 37.7^0 C is required throughout the life period of rice crop.

The practice of rice cultivation is greatly influenced by the pattern of rainfall distribution. In the eastern region of the country and peninsular India, the mean temperature throughout the year are favourable for rice cultivation. In the northern and central India rice is grown wherever rainfall conditions are favourable for cultivation of rice. In northern, central and western India where winter temperature is low, rice cultivation is limited mostly to one crop. In the peninsular India it is concentrated in the East and West Coastal areas of Tamil Nadu and Kerala. Photoperiodically, rice is a short-day plant, but there are varieties which are non-photoperiod sensitive. It is usually observed that late ripening varieties are sensitive and can be grown only during a specific season, while the early maturing varieties can be grown at any time of the year mostly during summer and kharif seasons.

Rice is grown in all types of soils ranging from pH 4.5 to 8.0. The soils most suited for the cultivation of rice crop are heavy neutral soils like clay, clay loam and loamy soils. Such soils are capable of holding water for long and sustain a good rice crop. The most important group of soils, under which rice can be successfully grown in India are laterite soils, alluvial, red soils, black soils, peat, marshy, saline and alkaline soils.

There are three main seasons for growing rice and these are named according to the season of harvest of the crop. The main rice growing season in the country is the *'kharif'* sown in June-July and harvested in November-December. It is known as winter rice as per the harvesting time. Winter rice is known as *aman* in West Bengal, *sali* in Assam, *sarrad* in Orissa, *agahani* in Bihar and Uttar Pradesh, *sarava* in Andhra Pradesh, *mundakan* in Kerala and *samba/thaladi* in Tamil Nadu. About 84% of the country's rice crop is grown in this season and generally medium to long duration varieties are grown in this season. Summer rice is called as *rabi* rice. It is known as *boro* in Assam and West Bengal, *dalua* in Odisha, *garma* in Bihar, *dalwa* in Andhra Pradesh, *punja* in Kerala and *navarai* in Tamil Nadu. The sowing time of summer rice is November to February and harvesting time is March to June. The area under summer rice is about 9-10% and early-maturing varieties are mostly grown in this season.

The yields are low in tropics, compared to temperate regions, may be due to high respiration rates. Thus there is a wide gap in the potential performance and actual performance and this provides a great challenge to the agriculturists and rice scientists.

Further, the diversity of rice ecosystems and varieties are governed by site-specific situations of drought, submergence and flooding, abiotic and biotic stresses etc. The human interventions have also contributed to diversity. Huke (1976) in the paper 'Geography and climate of rice' has shown part of West Bengal, part of Bihar, Odisha and the Western ghat region as the '*home area*' of rice where the crop is raised year after year without climatic modification by man. In other parts of India, at least one parameter climate (mainly rainfall) and some due to temperature is frequently less than the ideal for successful crop.

3.2. Rice agro-ecosystems

3.2.1 *Pluvial, phreatic* and *pluxial* rice lands

Hydrologically, according to Moorman and Breeman (1978) rice lands are classified earlier as: (i) *pluvial* rice lands where runoff and percolation occur due to excessive water, (ii) *phreatic* rice lands where the crop receive water from rainfall, or runoff water, or from shallow groundwater, at least a part of the season, and (iii) *pluxial* rice lands, which receive water wholly or partly from surface flow, mostly the runoff waters.

3.2.2 Rainfed, irrigated and flood-prone ecosystems

Based on the topography of lands and hydrologic conditions, rice ecosystems have been classified mainly as: rainfed upland, rainfed lowland, irrigated rice system, flood-prone system (Khush, 1984; Garrity, 1984; Pathak *et al.*, 2018; Kramer, 1969). Depending upon the depth of water during the growing period, rainfed lowlands include shallow water depth, semi-deep water, and deep water (Fig. 3.1 & 3.2). In addition, hill cold environments and coastal saline areas also include rice ecosystems.

The area under *rainfed upland ecosystems* is about 6.0 million ha, which is 13.5% of the total rice area in India. This type of rice ecosystem is prevalent in eastern Indian states *viz.* Odisha, West Bengal, Bihar, Jharkhand, Chhattisgarh, eastern Uttar Pradesh, Assam and north-eastern hill regions. This rice system is drought-prone.

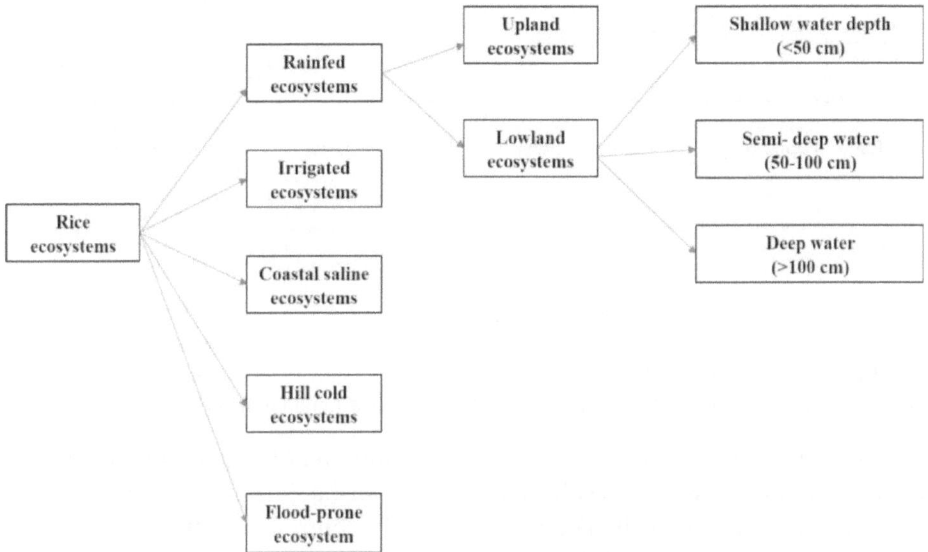

Fig. 3.1. Rice systems based on land situations, rainfall, temperature and water regime

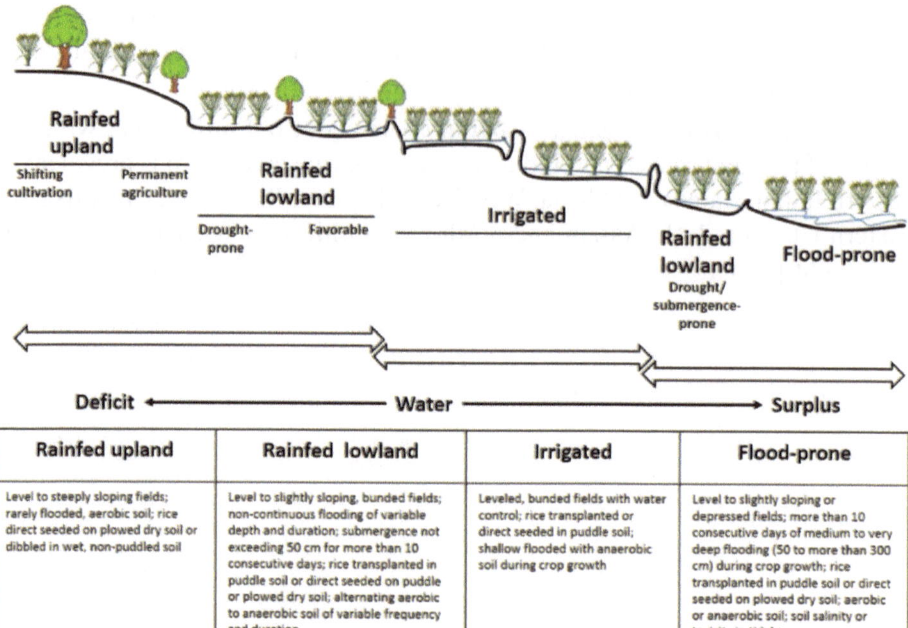

Rainfed upland	Rainfed lowland	Irrigated	Flood-prone
Level to steeply sloping fields; rarely flooded, aerobic soil; rice direct seeded on plowed dry soil or dibbled in wet, non-puddled soil	Level to slightly sloping, bunded fields; non-continuous flooding of variable depth and duration; submergence not exceeding 50 cm for more than 10 consecutive days; rice transplanted in puddle soil or direct seeded on puddle or plowed dry soil; alternating aerobic to anaerobic soil of variable frequency and duration	Leveled, bunded fields with water control; rice transplanted or direct seeded in puddle soil; shallow flooded with anaerobic soil during crop growth	Level to slightly sloping or depressed fields; more than 10 consecutive days of medium to very deep flooding (50 to more than 300 cm) during crop growth; rice transplanted in puddle soil or direct seeded on plowed dry soil; aerobic or anaerobic soil; soil salinity or toxicity in tidal areas

Fig. 3.2. Rice field systems depicting land situation, topography, water regimes and cultivation methods; *Source:* Greenland (1997) as adapted from Halwart and Gupta (2004)

Rainfed lowland ecosystems is about 14.0 million ha (32%) mainly located in eastern India *viz.* Assam, West Bengal, Bihar, Odisha, Eastern Madhya Pradesh and Eastern Uttar Pradesh. The probability of occurrence of flood or drought is more due to erratic rainfall pattern. In the rainfed lowland rice system, rice is traditionally grown in shallow water depth (up to 50 cm), semi-deep water (50-100 cm), and deep-water (>100 cm).

Irrigated ecosystem is about 26.0 million ha, which is about 60% of the total rice area in India. This system is prevalent in the states *viz.* Jammu & Kashmir, Haryana, Punjab, Himachal Pradesh, Uttar Pradesh, Telangana, Andhra Pradesh, Gujarat, Karnataka and Tamil Nadu.

The *flood-prone system* is about 2.5 million ha which occurs in West Bengal, Assam, Odisha, Bihar and eastern Uttar Pradesh. The hydrology of rainfed lowland rice system overlaps with those of upland, irrigated systems, deep water and the tidal wetland systems (Mackill *et al.*, 1996). However, a strong working knowledge of the rice ecosystem would be required. It has linkage with the hydrology, type of rice culture, climatic conditions especially the rainfall and temperature, soil-related constraints, biotic constraints, and socioeconomic factors.

Further rainfed lowlands comprise five environments (Khush, 1984) which are influenced by mainly the hydrological situations. The classifications have been made for development of rice varieties and management practices. These are shallow-favourable, shallow drought-prone, shallow drought-prone and submergence-prone, shallow submergence prone, medium deep waterlogged. The terminology of the rice-ecosystems has been furnished in Table 3.1.

Table 3.1. Terminologies for rice ecosystems & sub-ecosystems

Rice ecosystems	Sub-ecosystems/ water regime and crop growing condition
Rainfed upland	Favourable upland, long crop growing season
	Favourable upland, short crop growing season
	Unfavourable upland, long crop growing season
	Unfavourable upland, short crop growing season
Rainfed lowland	Favourable rainfed, shallow
	Drought-prone rainfed, shallow
	Drought-prone rainfed, shallow
	Drought and submergence-prone rainfed, shallow
	Submergence-prone rainfed, shallow
	Rainfed medium deep, waterlogged
Irrigated	Irrigated, favourable temperature
	Irrigated, low temperature, tropical
	Irrigated, low temperature, temperate
Deep-water	Deep water
	Very deep water
Tidal wetlands	Tidal wetlands, perennial fresh water
	Tidal wetlands, seasonal or perennial saline water
	Tidal wetlands, acid sulphate soils
	Tidal wetlands, peat soils

Adapted from Khush (1984), Mackill *et al.* (1996)

3.3. Rice cultivation systems

In India there are three principal systems of rice cultivation: dry, semi-dry, and wet cultivation

3.3.1. Dry cultivation system

The dry or upland rice are sown broadcast in the high-level lands with outbreak of the monsoons. This systems is followed in almost all rice growing states in

India but mainly confined to tracts which get either the south-west or north-east monsoon or both and do not have adequate irrigation facilities.

This system of cultivation is called *aus* in West Bengal, *aus* or *ahu* in Assam, *beali* in Odisha, *bhadai* in Bihar, *bhadi* or *kauri* in Uttar Pradesh, *virippu* in Kerala and *kuruvai/kar/ sornavari* in Tamil Nadu and so on, and it is of 90 to 100 days duration. The lands for this system of cultivation are ploughed a number of times in summer for necessary tilth and the available organic manure is applied @ 4-5 t ha^{-1}. As the sowing is usually done in the months of May-June, the crop is dependent on the south-west monsoon, and in September for the crop dependent on the north-east monsoon. The practice of opening up the soil in the summer and/ or in winter is adopted in order to ensure timely sowing and to utilize the monsoon rains to the best advantage of the crop. The seeds are sown broadcast at the rate of 60-80 kg ha^{-1}. Surface soil gets compacted using light-weight roller if the soil moisture is less than the optimum level. By this way, the seeds are brought into intimate contact with a large number of slightly moist soil particles and individual seeds are thus favoured to have a little more moisture for germination.

3.3.2. Semi-dry system of cultivation

Under this system of cultivation seeds are sown broadcast as a dry crop for about two months and when more water is available, after strengthening of monsoon, it is treated as wet crop. Usually, in July-August, when the south-west monsoon is active, the rainwater is impounded in the fields and the young crop of five to six weeks is ploughed crosswise with a light narrow wooden plough with about 5 cm of standing water in the field for partial weeding, thinning as well as stirring of soils to promote root formation.

3.3.3. Wet cultivation

The major portion of rice crops in India is grown under 'low land' conditions that is under 'flooded' or 'submerged' conditions. In this system, land is ploughed repeatedly three or four passes of country plough or mould-board plough giving an interval of four with 5-7 cm of standing water. After repeated ploughing the soil becomes puddled. Water for puddling may come from rainfall or from irrigation and for that the crop is named as rainfed or irrigated paddy. Before planting of seedlings or broadcasting of sprouted seed in the puddled land, inorganic and organic fertilizers are applied. The advantages of this system of cultivation over dry cultivation are: puddling operation buries all weeds which should otherwise compete with rice plants for nutrients in the soil, saves time by enabling the maturity of the crops to be so adjusted as to allow multiple cropping, assured plant population, increased availability of iron and phosphorus and nitrogen is conserved, germinated seeds or seedlings get established in soft soil initially to put forth their fresh roots in all directions and root system is thus extended, the infiltration rate of water is reduced and thus increases the water retaining capacity of soil, and the soft mud facilitates transplanting.

3.4. Rice seeding/ transplanting methods

3.4.1. Direct seeding in dry soil

This method requires dry land preparation. Land is tilled with plough and levelled properly before or at the onset of monsoon. Paddy seeds is sown directly in the field just like wheat. Seed rate varies from 60 to 80 kg ha^{-1} at row spacing of 25 to 30 cm depending on variety and time of planting. Seeds may be sown broadcast or behind ploughs or in line by seed drills. Immediately after sowing seeds should be covered by soil. The ideal depth of seeding is about 3 cm. If monsoon rain does not come after sowing of seed, arrangement may be made for irrigation. Method of planting was recommended earlier by Barker and Herdt (1979), and is presented in Table 3.2.

Table 3.2. Methods of planting according to the rice culture systems

Types of rice culture	Method of planting	Maximum water depth (cm)
Upland rice	Broadcast of drilled into dry soil	No standing water
Rainfed lowland rice	Transplant	0-50
Shallow rainfed low land rice	Transplant	5-15
Medium deep rainfed low land rice	Transplant	16-50
Deep water rice	Broadcast on to dry soil	51-100
Floating rice	Broadcast on to dry soil	101-600

Source: Barker and Herdt (1979)

3.4.2. Direct seeding in watered field

The land is prepared wet. The land is puddled by several ploughing and laddering in standing water. The land may be impounded with rain or irrigated water. Weeds present in the soil are incorporated into the soil and allowed to decompose for almost 3 weeks. The mud in the field is then levelled for seeding. During seeding excess water is drained from the field to increase good establishment of seedlings. From experiment it has been observed that excess water at seeding time affects germination of rice seed and seedling establishment. This is perhaps due to the low concentrations of dissolved oxygen in deep water. It is further noted that germination may be affected due to paucity of oxygen supply, in muddy water because settling mud may cover the seed. Where the depth of water is more the plumules elongate more than the radicles and the seedlings float on the impounded water and do not anchor properly.

Pre-germinated seed with radicle length of 1 to 2 cm long is broadcast by hand at the seeding rate of 90 to 100 kg ha^{-1}. The main disadvantages of this system are that, (a) seeds are exposed to rat and bird attacks and (b) seeds are carried or washed away during heavy rains resulting in poor seedling stand. To overcome the above two disadvantages pre-germinated seeds are sown in the puddled land with 5 cm standing water. If the water becomes muddy during seeding, soil particles are allowed to settle before seeding. This prevents soil from covering the seeds. The advantages over dry seeding are: (a) better control of grassy weeds in standing water, (b) better protection of seeds from bird attack. At the same time the disadvantages are (a) slow emergence and establishment of seedlings particularly in rainy season. (b) additional cost of construction of bunds to hold water in the field, (c) reduction in plant stand due to floating of seedlings, in case the water depth is beyond 10 cm high.

3.4.3. Transplanting

Planting of seedlings is done in two ways: random transplanting, straight row planting. In random transplanting rice seedlings are transplanted at random in the field, distance from plant to plant is not uniform and no definite pattern is followed. In this type of planting, it is difficult to operate a rotary weeder and/ or other machinery for intercultural operations. In the latter method of planting, seedlings are planted in straight rows. The advantages of this procedure are: (i) weeding operations are more efficiently done either by rotary weeder or by hand, (ii) proper plant spacing makes uniform growth of plants possible, and (iii) other field management operations, such as spraying and fertilizer top dressing are made easier.

Advantages and disadvantages of transplanting: i) Plant population becomes more assured, ii) since nurseries occupy only 1/10 of area of the main crop, the irrigation, control of insects and diseases and other management becomes easier to reduce the cost of raising seedlings as compared to a broadcast wet crop of raising seedlings as compared to a broadcast wet crop of the same age. Nurseries raised in advance enable the crop to be planted in the field at the right time, iii) the seedlings are growth up at the time of transplanting and they are able to compete with young weeds that spring up. Weed control and other inter-cultural operations are much easier, particularly in row planting. *Disadvantages* include i) high labour cost, ii) high water requirement, iii) seedlings are exposed to possible injury during handling, iv) plants tend to grow more slowly than direct seeding because of recovery time after transplanting, v) harvesting is delayed.

3.5. Agro-climatic zones (ACZs) of India & rice-based cropping systems

3.5.1. Zone-specific rice-based cropping

Rice-based is the major cropping system in India. It includes rotation of crops *viz.* rice, oilseeds, pulses, sugarcane, cotton, green manures, vegetables, etc. Rice is grown under so diverse soil and climatic conditions (ACZs) there is hardly any type of soil left where rice cannot be grown including saline, alkaline, and acidic soils. Therefore, it is grown from below sea level (Kuttanad area of Kerala) up to an elevation of 2000 m in Jammu & Kashmir, hills of Uttaranchal, Himachal Pradesh and North-Eastern Hills (NEH) areas (Table 3.3). Rice is the major crop in rice-based systems. Rice is followed by succeeding crops. Rice intercropped with other crops is also practiced in many regions.

Table 3.3. Rice-based cropping systems in different agro-climatic zones of India

Agro-climatic zones	Rainfall (mm)	Predominant soils	Prominent rice-based cropping systems
Western Himalayan (Himachal Pradesh, Jammu & Kashmir, Uttarakhand)	1650–2000	Hill and Sub-montane	Rice-wheat, Rice-potato-potato
Eastern Himalayan (Assam, North East states, West Bengal)	1840–3528	Red sandy, Laterite, hill, Alluvial	Rice-fallow, Rice-rice, Rice-pulses/oilseeds
Lower Gangetic Plain (West Bengal)	1302–1607	Alluvial, Red and yellow	Rice-rice, Rice-wheat, Rice-potato-jute/ vegetables
Middle Gangetic Plain (Bihar, eastern Uttar Pradesh)	1211–1470	Alluvial, Tarai and Calcareous	Rice-wheat Rice-maize, Rice-potato-sunflower
Upper Gangetic Plain (central and western Uttar Pradesh)	721–979	Alluvial, Tarai	Rice-wheat Sugarcane-ratoon-wheat
Trans Gangetic Plain (Punjab, Haryana)	360–890	Alluvial and Calcareous	Rice-wheat
Eastern plateau and Hills (Chhattisgarh, Jharkhand, Madhya Pradesh, Maharashtra)	1296–1436	Red, Yellow Laterite	Rice-blackgram Rice-niger/linseed Rice-vegetables

East coast Plain and Hills (Andhra Pradesh, Odisha, Tamil Nadu, Puducherry)	780–1287	Deltaic alluvium, Red, Laterite	Rice-groundnut-greengram, Rice-greengram/ blackgram, Rice-rice
West coast Plains and Hills (Goa, Maharashtra, Karnataka, Kerala, Tamil Nadu)	2226–3640	Coastal alluvium, Red, Laterite	Rice-rice
Andaman and Nicobar Islands	1600–3000	Red	Rice-fallow

Source: Adapted from Gill *et al.* (2008)

Broadly, rice growing areas in the country can be broadly grouped into five regions as discussed below

Eastern region: This region comprises of Bihar, Chhattisgarh, Madhya Pradesh, Orissa, Eastern Uttar Pradesh and West Bengal. Rice is grown in the basins of the Ganga and Mahanadi rivers and has the highest intensity of rice cultivation in the country. This region receives heavy rainfall and rice is grown mainly under rainfed conditions.

North-Eastern (NE) region: This region comprises of Assam and NE states. In Assam rice is grown in the basin of the Brahmaputra river. This region receives very heavy rainfall and rice is grown under rainfed conditions. Eastern and NE region combined have 75% of the rice area, and is under rainfed condition, accounting for about 45% of the total rice area and contributing about 43% of the rice production of the country.

Northern region: This region comprises of Haryana, Punjab, Western Uttar Pradesh, Uttaranchal, Himachal Pradesh and Jammu & Kashmir. The region experiences low winter temperature and single crop of rice from May-July to September-December.

Western region: This region comprises of Gujarat, Maharashtra and Rajasthan. Rice is largely grown under rainfed conditions during June-August to October-December.

Southern region: This region comprises of Andhra Pradesh, Karnataka, Kerala and Tamil Nadu. Rice is mainly grown in deltaic tracts of Godavari, Krishna and Cauvery rivers i.e. under irrigated conditions, and the non-deltaic rainfed areas of Tamil Nadu and Andhra Pradesh.

Rice in cold and hill ecosystems is described separately. Such rice area lies in the hill regions comprising of Jammu and Kashmir, hill areas of Himachal Pradesh, Uttarakhand and North-Eastern hill states. The total area under rice in cold and hill region is about 1 million ha which accounts for 2.3% of total area under rice in the country. The productivity is very less. The average yield is about 1.2 t ha^{-1} as against the national average yield. The major problems of these areas are cold injuries, blast, drought spell and very short span of cropping seasons. Because of the rolling topography in these areas bench terracing is being followed which limits the use of fertilizers and improved agronomical practices. In these areas the crop is sometimes affected due to low temperature in the early stage and sometimes at the flowering times which leads to sterility problems.

3.5.2. Intensified rice-based cropping systems in different locations

There is a need for intensified cropping systems for different locations in India. There is an ample scope of intensified cropping over the existing system, and has been presented for different regions in the country (Table 3.4).

Table 3.4. Recommended intensified cropping systems for different locations in India

Region	Existing system	Recommended system
Sandy clay loam soil of Varanasi (UP)	Rice-wheat	Rice-potato-blackgram, rice-maize (cob) + veg. pea (1:2) – green gram cropping
Karnal, Haryana	Rice-wheat	Rice - vegetable pea-wheat-greengram
Punjab	Rice-wheat	Maize-potato-onion
Clay loam soils of Jabalpur (MP)	Rice-wheat Rice-chickpea	Hybrid rice-gobhi sarson-okra
South Gujrat	Rice-wheat	Rice-sorghum-sorghum ratoon Rice-sweet corn-blackgram Rice-greengram-groundnut
High rainfall zone of south Gujarat	Rice-rice	Rice–fenugreek–okra
Krishna –Godavari delta areas of Andra Pradesh	Rice-rice	Rice-maize, Rice-mustard
North Coastal Zone of Andhra Pradesh	Rice-fallow	Sunnhemp-rice-blackgram
Low land rice ecosystem of Karnataka	Rice-maize	Rice-soybean Rice-marigold
Jharkhand (irrigated)	Rice-wheat Rice-fallow	Rice–potato+wheat(1:1)-greengram, Rice–potato–greengram
Irrigated medium land acid soil situation of Assam	Rice-rice	Winter rice–cabbage-greengram Rice–chilli–blackgram
Irrigated ecosystem in Bihar	Rice-rice	Rice–tomato–bottle gourd, Rice–potato–onion, Rice–coriander–lady's fingers, Rice–carrot–cowpea
Chhattisgarh plains	Rice-rice	Rice-potato-cowpea

Upper Brahmaputra valley zone of Assam	Rice-fallow	Rice– potato Rice –pea
Coastal plain zone of Odisha	Rice-fallow	Rice-ratoon-greengram, Rice-greengram/blackgram
Khurda District of Odisha	Rice-fallow	Rice-utera blackgram

Adapted from Panda *et al.* (2018)

3.6. State-wise rice-based systems

Rice ecosystems in Odisha

Agriculturally, the state of Odisha is one of the dominant states in rice cultivation. The main food for the people and livelihood is dependent on rice. It supplies considerable amount to central rice grain stocks. In the year 2021-22, the rice area in the state is 4.97 million ha, which is 3.81% to all India area, with the production of 9.95 million tonnes, which is 3.15% to all India production, and its yield is 2001 kg ha^{-1} (DES, 2023).

Like other states of eastern India, farmers of Odisha have classified their lands into three main land types- upland, medium and lowland based on the topography and water retention i.e., water regime in the fields. In the year 2017-18, total cultivated area is 6180 thousand ha, net sown area of 5356 thousand ha, gross cropped area of 8361 thousand ha of cultivated area with the cropping intensity of 156% (OAS, 2020). Out of the total cultivated area, 47.15% is upland, 28.40% medium land and 24.45% lowland; rice area is 3544 thousand ha during kharif season (2017-18), which is spread as: 16.03% in upland, 42.30% in medium land and 41.68% in lowlands (Table 3.5). Other category of field crops and horticultural crops are also grown in the state (Table 3.6). Major cereal other than rice is maize and ragi, major pulses area green gram (moong), black gram (*biri*), arhar, chickpea; and major oilseeds are groundnut, mustard, sesamum and sunflower. A few fruit crops are also predominant *viz.* mango, banana, guava, citrus etc.

Table 3.5. High, medium and low lands in Odisha and rice area during *kharif* season (2017-18)

Land types	**Cultivated area ('000 ha)**	**Rice area in *kharif* ('000 ha)**
High	2914 (47.15%)	568 (16.03%)
Medium	1755 (28.40%)	1499 (42.30%)
Low	1511 (24.45%)	1477 (41.68%)
Total	6180 (100%)	3544 (100%)

Source: OAS (2020)

Table 3.6. Major cropped area during *kharif* and *rabi* seasons in Odisha (2017-18)

Crops	*Kharif* season ('000 ha)	*Rabi* season ('000 ha)
Rice	3544.45	221.94
Cereals (other than rice)	376.68	24.74
Total cereals	3921.13	246.68
Total pulses	708.09	1339.26
Total oilseeds	286.50	316.87
Total vegetables	275.67	375.85
Total fibre	160.17	-
Total spices	76.72	84.90
Sugarcane	-	27.00
Total cropped area	5428.28	2390.87

Source: OAS (2020)

Rice ecosystems are very diverse in the state because of varied land situations, climatic conditions and water regimes. Because of this variability, rice is grown in all three systems of cultivation: dry, semi-dry and wet cultivation. About 18% of the rice area is under dry ecosystems, and the remaining is under semi-dry and wet ecosystems (Pani and Patra, 2004).

According to a report in IRRI Rice Knowledge Bank, Odisha hub office (IRRI, 2021), rice is cultivated in seven different ecosystems: irrigated kharif (27.4%), rainfed upland (19.1%), medium land (12.4%), shallow lowland (22.5%), semi-deep (7.9%), deep (3.4%), and irrigated rabi (7.4%) (Table 3.7). Sometimes classification of rice ecosystems becomes difficult because of varied immense variability. Of course, over the past several years of efforts, rice researchers at the international, national and state level have developed many varieties which are suitable to different ecosystems, and farmers have also identified varieties which are well-adapted to actual situations of land and hydrology.

The secondary centre of origin of cultivated rice is in Jeypore area of southern Odisha (Ramiah and Ghose, 1951, Ramiah and Rao, 1953). The centre of origin of '*aus*' ecotypes is in the regions *viz.* Odisha, Jharkhand and Chhattisgarh (Sharma *et al.*, 2000). As indicated in the introduction chapter, there are several festival and rituals prevalent in Odisha, which are related to rice seeding and its different growth stages (Das, 2012). The '*Akhaya tritiya*' coincides with rice seeding in the month of May-June, '*Raja sankranti*' is celebrated in the mid-June to mark the completion of rice sowing, '*Garbhana sankranti*' in October signifies the reproductive stage of rice, '*Nuakhai*' is linked to harvesting of upland paddy, and '*Laxmi puja*' to harvesting lowland paddy. Some tribal people celebrate '*Chaita parab*' in mid-January when rice threshing and grain storage is completed. This festival matches with the celebration of '*Makar sankranti*' in Odisha and elsewhere in India.

Table 3.7. The rice ecosystems in Odisha and the approximate land area under each category

Ecosystems	Area under rice (m ha)	Major stress
I. Rainfed ecosystems	2.90 (65%)	
Rainfed upland	0.85 (19.1%)	Drought
Rainfed medium land	0.55 (12.4%)	Fairly stress free
Rainfed lowland	1.50	
Rainfed lowland- shallow water	1.00 (22.5%)	Flood and waterlogging and
Rainfed lowland- semi deep water	0.35 (7.9%)	Flood and waterlogging
Rainfed lowland- deep water	0.15 (3.4%)	Early drought, flood and waterlogging
II. Irrigated ecosystems	1.55 (35%)	
Irrigated kharif (bunded; upland, medium land, lowland)	1.22 (27.4%)	Waterlogging: too much irrigation and rainwater
Irrigated rabi	0.33 (7.4%)	Low temperature at early seedling stage

Adapted from Das (2012)

Farmers practice the ***dry-system of cultivation*** mostly in rainfed uplands. The soil remains aerobic. The land is ploughed 2-3 times in the summer i.e., summer ploughing which favours interception of sunlight by the ploughed soil and killing of soil borne insect-pests, pathogens and weeds. Before on-set of monsoon rain, seeds of short duration early maturing rice are usually sown broadcast in dry-system. The seeds are then covered with surface soil by light ploughing with a country-plough and also by laddering. A few furrows are also made in the field to drain out excess water, if occurs during the on-set of monsoons. A blind tillage, locally called as '*mendha*', is followed cross-wise after 3-4 days of broadcasting to break soil crusting and facilitating germination of rice seeds and aeration, destroy germinating weed seeds and redistribute rice seeds. Again, mechanical weeding, locally known as '*bida*', is done after about two weeks of germination. Other agronomic practices are also followed. The ***semi-dry system*** is generally followed in rainfed ecosystems: medium and shallow lowlands. It is mostly dry-seeding and then wet cropping period. This is basically the direct-seeding of rice. After land preparation, seeding in May-June and aerobic condition persists for about 30-45 days before monsoon rains. On receipt of rainfall, field is lightly ploughed with a bullock-drawn country plough in the standing water (5-10 cm). Weeds are managed and crop stand is made almost uniform manually. This particular tillage operation is locally called as '*beushening*' or '*beushen*', and gap filling with uprooted seedlings, thinning of dense areas and making the crop stand uniform is locally called as '*khelua*'. Sometimes '*khelua*' is done 7-10 days after '*beushening*'. ***Wet system of rice cultivation*** is the traditional practice of saturated soil conditions almost throughout the crop growing period, except at the maturity stage. Availability of water is assured. With the monsoon rain or by irrigation, the saturated field is cultivated and puddled by repeated ploughing with 3-5 cm depth of water, then rice seedlings are transplanted, or sometimes germinated seedlings are placed on the wet-puddled and levelled field with no standing water. In this system, the field is soil is mostly anaerobic. Weeds are controlled by hand weeding once or twice. Other practices *viz.* water management, fertilizer application and need-based plant protection measures are taken. The constraints and potential technological solutions have been discussed in the chapter 4.

Rice-based systems in different agro-climatic zones (state) of Odisha

Broadly two zones are distinct- the plateau region and the coastal region. About 85% of the geographical area is under plateau region with about 2.75 million ha of rice, and 15% is under the coastal region with about 1.70 million ha of rice area (Das, 2012). The districts *viz.* Balasore, Bhadrak, Kendrapara, Jajpur, Jagatsinghpur, Cuttack, Puri and Ganjam belong to coastal alluvial plains. The coastline varies with its width from 24 to 72 km from the sea coast. Flood and waterlogging are the frequent occurrences in the coastal districts, and lowland rice ecosystem is prevalent in the coastal region.

The state is divided into ten distinct agro-climatic zones based on topography of land, soil characteristics, rainfall, atmospheric temperature etc. (Fig. 3.3), and rice is the major crop in all the agro-climatic zones (Gangopadhayay, 1991). Rice-based systems have been presented in the following section under different agroclimatic zones in brief with the source of information as reported by Das (2012).

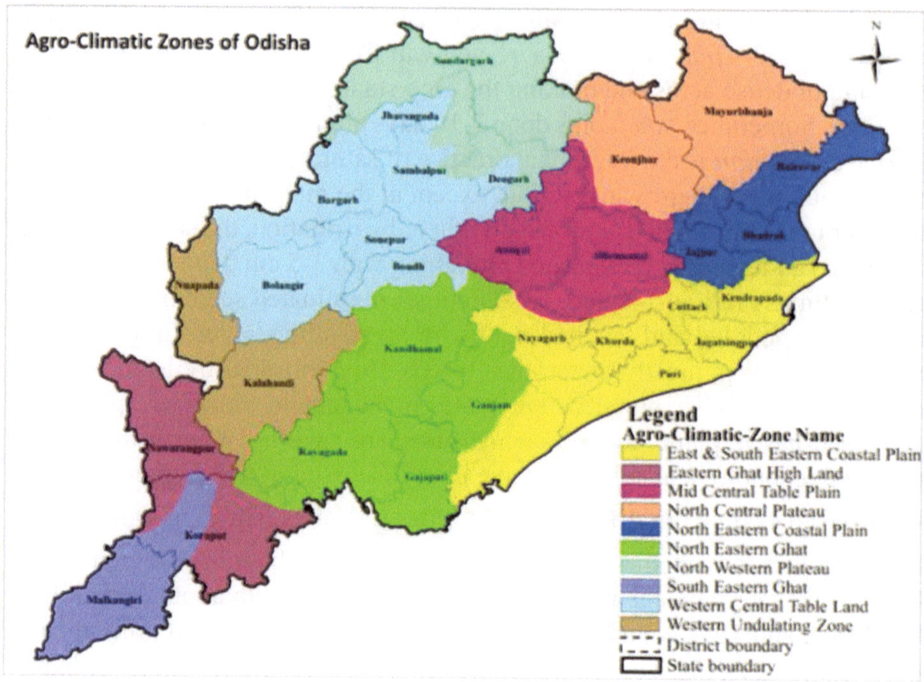

Fig. 3.3. Agro-climatic zones of Odisha

North-western plateau: comprising districts of Sundargarh, northern parts of Sambalpur, Deogarh and Jharsuguda, 1,291 thousand ha of geographical area with 457 thousand ha net sown area, 581 thousand ha gross cropped area and 127% cropping intensity.

Major rice-based systems:

Rice/ pigeonpea intercropping, Rice-horsegram, Rice-castor, Rice-greengram in uplands

Rice-rapeseed & mustard, Rice-linseed, Rice-chickpea, Rice-sesamum, Rice-vegetables in medium lands

Rice-wheat, Rice-chickpea, Rice-pea, Rice-lentil, Rice-linseed, Rice-sesamum, Rice-vegetables in lowlands.

North-central plateau: comprising districts of Mayurbhanj and major of Keonjhar, 1,726 thousand ha of geographical area with 715 thousand ha net sown area, 818 thousand ha gross cropped area and 114.4% cropping intensity.

Major rice-based systems:

Rice/pigeonpea intercropping, Rice-horsegram, Rice-blackgram, Rice-greengram Rice-vegetables in uplands

Rice-rapeseed & mustard, Rice-greengram, Rice-blackgram, Rice-linseed, Rice-lentil, Rice-safflower, Rice-vegetables in medium lands

Rice-chickpea, Rice-greengram, Rice-blackgram, Rice-pea, Rice-vegetables in lowlands.

North-eastern coastal plains: comprising districts of Baleswar, Bhadrak, major part of Jajpur, south-eastern part of Keonjhar, 884 thousand ha of geographical area with 642 thousand ha net sown area, 889 thousand ha gross cropped area and 138.4% cropping intensity.

Major rice-based systems:

Rice/ pigeonpea intercropping, Rice-horsegram, Rice-blackgram, Rice-greengram Rice-vegetables in uplands.

Rice-rapeseed & mustard, Rice-greengram, Rice-blackgram, Rice-pea, Rice-castor, Rice-groundnut, Rice-vegetables in medium lands.

Rice-chickpea, Rice-greengram, Rice-blackgram, Rice-pea, Rice-vegetables in lowlands.

East and south-eastern coastal plains: Comprising districts of Puri, Jagatsinghpur, Kendrapara, Khorda, mid southern part of Cuttack and Nayagarh, southern part of Ganjam, 1,685 thousand ha of geographical area with 1,113 thousand ha net sown area, 1,932 thousand ha gross cropped area and 173.6% cropping intensity.

Major rice-based systems:

Rice/ pigeonpea intercropping, Rice-horsegram, Rice-blackgram, Rice-greengram, Rice-vegetables in uplands

Rice-rice, Rice-rapeseed & mustard, Rice-greengram, Rice-horsegram, Rice-blackgram, Rice-chickpea, Rice-potato-sesame, Rice-groundnut, Rice-vegetables in medium lands;

Rice-rice, Rice-greengram, Rice-blackgram, Rice-groundnut, Jute-rice-pulses, Rice-vegetables in lowlands.

North-eastern ghats: comprising districts of Kandhamal, Gajapati, Rayagada, northern part of Ganjam, northern fringe of Koraput, southern part of Boudh and small part of Kalahandi, 2,305 thousand ha of geographical area with 606 thousand ha net sown area, 898 thousand ha gross cropped area and 148.2% cropping intensity.

Major rice-based systems:

Rice-horsegram, Rice-blackgram, Rice-greengram in uplands

Rice-greengram, Rice-rapeseed & mustard, Rice-lathyrus, Rice-groundnut, Rice-vegetables in medium lands;

Rice-groundnut, Rice-greengram, Rice-blackgram, Rice-lathyrus, Rice-potato, Rice-vegetables in lowlands.

Eastern ghat highlands: comprising districts of Nabarangpur and major part of Koraput, 995 thousand ha of geographical area with 400 thousand ha net sown area, 505 thousand ha gross cropped area and 126.3% cropping intensity.

Major rice-based systems:

Rice monocropping, Rice/pigeonpea intercropping, Rice-horsegram, Rice-blackgram, Rice-greengram, in uplands

Rice-rapeseed & mustard, Rice-linseed, Rice-chickpea, Rice-ragi, Rice-lentil, Rice-maize, Rice-vegetables in medium lands;

Rice-rice, Rice-lentil, Rice-greengram, Rice-blackgram, Rice-rapeseed & mustard, Rice-vegetables in lowlands.

South-eastern ghats: comprising districts of Malkangiri and a part of Koraput, 659 thousand ha of geographical area with 242 thousand ha net sown area, 296 thousand ha gross cropped area and 122.4% cropping intensity.

Major rice-based systems:

Rice monocropping, Rice/pigeonpea intercropping, Rice-horsegram, Rice-blackgram, Rice-greengram in uplands

Rice-chickpea, Rice-safflower, Rice-lentil, Rice-blackgram, Rice-vegetables in medium lands;

Rice-chickpea, Rice-greengram, Rice-blackgram, Rice-vegetables in lowlands.

Western undulating: comprising districts of Nuapada, Kalahandi and part of Rayagada, 1,259 thousand ha of geographical area with 586 thousand ha net sown area, 816 thousand ha gross cropped area and 139.3% cropping intensity.

Major rice-based systems:

Rice monocropping, Rice/pigeonpea intercropping, Rice-horsegram, Rice-blackgram, Rice-greengram in uplands

Rice-chickpea, Rice-horsegram, Rice-greengram, Rice-blackgram, Rice-vegetables in medium lands;

Rice-rice, Rice-wheat, Rice-chickpea, Rice-greengram, Rice-blackgram, Rice-rapeseed&mustard, Rice-niger, Rice-sesamum, Rice-vegetables in lowlands.

West-central table land: comprising districts of Balangir, Subarnapur, Bargarh, southern part of Boudh, northern part of Deogarh, southern part of Sambalpur and western part of Jharsuguda, 1,719 thousand ha of geographical area with 978 thousand ha net sown area, 1,331 thousand ha gross cropped area and 136.1% cropping intensity.

Major rice-based systems:

Rice monocropping, Rice/pigeonpea intercropping, Rice-horsegram, Rice-blackgram, Rice-greengram in uplands

Rice- rapeseed & mustard, Rice-chickpea, Rice-rice, Rice-blackgram, Rice-greengram, Rice-groundnut, Rice-sesamum, Rice-vegetables in medium lands;

Rice-rice, Rice-chickpea, Rice-greengram, Rice-blackgram, Rice-groundnut, Rice-sesamum, Rice-vegetables in lowlands.

Mid-central table land: comprising districts of Angul, Dhenkanal, parts of Cuttack, Nayagarh, Jajpur and Keonjhar, 1,364 thousand ha of geographical area with 508 thousand ha net sown area, 734 thousand ha gross cropped area and 144.5% cropping intensity.

Major rice-based systems:

Rice monocropping, Rice/pigeonpea intercropping, Rice-horsegram, Rice-blackgram, Rice-greengram, Rice-rapeseed & mustard, Rice-sesamum in uplands

Rice- rapeseed & mustard, Rice-horsegram, Rice-blackgram, Rice-greengram, Rice-groundnut, Rice-sesamum, Rice-vegetables in medium lands;

Rice-chickpea, Rice-pea, Rice-blackgram, Rice-groundnut, Rice-vegetables in lowlands.

Rice ecosystems in West Bengal

The categorization of rice lands, especially the depth of water, was made by Biswas and Mallick (1983) for West Bengal as indicated in Table 3.8. The lowlands are subjected also to flash floods, where water depth and duration of flooding may vary. Chatterjee and Maiti (1985) classified lowland, which is followed widely for varietal improvement and management practices as:

i) Shallow, 0-25 cm flooding depth,

ii) Medium, 26-50 cm flooding depth,

iii) Deepwater, 51-100 cm flooding depth, and

iv) Floating, above 100 cm flooding depth.

Deepwater and floating rice culture is observed in Assam, Bihar, West Bengal, Andhra Pradesh and Uttar Pradesh.

Table 3.8. Rice ecosystems during *kharif* season in West Bengal

Type of land	Percentage of total	Area in m ha
Upland unbunded	10	0.50
Rainfed shallow (0-30 cm)	28	1.40
Rainfed medium (31-50 cm)	20	1.00
Rainfed lowland (51-100 cm)	5	0.25
Deepwater (above 100 cm)	5	0.25
Irrigated	32	1.60

Source: Biswas and Mallick (1983)

In West Bengal rice is grown under varying eco-systems with varying climatic and hydrological conditions ranging from waterlogged and poorly drained areas to well-drained soils. Along with rice, various other crops like pulses, oil seeds, vegetables and jute are grown in most suitable areas compatible with various seasons. The major systems are: rice- potato/ vegetable- vegetable, rice-potato/ vegetable-maize, rice- potato/vegetable- rice, rice-potato/vegetable-sesame, rice-mustard-vegetable, rice-potato/vegetable-jute, rice-wheat-maize, rice-wheat-rice, rice-wheat-sesame, rice-mustard-maize, rice-wheat-greengram, rice-mustard-rice, rice-wheat-jute, rice-mustard-greengram, rice-mustard-jute, rice-rice-greengram/ jute, rice-blackgram-potato-vegetable, rice-potato-maize/jute, rice-blackgram-potato/ vegetable, rice-potato-wheat/greengram, rice-fish system.

The major rice-based systems under rainfed conditions of West Bengal are: rice-pulses, rice-rapeseed-mustard, rice (aus)-aman rice, jute-aman rice, aman rice-lathyrus, aman rice-greengram/ vegetable, aman rice-maize, aman rice-barley.

The climate of West Bengal provides scope for growing jute in pre-*kharif* season (March-July) and it is an important cash crop of the farmers. After harvesting of 120 days old jute the fields can be put to transplanted rice. In upland situations after jute, pulses or oilseeds or potato or wheat is grown.

Rice-based cropping systems in Kerala

The major rice-systems are: rice-rice-pulse/oil seeds, rice-pulse-vegetable, rice-rice-fallow and rice-vegetable pattern, rice-fish system. Rice intercropping with millets and pulses (red gram, finger millet, little millet, maize and cowpea) are mainly practiced by some tribal farmers in certain parts of hilly areas in the state. The seeds of these crops are broadcasted during the months of April-May, and harvested during September-October. Little millet, cucumber and pumpkin are also grown as border crops.

In Kerala, typically rice-rice-water fallow or water fallow- water fallow-rice is practiced in Kuttanad; rice-rice-oilseeds/ greengram in Onattukara; water fallow- rice-rice in Kole or Kaippad; rice-fish/ prawn in Pokkali; rice-rice-fallow in high ranges.

In dry seeded low land (semi-dry) rice system, cowpea may be raised as an intercrop. There is the practice of mixed varietal combination of rice (3:1 ratio of autumn rice and winter rice), locally called as *koottumundkan*.

Relay cropping i.e. broadcasting of pulses seeds is also practiced in standing paddy field about 21 days before the harvesting of rice, after draining out of standing water. Crops like lentil, gram, pea, etc. continue to grow by utilizing residual moisture and residual nutrients. This practice is common in both upland and lowland rice culture.

Rice-based cropping systems in Tamil Nadu

Tamil Nadu is a major rice growing state. There has been a long and rich history of rice cultivation in the state because of favourable climatic, soil and hydrological conditions. The major growing districts are Thanjavoor, Thiruvaroor and Nagapatnam. Major rice-systems are: rice-rice-pulse/oil seeds, rice-pulse-green manure, rice-pulse/oilseeds, rice-rice-rice.

The first rice crop (locally known as *Kuruvai*) starts during June-July and is harvested in October. The second crop (locally known as *Thaladi*) usually starts in October-November and is harvested during January-February. Farmers grow pulses like black gram, cow pea, horse gram, and oil seeds such as sesame, etc in the paddy lands.

Rice based cropping systems in Karnataka

In Karnataka, rice is grown in a range of agro-ecological zones like arid lands, hilly areas, wet lands and coastal plains. The major rice-based farming systems are followed Davangare, Haveri, Shimoga, Mysore, Mandya districts: rice-rice, rice-pulse/green manure, rice-pulse/millets, rice-other cereals/millets.

In the southern transitional zone of Karnataka: rice-rice, rice-groundnut, rice-pulses, rice-fallow, rice-vegetables and rice-maize. Tungabhadra project area is dominated by rice-rice mono-cropping system.

Rice ratoon cropping system is also practiced in the hilly region of Karnataka *viz.* Malnad region.

3.7. Rice-pulses/ oilseeds ecosystems

Pulses and oilseeds are grown in the rice-based systems in the form of sequential cropping on the same field, intercropping, relay cropping, catch cropping etc.

Rice-chickpea system in eastern Uttar Pradesh and Bihar is very prominent. This is very emerging in stead of rice-wheat due to the constraints of irrigation water, especially in the tail-ends of canal commands. Chickpea is more remunerative than wheat under resource constraints situations.

Rice-lentil is better than chickpea in low land areas of eastern region because of pest problems in chickpea.

In the northern, western and central India (Maharashtra, Uttar Pradesh, Madhya Pradesh, Karnataka and Gujarat), pigeon pea-wheat systems is emerged better than rice-wheat system because of changing weather conditions *viz.* temperature, better price and limited irrigation facilities. Availability of short duration varieties of pigeonpea, which takes about 120-160 days to mature has enabled their introduction in rice wheat systems in irrigated area of western U.P., Punjab and Haryana, Delhi and North-East Rajasthan. This has provided desired stability and sustainability to productivity of cereal based cropping system.

Rice- blackgram/ green gram/ lathyrus is popular in coastal regions of South India. It is also suitable for rice-fallow areas in Andhra Pradesh, Odisha, Karnataka; while under irrigated conditions rice-rice-green gram/ black gram or cowpea is common.

In addition to sequence cropping, there are practices of intercropping with pulses and oilseeds under rice-based systems. More than 70% area of pulses is covered under intercropping systems in India.

Relay or catch cropping is commonly practiced in Bihar, Eastern Uttar Pradesh, West Bengal, Chhattisgarh and Odisha is a kind of relay method of sowing in which lentil/ lathyrus/ urdbean/ mungbean seeds are broadcast in the standing crop of rice about 2 week before its harvest. This practice enables to use better soil moisture available at the time of harvesting of rice crop, which could otherwise be lost quickly. Experimental evidences showed that paira cropping produced more yield of lentil than planting with tillage after harvesting of the rice crop.

Rice-fish farming system

The field with sufficient water retaining capacity for a long period and free from heavy flooding are suitable for rice-fish farming system. This system is being followed by the small and marginal poor farmers in rainfed lowland rice areas in Odisha and West Bengal. Farmers raise a modest crop of traditional low-yielding rice varieties, and fish co-culture. Rice-fish farming have been described in details in a separate chapter.

3.8. Deep water rice culture

The deep water rice culture differs in different areas with respect to depth of water, duration of flooding, the rate of increase in water level, temperature, turbidity and time of occurrence of flooding. There are certain conditions: varieties should have good seedling vigour, ability to elongate faster, submergence tolerance, kneeing ability for specific situation, varieties should be photoperiod sensitive and should be able to flower when it is least vulnerable to submergence.

At places double transplanting is also practiced. Sowing of deep-water rice should be done at least 4-5 weeks before the time of stagnation, so that the crop may have established quite well and would be in a position to grow with the rise of water. The older seedlings show better resistance than younger ones. The initial height of seedlings is also a good index of their capacity to resist submergence. Such seedlings firmly establish themselves by the time the water rises. As the water level increases, the culms, leaves and leaf sheaths elongate. The plant height varies from 2.5-5.0 m depending upon the depth of water where it is grown. Daily rate of growth of as much as 25-30 cm in height is not uncommon, when the water level rises rapidly. The total number of nodes and internodes became 15-20. The main (basal) tillers under deep water conditions are remarkably few. The deep-water varieties produce significantly more nodal (aquatic) tillers. Most of the nodal tillers initiate and produce shoots from the active region of the nodes when water level decreases to some extent, and the culm floats horizontally over the water surface. Actually a relationship exists between the growth of culm and level of water for the formation of nodal tillers. With a decrease in the water level, vigorous nodal branching starts. As a result there is no further elongation of next closer internodes, and the diameter of the internodes shows no remarkable change. At the pre-emerging reproductive stage, only 3 or 4 uppermost internodes elongate with usual rapidity.

Nodal tillering begins about 3 months after sowing and continues for 4-6 weeks thereafter. Nodal tillering is a characteristic feature of the deep water varieties which produce a considerable amount of grain and straw and contribute towards the final yield. Besides 4-5 main tillers per hill, there are about 7-8 nodal tillers also, borne on the nodes under water. These nodal tillers reach the height of the main shoot and bear ear heads. Most of the nodal tillers arise from some specific active nodes (*viz.* 7th, 8th and 9th) and their formation depend upon the level of water. In general nodal tillers grow strictly from the submerged nodes and contribute 20-30% of the grain yield. Anatomic studies suggest that air sac formation in stem is a characteristic feature of deep water varieties which helps in floating and tissue respiration under the hydrologic condition i.e. up to 4-5 cm depth of water. Deep water rice varieties develop 3-4 times more area of air sac than in the stem of normal rice. Presence of air sac is noticed in the higher nodes i.e. up to 19th-20th internode (250-300 cm) above soil surface under deep water condition. But under normal rice culture air sacs are present only up to the 4th-6th internode.

Aged seedlings (2-2 ½ months old and 100-200 cm height) are transplanted late in 50-60 cm depth of water. Sometimes, the abundant multiplication of filamentous algae in the deep-water rice fields is a serious problem. These floating algae (thick scum), particularly at high water levels, can completely suppress crop growth by preventing the young tender plants from rising above water so that they choke below.

High storm and strong water current particularly during peak monsoon season, sometimes uproot the deep water rice plants. The detached rice plants may still produce some grain because they form profuse adventitious roots at the upper nodes of the stem. Submergence under water induces root formation. Like some typical hydrophytes they can absorb nutrients from silty water through these physiologically active adventitious roots.

As the level of water in the field rises gradually plants grow, keeping pace with the rise of the level of water.

Rice plants usually lodge after the internodes have elongated and flooded water receded. Deepwater varieties have the kneeing ability to keep the first 3 leaves above the water level, preventing leaf decay and providing better leaf arrangement. Floating rice is grown where maximum depth ranges between 1 and 6 m for more than half the growth duration. These areas cover the flood plains of the Ganges and Brahmaputra.

The problems of deep water rice culture are (i) drought damage at germination and the seedling stage because the crop is direct-seeded into dry soil, (ii) weed competition at the seedling stage, (iii) poor stand establishment, (iv) high seedling mortality due to sudden flood, (b) submergence of rice at various growth stages and lodging susceptibility.

3.9. Rice ecosystem in coastal saline areas

Rice is grown under saline/ alkaline conditions, nearly in 4 million ha where yield returns are fairly low. About 2 million ha of land is under coastal regions of West Bengal, Odisha, Andhra Pradesh, Puducherry, Tamil Nadu, Kerala, Karnataka, Maharashtra, Goa and Gujarat. The coastal areas are subjected to saltwater inundation/ tides from the adjoining sea and the crop yields are adversely affected. Rice is the only crop grown under rainfed conditions i.e. during the monsoon season, in the coastal saline soils of those states. Where irrigation water is available second crop of rice is taken in Andhra Pradesh and Tamil Nadu. Under limited or no irrigation facilities watermelon, vegetables, barley are grown in limited areas in coastal areas of West Bengal. The rest of the land is kept fallow in West Bengal and Odisha. Where irrigation water is available in Andhra Pradesh, rice, groundnut, chickpea, sugarcane are grown. Mustard could be grown on those upland areas, where soil salinity does not exceed 7.0 dS m^{-1} and the monsoon rice is harvested earlier. Sunflower or sugar beet can also be grown when salinity does not exceed 9.0 dS m^{-1}.

The saline soils are economically important because those areas are located in deltas, estuaries, coastal fringes or river basins that are climatically, physiographical and hydrologically suited to rice; rainfall is adequate for leaching the soil and growing at least one rice crop/year; the lands are close to densely populated areas, where the demand for both land and food is high; and the labour and the know-how in rice cultivation are available. The coastal saline soils are often affected with deficiency of ferrous and zinc which causes chlorosis and reduced tillering. The effect of salinity on the growth of rice depends on the stage of development at which salinity occurs. Sensitive stages are transplanting stage, again on panicle initiation to flowering. The use of ammonium sulphate as fertilizer is better than urea. Salt injury may be due to osmotic stress, specific ion effects, ionic antagonism, toxicities caused by ions released by cation exchange and effects of ionic activities. In flooded soils, salt displaces K+, NH_4^+, Fe^{2+}, Ca^{2+}, Mg^{2+} from exchange sites into soil solutions.

Coastal saline soils are subjected to tides of different nature: (i) the strip nearest to sea subject to daily tides (ii) a strip several km wide subject to only exceptionally high tides (iii) an area where tides exercise their effect through rivers and creeks and (iv) a wide zone not influenced by tides but by the upward thrust of saline water. Zones (iii) and (iv) are suited to rice production without high inputs on reclamation.

West Bengal: The coastal saline soils of West Bengal, comprising about 0.85 m ha, are situated in the Gangetic plains and gradually slopes towards the Bay of Bengal. In Sundarban condition roughly soils containing 0.15% or more salt are referred as saline soils. The Sodium and Magnesium are the dominant cations on the exchange phase. This is partly responsible for stickiness of the soil under wet condition and very hard under dry condition. Sunderban soils under rice cultivation undergoes the following five operations in sequences (i) accumulation of rainwater on the soil surface (ii) puddling (iii) maintenance of a depth of standing water and soil throughout the growth period. In most of the places the depth of standing water is quite high even beyond 50 cm, (iv) drying of field during harvest and (v) reflooding the soil after 7-8 months for the cultivation of next rice crop. Major problem in raising seedlings is that initially high soil salinity affects germination in dry seed-bed and on the other hand sufficient water is not available for wet seedbed.

Odisha: About 0.4 m ha of area is affected by salinity. Water table is shallow at 1.5 m even in summer. Three distinct physiographical features could be recognized in the coastal districts of Balasore, Cuttack and Puri.

i) Narrow / saline marshy tract: This is a narrow strip of land running parallel to the coast varying from 3 to 10 km from the coastal line with essentially low relief. Along this tract, there are sand bars running parallel to the coast.

ii) Coastal tract: This tract lies between the narrow marshy land and the hilly upland tract and has gently undulating topography. This has an average width of about 50 km in the Subarnarekha basin region, which is gradually reduced to 15 km around Soro (Salandi Basin). This again gradually widens and attains maximum width of 80 km in the region of Baitarani, Brahmani and Mahanadi delta system and tapered again gradually towards Chilika lake region.

iii) Upland tract: These soils, distributed over the deltaic plains and down slope of the hill ranges constituting sedimentary and igneous rocks, and often characterized by lateritic capping of uneven thickness.

Andhra Pradesh: About 0.24 m ha deltaic alluvium and medium deep black soils are saline and saline-alkali soils. The electrical conductivity values vary from 0.4 to 16.8 dS m^{-1}. The rainfall varies from 763 mm in Prakasan district to 1138 mm in East Godavari district.

Tamil Nadu: Area affected is estimated to be 0.3 m ha. In the coastal areas of South Arcot, Thanjavur and east coast of Tirunelveli district, vast areas are affected by salinity. In the South of Cauvery delta alkaline soils occur mainly due to poor drainage. In the command areas of Periyar irrigation systems and Krishnagiri reservoir project soils are affected by alkalinity.

Kerala: In Kerala special type of saline soils are present. These soils are highly acidic and hence classed as acid saline (pH 3.3 to 5.5). These soils are periodically subjected to inundation by sea water. A special type of paddy cultivation is followed here. Big bunds strengthened by mangrove plants or bamboo and plaited coconut plants are formed to prevent the tides. During February-March the top soils is made into mounds one meter in diameter and ¼ to ¾ of a meter in height. The height of the mound will vary depending upon the level of the land in relation to sea level, lower the level greater will be the height. With the onset of monsoon the salinity in mounds will be washed down, and water in the field will not be saline. On flat top of the mounds the specially germinated seeds are sown. For sprouting the seeds it is first wrapped in green banana leaves and tightly covered by plaited coconut fronds. These bundles are immersed in fresh water for 12 hours after which they are stored in shade. The seeds will be stored to a maximum of 30 days. During the season of planting the seed bundles are resoaked for 6 hours. This process imparts salt tolerance to the crop. When the seedlings are 30-35 days old the mounds are dismantled and soil pieces with 3-4 seedlings will be distributed throughout the field giving an appearance of transplanted area. Thus crop will be ready for harvest in October-November.

Karnataka: About 4 million hectares are salt affected in black soil (ECe 3-4 dS m^{-1}), red soils (2-3 dS m^{-1}) and alluvial soils (0-16 dS m^{-1}). In the coastal area the early inundation and further flooding with saline water cause damage to the standing rice crop resulting in poor yield. At other places the rise in water table with increase in salts, particularly in low-lying areas, as a result of irrigation is the main reason of salinization.

Maharashtra: The state has 720 km of coastal belts. The average rainfall receipts in the area arise from 2540 to 3160 mm at different places. Except some parts of Ratnagiri district soils are usually clay loam to clayey loam in texture. pH ranges from 7.0 to 8.5. Total soluble salts vary from 3 to 6% and ECe may be as high as 14 dS m^{-1}. The recommendations for increasing rice yields in these area are (i) levelling of paddy land (ii) monsoon showers should be impounded for maximum possible time (iii) puddling should never be done (iv) aged seedlings should be planted closely (v) FYM and compost should adequately be used (vi) salt tolerant varieties should be planted.

Gujarat: Gujarat state has about 1600 km of sea coast. Saline soils are extensively distributed on the coastal areas as well as inlands. About 0.9 m ha are affected by salinity. In Surat district, where irrigation facility is available, rice is recommended to be grown in *kharif.* Rainfall ranges from 1000-1500 mm.

Goa: In Goa out of 17,600 ha of alluvial stretch of land, 15,000 ha are subjected to inundation of saline water and hence problematic. These soils are classified as '*Zuari*' series and locally known as '*Khazan* lands'. Rainfall is very high 3000-5000 mm. pH ranges from 5.0 to 5.2 cm. ECe may go as high as 10-15 dS m^{-1}. *Khazan* lands are utilized for rice cultivation during *kharif* only.

Mitigating the evils of total water inundation through construction of embankments, accelerating leaching of excess salts by application of organic manures in *kharif* and mulches in the dry season, lowering of water tables through improvement of drainage system, water available for irrigation, evaluation of water quality, suitable agronomic practices- land levelling, combined application of gypsum and compost @ 10 t ha^{-1} for particularly inland saline-alkali soils, no puddling, close planting (15 x 15 cm) with 4-5 plants/ hill, growing of suitable varieties, providing adequate amount of fertilizer including Zn, Fe etc. Besides, this increases the availability of trace elements mainly Fe. Salt injury is likely to be more severe in soils low in Ca and Mg, such as acid sulphate soils and peat soils than calcareous soils. Salt injury in rice depends on both salt concentration and sodium absorption ratio (SAR). Salt tolerance appeared to be associated with K absorption in plants. The decrease in concentration of K in the shoots and the increase of Na, Ca and Mg. Salt depressed concentration of Fe in the tolerant variety as compared to salt susceptible varieties.

3.10. Rice-ecosystem in acid soils

The total area under red, laterite and lateritic soils in India is about 1.08 million square km. These soils are generally light textured, acid to moderately acid, low in C, N, available P, cation exchange capacity, exchangeable cations, water holding capacity, and high in Fe and Mn. Excepting soils from Pattambi, Kaolinite is the dominant mineral in most of the soils while illite in some soils. Laterite soils are often deficient in K. Soils of Chotanagpur plateau are normal with respect to Zn, high in Cu and normal to deficient in B. Acid soils have been reported to give low yields due to inherent soil problems: (i) moderate to high acidity causing nutrient imbalance, (ii) lack of adequate native nutrients, (iii) high Fe and Al content resulting in high P fixation, (iv) toxicity of Fe, (v) impeded drainage in certain areas causing acute reduction of the soils, and (vi) terraced topography, loose texture and poor physical conditions encouraging the loss of soil and applied nutrients due to leaching and erosion.

The upland rainfed crop on these acid soils faces the problem of severe weed infestation, which when controlled can provide good yields with right choice of varieties and fertilizer management practices. In lower terraces where the long-duration varieties are grown, the crop suffers from bronzing disease due to Fe-toxicity, which can be overcome by the provision of drainage, soil application of lime, phosphate and potash. In P deficient soils and soils having high P fixing capacity, it is necessary to saturate the P fixing capacity while formulating P application. Application of carbonate dominant rock phosphate is recommended to moist aerobic P-deficient acid soils, 2-3 weeks before flooding and transplanting. Mixture of rock phosphate and superphosphate is advisable in acid lateritic soils.

Basic materials like lime, calcium silicate etc. in acid soils help in correcting acidity, increasing microbial activity, decreasing phosphate fixation and non-release of toxic elements like Fe, Al and Mn into soil solution. Low response to liming might be due to the stabilization of pH around 6.0 -6.5 in the flooded soils. Hence, liming might not have a direct beneficial effect on the rice crop but might have an indirect effect by dispersing soil organic matter due to rise in pH for rapid microbial decomposition to release more of ammonium nitrogen equivalent to 20-30 kg N ha^{-1}.

3.11. Conclusion

Main reasons of low rice yields particularly in the eastern regions of the country are: in-efficient utilization of applied nitrogen, limited scope for water management in heavy rainfall areas, cloudiness and its ill effects on the photosynthetic activity of rice plant in monsoon season, susceptibility to and heavy incidence of disease and pests, and adverse soil conditions such as soil salinity management problems. Rice and water management is comparatively complex and specific. Water management and crop establishment practices produce beneficial outcomes for any rice system. The saturated soil culture and ponding with water facilitate rice systems even in sloppy areas. Over time, submerged rice ecosystems have been evolved and designed to enable integrated farming systems with the components *viz.* fisheries, other cash crops, and livestock. There is the need for developing strategies for yield maximization in different land situations and soil conditions. Water management and associated technological options have been elaborately described in the subsequent chapters in this book.

References

Barker, R. and Herdt, R.W. 1979. Rainfed lowland rice as a research priority- an economist's view. IRRI 1978 Rice Research Conference, International Rice Research Institute (IRRI), Philippines, pp. 3-50.

Biswas, S. and Mallick, S. 1983. Strategy for yield improvement-varieties for lowland waterlogged situation and flood-prone areas in West Bengal. Workshop on All India Coordinated Research Project, pp. 12-18.

Chatterjee, B.N. and Maiti, S. 1985. Principles and Practices of Rice Growing. Oxford & IBH Publishing Co. Pvt. Ltd., New Delhi, 419 p.

Das, S.R. 2012. Rice in Odisha. Technical Bulletin No. 16, International Rice Research Institute (IRRI), Los Baños, The Philippines, 31p.

DES, 2020. Area, production and yield along with coverage under irrigation. In: Agricultural Statistics at a Glance 2019. Directorate of Economics and Statistics, Ministry of Agriculture & Farmers' Welfare, Govt. of India, New Delhi. p. 46-92.

DES, 2023. Agricultural Statistics at a Glance 2022. Directorate of Economics and Statistics, Economics and Statistics Division, Ministry of Agriculture & Farmers' Welfare, Govt. of India, New Delhi. 280p.

Gangopadhyay, S. 1991. Agricultural characteristics in the agro-climatic zones of Indian states: Odisha. In: Ghose SP, editor. Agroclimatic Zones in Specific Research, Indian perspective under NARP. Indian Council of Agricultural Research, New Delhi.

Garrity, D.P. 1984. Rice environmental classifications: a comparative review. In: Terminology for rice-growing environments. International Rice Research Institute, Manila, Philippines, Pp. 11-35

Gill, M.S., Shukla, A.K., Pandey, P.S. 2008. Yield, nutrient response and economic analysis of important cropping systems in India. Indian Journal of Fertilisers. 4:11-48

Greenland, D.J. 1997. The sustainability of rice farming. CAB International, UK, 273 p.

Halwart, M. and Gupta, M.V. (eds.). 2004. The Rice Field Ecosystem. In: Culture of Fish in Rice Fields. FAO, Rome and The World Fish Centre, Malaysia, p.5.

Huke, R. 1976. Geography and climate of rice. In: Proc. Symposium on Climate and Rice. International Rice Research Institute (IRRI), Philippines, 176p.

IRRI, 2021. Rice in Odisha. In: Rice-based cropping systems knowledge bank, Odisha. Rice Knowledge Bank, Information source, Govt of Odisha and ICAR. International Rice Research Institute (IRRI), Odisha Hub Office, Bhubaneswar, https://www.rkbodisha.in/rice#rice-colsp2.

Khush, G.S. 1984. Terminology for rice growing environments. International Rice Research Institute (IRRI), Los Baños, Philippines, pp. 5-10.

Kramer, P.J. 1969. Plant and soil water relationship- a modern synthesis. McGraw- Hill Book Co., New York, 482p.

Mackill, D.J., Coffman, W.R., Garrity, D.P. 1996. Rainfed lowland rice improvement. International Rice Research Institute, P.O. Box 933, Manila, Philippines, pp. 3-24.

Moorman, R. and Van Breeman, N. 1978. Rice, soil, water and land. IRRI Publication, International Rice Research Institute (IRRI), Philippines, 185p.

OAS, 2020. Odisha Agriculture Statistics 2017-18, Directorate of Agriculture & Food Production, Odisha, Govt. of Odisha, Agriculture Information Wing, Bhubaneswar, August 2020, p. 11-20

Panda, B.B., Satapathy, B.S., Nayak, A.K., Tripathi, R., Shahid, M., Mohanty, S., Bhaduri, D., Khanam, R. and Nayak, P.K. 2018. Agro Ecological Intensification of Rice Based Cropping System. In: (Eds) Pathak, H., Nayak, A.K., Jena, M., Singh, O.N., Samal, P. and Sharma, S.G. Rice Research for Enhancing Productivity, Profitability and Climate Resilience. ICAR- National Rice Research Institute, Cuttack, India, pp. 256-272.

Pani, D., Patra, B.C. 2004. Genetic improvement of rice varieties in Odisha. In: Sharma SD, Prasada Rao U, editors. Genetic improvement of rice varieties of India. Today and Tomorrow's Printers and Publ., New Delhi, India. p 793-833.

Pathak, H., Samal, P. and Shahid, M. 2018. Revitalizing Rice-Systems for Enhancing Productivity, Profitability and Climate Resilience. In: (Eds) Pathak, H., Nayak, A.K., Jena, M., Singh, O.N., Samal, P. and Sharma, S.G. Rice Research for Enhancing Productivity, Profitability and Climate Resilience. ICAR- National Rice Research Institute, Cuttack, India, pp. 1-17.

Ramiah, K., Ghose, R.L.M. 1951. Origin and distribution of cultivated plants of South Asia: Rice. Indian J Gent Plant Breed. 11: 7-13.

Ramiah. K, Rao, M.B.V.N. 1953. Rice Breeding and Genetics. Indian Council of Agricultural Research, New Delhi.

Sharma, S.D., Tripathy, S. and Biswal, J. 2000. Origin of Oryza sativa and its ecotypes. In: Nanda, J S (ed) Rice Breeding and Genetics: Research Priorities and Challenges. Science Publishers, Enfield and Oxford and IBH, New Delhi, pp. 349-369.

4

Water Management Strategies and Technologies in Rice

4.1. Introduction

Globally, rice is cultivated on about 165 million ha of land, which provides around 787 million tons of paddy annually (FAO, 2023). In Asia, 708 million tons of paddy is produced from nearly 143 million rice-cultivated areas. Rice crop is grown in a wide range of climates, varied soil types, and hydrological situations. Rice is the lone cereal crop that grows well in floods.

The rice habitat can be divided into four groups based on the hydrology of the rice field: irrigated lowland rice (79~80 million ha), rainfed lowland rice (54~55 million ha), upland rice (14~15 million ha), and flood-prone rice (11~12 million ha) (Table 4.1) (Bouman *et al.*, 2006).

Table 4.1. Characteristics of rice field environments

Category	Sub-category	Characteristics
Lowland rice		Known as "paddy rice," these fields have standing water for a minimum of 20% of the crop's growing season and anaerobic (saturated) soil conditions.
	Irrigated lowlands	Guaranteed irrigation to sustain stagnant water for a minimum of 80% of the crop's life.
	Rainfed lowlands	Rainfall is the only source of water, and the duration of stagnant water in the rice field cannot be guaranteed due to the unpredictable nature of rainfall.
Flood-prone environments	-	Rice fields with periodically excess and uncontrolled water, deep flooding (> 25 cm for 10 days or more). For example- deepwater rice and floating rice
Upland rice	-	Well-drained, aerobic (unsaturated) soil conditions without any stagnant water for > 80% of the crop's duration.

75% of the world's rice production comes from 80 million hectares (50%) of irrigated lowlands (Maclean *et al.*, 2002). Lowland irrigated rice is primarily grown with an additional supply of irrigation during the rainy season if needed, however, it depends on irrigation during the dry season. In Asia, average yields of rice grown under irrigation varied from 3 to 9 t ha^{-1}, with an average of roughly 5 t ha^{-1}. Large tracts of rice are planted in systems of rice-wheat, rice-maize, rice-groundnut, and rice-potato rotation alongside a variety of other crops.

In the deepwater rice system, paddy is grown in flooded conditions with water > 50 cm deep for a minimum of a month. Traditional tall varieties are grown at water depths between 50 and 100 cm. Floating rice grows in water deeper than 100 cm through advanced elongation ability. Deepwater rice is grown in roughly 90,000 km^2 of area, mainly in South East Asia including North Eastern India, where more than 100 million people depend on it for their livelihood. In Southeast Asia, the main area of deepwater rice is grown in the Brahmaputra valley in Assam and other Northeast Indian states. Other regions are in Burma in the Irrawaddy Delta, in Thailand in the Chao Phraya, and in the Mekong in Vietnam and Cambodia. In West Africa, deep-water rice is grown in the Niger River basin and some areas in Ecuador.

For deep-water rice cultivation, seeds are sown directly into the soil, and when flood or water comes the plants grow and elongate with the water level rise. Overall grain yield is very low in deep water or floating rice. Sometimes sudden flooding, where a large volume of water enters the field in a short time, can lead to a high level of seedling death. It also faces problems due to the depth and time of the flood water in which it grows. Water conditions such as turbulence and temperature can adversely affect the crop. Natural disasters like cyclones can damage or destroy deep-water rice crops, particularly in Asia.

A cultivar 'Swarna Sub1' was developed with the ability to withstand submergence of 14 days. The submergence tolerance ability of this variety is conferred by the presence of the Sub1A gene, introgressed from the Indian cultivar 'FR13A' into the flood-vulnerable, but high-yielding cultivar 'Swarna'.

4.2. The water balance in the rice field

Generally, when the seedlings are 2-3 weeks old, rice is transplanted into the main field after being raised in a seedbed, which is one-tenth of the original field. In the main field, it can also be grown by direct wet seeding, which involves broadcasting pre-germinated seeds onto puddled soil free of standing water, or direct dry sowing, which involves broadcasting (manually or by using a machine) dried seeds onto moist or dry soil. Direct seeding covers around one-fifth of Asia's rice-growing land, yet production levels are still lower than with transplanted rice and seed requirements are also high.

The main rice field is prepared under wet/flooded conditions before transplanting. It consists of dry ploughing, soil soaking with water, ploughing, and puddling. Puddling is used to manage weeds, lessen soil permeability, and make seedling transplantation easier. Puddling leads to the complete or partial destruction of soil aggregates and macro-pore volume, increasing micro-pores. A vertical cross-section of a puddled rice field reveals- a top ponded water layer (0-10 cm), a puddled and muddy topsoil (10-30 cm), a hard plough pan that is formed due to puddling by decades or centuries, and a bottom undisturbed subsurface soil (Fig. 4.1). Most of the rice roots are shallow and limited within the puddled layer. The plough pan reduces the percolation rate (downward movement of water) and hydraulic conductivity of rice fields. After initial crop establishment, the rice field is usually kept flooded with a 5-10 cm depth of water throughout the crop growth to control weeds and also some pests. Ponded water is drained 7-15 days before crop harvest from the rice field.

Because the rice field remains flooded, it has a different water balance than the dryland crops like wheat (*Triticum aestivum*) or maize (*Zea mays*). The water balance in a typical paddy field consists of rainfall, irrigation, and capillary rise as an inflow and transpiration, evaporation, over-bund flow or runoff, seepage, and percolation as outflows (Fig. 4.1). In a rice field, a main source of water inflow is either rainfall or irrigation. Capillary rise is the upward water movement from the groundwater. In aerobic or non-flooded soil conditions, water through capillary rise moves into the root zone and provides water to the crop. However, "percolation," or the downward movement of water from the puddled layer down below the plough pan, happens in flooded rice fields, preventing capillary rise into the root zone. Therefore, the capillary rise is usually neglected in the water balance of typical rice fields.

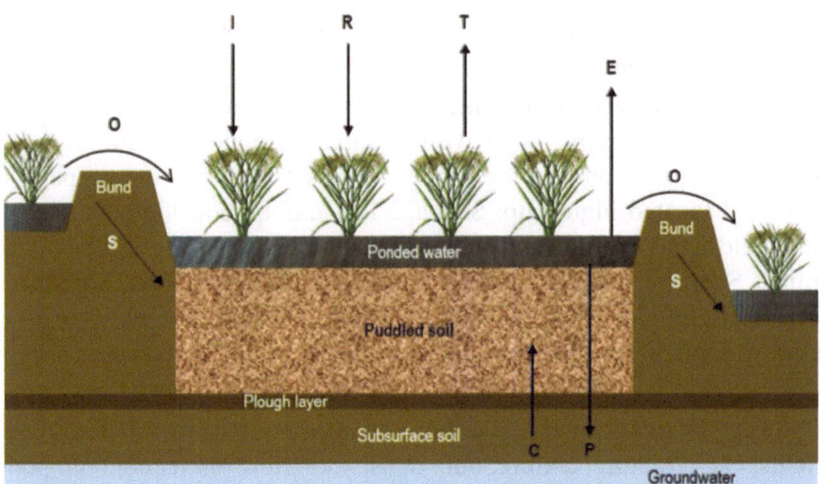

Fig. 4.1. Water balance of a lowland rice field. C = capillary rise, E = evaporation, I = irrigation, O = over-bund flow, P = percolation, R = rainfall, S = seepage, T = transpiration (*Source*: Bouman *et al.*, 2007)

All water outflows from a rice field-runoff or over-bund flow, evaporation, seepage, and percolation-are regarded as losses from the crop since they are unproductive. The only water flow that is productive since it aids in crop growth and development is transpiration.

Surface runoff or over-bund flow occurs when the stagnant water level rises above the bund height due to rainfall and excess rainwater leaving the rice field. This runoff can flow into a neighbouring field, but, in a sequence of fields, neighbouring fields will pass on the runoff until it is lost in a drain or ditch. Interestingly, it also causes loss of nutrients applied through fertilizer and could lead to environmental pollution.

Evaporation (E) - water losses directly from the stagnant water of the rice field.

Transpiration (T) - water is released from plant leaves. It is the process of water release from aerial parts of the plants, such as leaves, stems, and flowers. Water is essential for plants but only a small proportion of it is taken up by the roots and is used for growth, development and metabolism, the remaining 97-99.5% is lost by transpiration. Rice plants withdraw water from the puddled layer, as its roots generally don't penetrate the plough or compacted layer, the contribution to transpiration from the subsoil is almost nil. Since evaporation and transpiration are difficult to measure separately in the field, they are usually measured together as 'evapotranspiration' (ET). During the active crop growth phases, nearly 30-40% of ET is evaporation. Typically, ET rates of rice fields in the wet season (rainy days) are 4-5 mm d^{-1} and 6-7 mm d^{-1} in the dry (*rabi*) season but on hot days ET can be as high as 10-11 mm d^{-1}, especially in subtropical regions (Tabbal *et al.*, 2002).

Seepage (S) – It refers to the water that flows beneath a rice field's bunds. Seepage can be negligible with well-maintained bunds. In continuous rice fields, seepage loss from one field may be offset by incoming seepage from another field located higher up. Seepage loss can be considerably high from top-end fields and bottom-end fields, those boundaries are ditches, cracks, or drains. Seepage rates are affected by several factors, *viz.* soil physical characteristics of the field and bunds, bunds maintenance and length, and water table depth in the field and the surrounding drains, ditches, or creeks.

Percolation (P) – It is the vertical downward water flow below the crop's root zone. Numerous soil characteristics have an impact on the rate of percolation in rice fields, viz. structure, texture, bulk density, organic matter content, ponding depth, groundwater depth, mineralogy, and salt type and its concentration (Bouman *et al.*, 2007). Puddling changes the soil structure, and heavy-textured clay soil having high bulk density is favourable for reducing percolation rates.

However, light-textured sandy soil has a high percolation rate. Higher depths of ponded (stagnant) water favour high percolation rates, however, rice fields with no stagnant water have a very low percolation rate. Percolation rates were found to be higher in fields with deep groundwater tables (>2 m depth) than in fields with shallow groundwater tables (0.5–2 m depth). Usually, seepage and percolation flows are not easily separated because of transition flows that cannot be classified as either percolation or seepage, therefore, mostly it is measured as seepage and percolation (SP) rates. In heavy clay soils, SP rates fluctuate from 1-5 mm d^{-1}, while it goes to 25-30 mm d^{-1} in sandy and sandy loam soils. SP losses of water account for about 25-50% of all water inputs in heavy soils with shallow groundwater tables of 20-50 cm depth (Cabangon *et al*., 2004; Dong *et al.,* 2004), and 50-85% in coarse-textured soils with deep groundwater tables of 1.5-m depth or more (Sharma *et al*., 2002; Singh *et al*., 2002).

It is helpful to identify the water outflows that are recyclable or not from the rice fields. Non-reusable outflows are called 'depleted water' for example, evaporation, and transpiration. At the field level, SP losses and over-bund flow are often reused at the downstream side. Only when these flows enter very deep or saline groundwater (or saline surface water) from where they cannot be recovered are they not reusable anymore and they become depletion flows as well. The water reused in rice-based irrigation systems is difficult to know and varies widely amongst irrigation systems.

Generally, in dryland crops like wheat and maize, daily SP losses do not occur as in rice because of the ponded water. Also, evaporation from surfaces of ponded water is higher than from soil surfaces (as in dryland crops). Therefore, lowland rice fields are considered "heavy water users" due to the comparatively substantial water fluxes caused by seepage, percolation, and evaporation. Rainfall + irrigation can result in a rice crop that receives up to 2-3 times the total amount of water input compared to other crops like wheat or maize. It ranges from > 2,000 mm in sandy or loamy coarse-textured soils having deep groundwater levels to as little as 400 mm in clay and heavy textured soils with shallow groundwater tables (mostly it is used for crop transpiration) (Bouman and Tuong, 2001; Cabangon *et al*., 2004).

Table 4.2 gives typical values of water use from a rice field. For a crop of 120 days duration (modern high-yielding varieties), total water uses vary from 775 to 5,190 mm, depending on the crop season and soil characteristics. In Asia, the average water input for irrigated rice is between 1,500 and 2,000 mm.

Table 4.2. Typical daily rates and seasonal water use in rice production in the tropics

Item	**Daily rates (mm d^{-1})**	**Seasonal use (mm)**
Land preparation (wet)	-	175-750
Evapotranspiration (ET)		
Wet season	4-5	480-600
Dry season	6-7	720-840
Seepage and percolation (SP)		
Heavy clay soil	1-5	120-600
Loamy / sandy soils	25-30	3000-3600

Water is required for rice before the crop begins to prepare wetland. After puddling, the field is usually left fallow and flooded for a few days (or 1 to 4 weeks) before the seedlings are transplanted called an 'idle period'. When the rice is directly wet-seeded or if there is a short period between soaking the field and transplanting, the water required for land preparation/puddling can be as low as 150–200 mm. However, sometimes in large-scale irrigation systems with poor water control, the time between land soaking and transplanting can go up to 2 months, and water requirement during this period can be very high (up to 940 mm), a wastage of this precious input (Tabbal *et al.*, 2002).

4.3. Global water use in rice cultivation

The amount of irrigation water utilized worldwide by rice fields is not yet well-documented. It can be approximated, therefore, using the total amount of water extracted globally for irrigation, the proportion of irrigated rice land (in comparison to other crops), and the amount of water that rice fields utilize. Total worldwide withdrawals of fresh water are estimated at 3,999 billion m^3 annually, of which 2,799 billion m^3 is used to irrigate crops, i.e. 70% of total withdrawal (FAO, 2021). The rest is used in industry (760 billion m^3) and for domestic purposes (440 billion m^3). Global freshwater use (%) by different sectors is presented in Fig. 4.2.

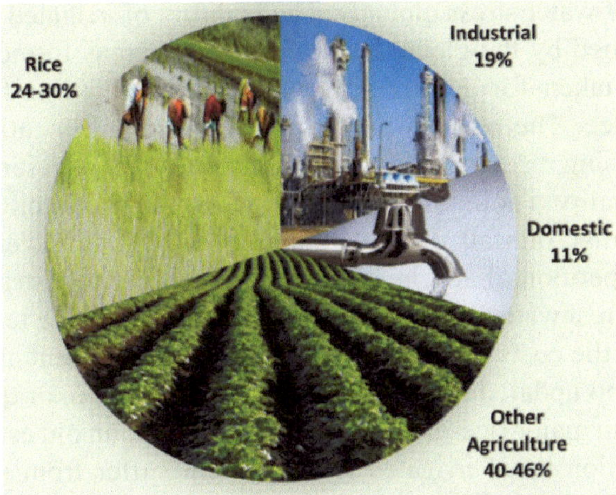

Fig. 4.2. Global freshwater use by the sectors

Asia is home to over 56% of the 275 million hectares of irrigated land worldwide, with 40–46% of this net irrigated area of rice cultivated (FAO, 2021). Rice can receive 2-3 times as much water at the field level as other irrigated crops, but an undetermined percentage of the water lost from one field is recycled by other fields downstream. Based on a 25% reuse percentage, irrigated rice is predicted to get 34–43% of the world's irrigation water, or 24–30% of all freshwater withdrawals worldwide. The amounts of irrigation water used for agricultural and rice production are shown in Fig. 4.3.

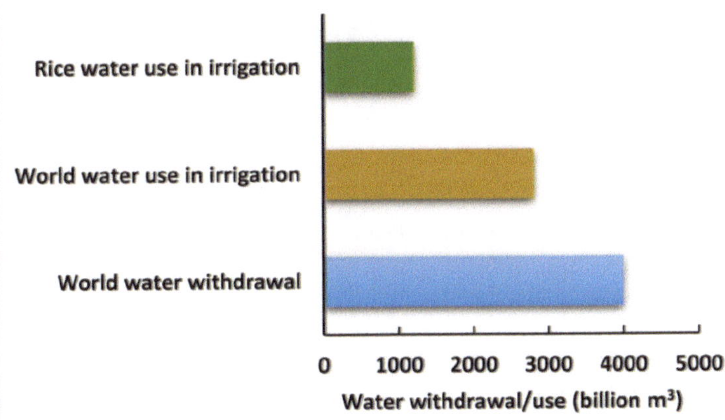

Fig. 4.3. Volumes of water withdrawal and used for agriculture and rice production annually.

4.4. Water scarcity in rice-cultivated areas

Worldwide, several countries are experiencing water stress due to greater freshwater withdrawals than it is being replenished. Fig. 4.4 provides a visualization of water stress globally. The viability of irrigated rice production is now threatened by the global "water crisis," whereas formerly the amount of water was taken for granted. The availability of water for irrigation is becoming scarce. The reasons for this are diverse and location-specific but include decreasing resources (falling groundwater tables, reservoirs siltation), decreasing quality (pollution, salinization), malfunctioning of irrigation systems, irregular rainfall due to climate change, and several others. The increased competition of accelerated urbanization and industrial development further limits freshwater resources for rice production. Due to the increase in water scarcity, the costs of its use and resource development are increasing as well. There is no updated systematic inventory, definition, or quantification of water scarcity in major rice-growing areas. Moreover, an old estimate indicates that 15-20 million ha of irrigated rice area will suffer from some degree of water scarcity by 2025 (Tuong and Bouman, 2003).

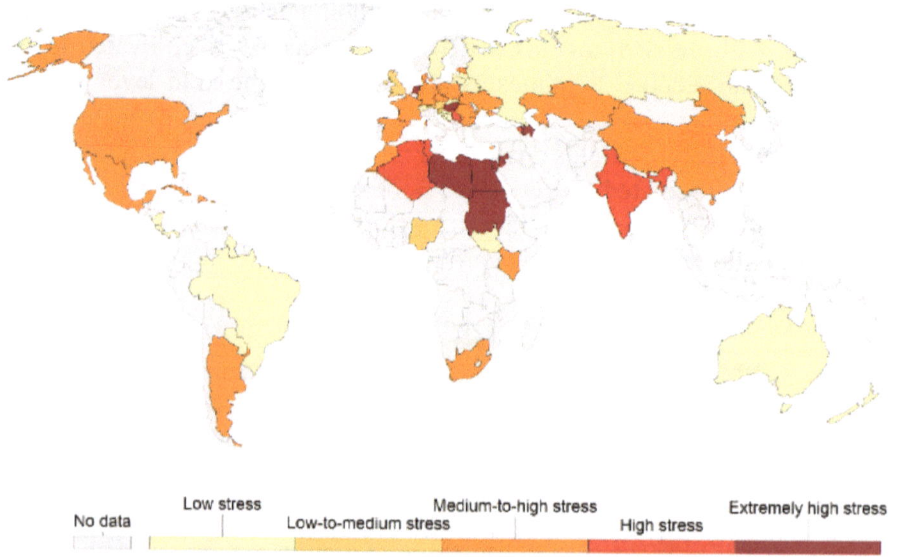

Fig. 4.4. Levels of water stress across the world (*Source:* FAO, 2021)

In South Asia, several rivers, *viz*. the Ganges, Indus, Mahanadi, Godavari, and Krishna have small outflow to the sea during the post-rainy season affecting rice-growing areas downstream. During recent decades, overexploitation of groundwater has caused serious problems in South Asia and northern China, affecting major rice-wheat-growing areas. In the Indian states of Punjab, Haryana, Rajasthan, northern Gujarat, Maharashtra, and Karnataka, groundwater tables have decreased by an average of 0.5–0.7 m yr^{-1}. In Tamil

Nadu and southern India, they have decreased by around 1 m yr^{-1}, and in the North China plain, where flooded rice is the main crop, they have decreased by 1-3 m yr^{-1}. The extensive usage of groundwater in West Bengal, India, and Bangladesh has resulted in serious issues with arsenic pollution in rice-growing regions. The Cauvery Delta in southern India and the Chao Phraya Delta in Thailand, a significant regional rice bowl, are experiencing water scarcity due to intense competition for river water between states and various sectors (domestic, industry). Numerous studies indicate that there may be localized hotspots of water scarcity due to heightened competition among various water users, even in places that are typically not thought to be water-scarce, such as the Zanghe Irrigation System in the middle reaches of the Yangtze and Angat reservoir near Manila, Philippines. In general, during the dry season, when farming is impossible without irrigation due to a lack of rainfall, water is always in short supply.

4.5. Approaches to managing water scarcity

In this section, technological options to deal with water scarcity at the field level have been presented. Recently, some of these technological interventions have been described by researchers (Surendran *et al.*, 2021). A major portion of water input can be reduced by minimizing unproductive outflows like seepage, percolation, or evaporation while maintaining transpiration flows. There are two ways to do so: i) by enhancing the resistance to water flow in the soil, and ii) by decreasing the hydrostatic pressure (i.e., depth) of the ponded water. This can be done starting from land preparation to crop establishment, and also throughout the entire crop growth period.

4.5.1. Soil and field management

Soil and field management strategies to save water are field channels for irrigation, land levelling, tillage practices (puddling, bund preparation, and maintenance), and mulching in the rice field. These techniques are done while the land is prepared or after transplanting. In addition to being crucial for effective water management, land preparation sets the stage for the entire cropping season.

4.5.1.1. Field channels

The most common way for irrigation and drainage of rice fields in Asia is 'plot-to-plot' irrigation, where water flows from one to another field either through opening or over the bunds. Furthermore, a lot of irrigation systems lack field channels, also known as "tertiary" irrigation or drainage channels, which makes it impossible to regulate the amount of water entering and leaving a rice field and makes field-specific water management impossible.

Water continuously flowing from one rice field to another removes valuable (fertilizer) nutrients.

Farmers frequently find that water from other fields keeps seeping into their fields, making it impossible to drain them before harvest. Several water scarcity technologies necessitate effective water management for certain areas. Building distinct field channels to transport water to and from each field (or to a small group of fields) is the preferred practice in any kind of irrigation system because it significantly enhances individual control over water and conserves a significant amount of water.

Plot-to-plot irrigation in rice

Field channels in the rice field

4.5.1.2. Land leveling

A well-leveled field is a prerequisite for a good and uniform rice crop. When paddy fields are not properly levelled, water may stagnate in depressions, whereas higher parts may become dry. This results in uneven crop emergence and uneven early growth, uneven fertilizer distribution, and maybe extra weed problems. Properly levelled land can enhance yield, reduce water use, improve nutrient-use efficiency and help in weed control.

Field trials were conducted in rice to evaluate the interactive effects of rice planting methods and land levelling on 16 farmers' fields in Western Uttar Pradesh, India. Researchers found that precision land levelling improves the grain yield and water productivity of rice grown as puddled transplanted or unpuddled direct-seeded rice (Table 4.3). The total saving in irrigation water in DSR rice over the puddled transplanted rice was nearly 250 ha-mm shared equally between direct seeding and precision land leveling. The water productivity of rice cultivation was increased from 0.41 to 0.47 under DSR and from 0.33 to 0.39 kg grain m^{-3} water under transplanted rice (Jat *et al.*, 2006).

Table 4.3 Effect of planting methods and precision levelling on rice yield and water productivity

Crop establishment method	Rice grain yield (t ha^{-1})		Water productivity (kg grain m^{-3})	
	Laser leveling	Traditional leveling	Laser leveling	Traditional leveling
DSR	5.25	5.10	0.468	0.409
TPR	5.41	4.98	0.394	0.331

DSR-Direct seeded rice; TPR-Transplanted rice
Source: Jat *et al.* (2006)

4.5.1.3. Tillage to reduce soil permeability

SP losses from paddy fields are determined by the permeability (hydraulic conductivity) of the soils: their capacity to move water downward and sideward. Rice fields with few rat holes or leaky spots can cause a rapid loss of water by seepage and percolation.

While land soaking before puddling, large amounts of water can be lost if there are large and deep cracks in the field, hence additional shallow tillage before land soaking could save a huge amount of water. Before land soaking, additional shallow soil tillage closes the cracks of the field and the quantity of water required for wet land preparation is reduced from about 350 mm to about 250 mm, a 100 mm saving (Cabangon and Tuong, 2000).

A compacted hard plough pan produced by thorough puddling lowers permeability and percolation rates during the crop-growing season. Puddling can decrease percolation, but its effectiveness is mostly dependent on the characteristics of the soil. Puddling, for instance, might not be the best method for lowering percolation losses in coarse soils because there aren't enough fine clay particles in these soils for the particles to move downhill and fill in the pores and cracks in the plough pan. Conversely, puddling works particularly well on clay soils where the fallow period creates cracks that penetrate the plough pan. It is evident that puddling consumes lots of water but also reduces percolation rates of the soil, and there should be a trade-off between the amount of water used for puddling and the amount of water that could be 'saved' by reduced percolation rates during the entire crop growth period. In heavy clay soils having low vertical permeability or limited internal drainage, puddling may not be necessary. In such soils, direct dry seeding on tilled land (without puddling) is very well possible with minimal percolation losses.

In sandy soil with loamy subsoils containing a minimum of 5% clay, the application of heavy machinery for soil compaction has been demonstrated to reduce soil permeability; nevertheless, most farmers are unable to afford to compact their soils. This technology may only be feasible with governmental support.

4.5.1.4. Bund preparation/maintenance

To limit seepage losses and under-bund flows, good and compact bunds are essential. Bund should be well plastered with mud before the crop season starts to seal cracks or rat holes.

Bund plastering with mud to check seepage and under-bund losses of water

The height of the bund is important to avoid over-bund flow during heavy rainfall. In general, a bund height of 20 cm is sufficient to avoid over-bund flow, but it depends on the amount and intensity of rainfall in a particular area. The use of plastic sheets in bunds reduces seepage losses. Bouman *et al.* (2005) demonstrated a reduction of 450 mm in total water use in a rice field by lining the bunds with plastic.

Bund lining with plastics

4.5.1.5. Mulching in rice

Different methods/ materials of mulching (e.g., dry soil, rice straw, and plastic sheets) have been experimented with non-flooded rice systems in various countries and shown to reduce evaporation as well as percolation losses and give high yields (Yan *et al.*, 2018). In hilly areas in Shiyan, Hubei Province, in China, farmers are using plastic sheets to cover rice fields where the soil is kept below the saturation level. The reported advantages are earlier crop establishment by 3 weeks (rice is established at low temperatures at the same time soil temperature rises due to plastic sheets), more yields, suppression of weed growth, and saving water.

Mohammed *et al.* (2016) used rice straw mulch (2 mm thick) in the SRI field and found it could control 98.5% of weeds. Approximately 4 tonnes of rice straw can be used to cover soil on 1 ha of land (Devasinghe *et al.*, 2011). Rice straw can remain for a longer period in the field due to higher lignin and silicon contents as well as low protein and digestibility (Hanafi *et al.*, 2012). It has also been recognized as a natural herbicide because, during its degradation process (El-Shahawy *et al.*, 2006; Kato-Noguchi and Ino, 2005), it releases phenolic compounds (caffeic, cinnamic, ferulic, pcoumaric, o-coumaric, and p-hydroxybenzoic acids), which acts as the allelopathic compound (Chung *et al.*, 2003; El-Shahawy and Zydenbos, 2010) for weeds suppression (Devasinghe *et al.*, 2011). At maturity, the vegetative part of rice crops contains about 40% of N, 30-35% of P, 80-85% of K, and 40-50% of S, of the total absorbed N, P, K, and S from the soils (Hanafi *et al.*, 2012). Thus, rice straw can be used as organic fertilizer for recycling (Nader and Robinsons, 2010) and it can provide feeds for the soil microbes. Therefore, it leads to less application of nitrogen, lowers production costs, and reduces water pollution (Bird *et al.*, 2002). Wheat straw mulch improves root growth, dry matter accumulation, grain yield, rice quality, and nutrient uptake (Yan *et al.*, 2018; 2019). They increased total N, P, and K accumulation in rice plants by 1.8-10.8%, 2.7-42.2%, and 16.4-17.9%, respectively, thus improving utilization of N, P, and K. However, extensive research is required to verify these benefits. Another issue is the residual plastic during harvest, which, if improperly disposed of, could pose a risk to the environment.

Rice crop with plastic mulching

4.5.2. Reducing turnaround time

The time interval between land soaking for wet soil preparation and transplanting determines the amount of water waste in the rice production system. The main rice field is not producing any crops during this time, and field water outflows are not helping to produce any crops either. Ideally, there should be as little time as possible between soaking the field and transplanting it to reduce this wasteful water loss. Plot-to-plot irrigation is practised in many large-scale irrigation systems, yet water losses remain very high during the turnaround period. Farmers usually raise seedlings in one part of their main field. Therefore, the main field remains completely submerged during seedbed preparation and it remains submerged for the entire period of the nursery raising because of the absence of tertiary field channels. Making field channels, the use of shared seedbeds, or direct wet or dry seeding can all reduce the turnaround time in such systems. Field channels eliminate the need to flood the main field by allowing water supply to the individual seedbeds independently. Common seedbeds can be watered as a single block and placed strategically next to irrigation canals.

In direct dry sowing, the crop begins using water when it is established. Moreover, direct dry seeding lessens the demand for irrigation by efficiently utilizing rainwater. Dry seeding followed by flooding, however, is only feasible in heavy (clayey) soils that have poor internal drainage and low permeability. Since labour is not needed for transplanting and can be done mechanically, direct seeding has become increasingly popular in Asia due to labour shortages.

4.5.3. Irrigation methods

4.5.3.1. Furrow irrigation

To address the water shortage, one solution is the raised bed systems, in which rice is planted on beds divided by irrigation water-applying furrows. In the Indo-Gangetic Plain, this method is widely used in the rice-wheat system. Raised bed systems are quite similar to "furrow irrigation" systems. Since the beds' soil is mostly aerobic, the method can be referred to as an aerobic rice system. Intermittent irrigation is administered through a furrow.

Raised bed systems (rice on bed and furrow for irrigation)

The dimensions of beds and furrows may vary, usually, beds are around 35 cm wide, separated by 30 cm wide furrows and 25 cm deep. On the beds, rice can be directly seeded or transplanted. It is now possible to create beds and furrows concurrently with tractor-pulled equipment and to drill seeds simultaneously (often in conjunction with fertilizer). Raised bed systems had several benefits, *viz.* improved water and nutrient use efficiency, water-saving, etc. When it is constructed with the machine, reduces labor requirements.

Numerous reports indicate that rice cultivated on beds yields less than rice transplanted. According to Balasubramanian *et al.* (2003) and Hobbs and Gupta (2003), the yield of rice that was transplanted or directly sown on beds was ± 5-6% higher than that of transplanted rice that was puddled, although there were roughly 37-40% irrigation water savings. According to Kukal *et al.* (2006), rice yields on beds are 20-40% lower than those of puddled transplanted rice, even with equal irrigation scheduling. Weed infestation, iron deficiency, and occasionally nematode problems are also present. According to Choudhury *et al.* (2007), yields on raised beds (maintained at field capacity) were likewise shown to be 21% lower than in flooded wet-seeded circumstances and 32-42% lower than in flooded transplanted conditions. The overall water input for rice grown on raised beds was 32-37% less than that of flooded wet-seeded rice and 38-42% less than that of flooded transplanted rice. Raised bed systems and flooded transplanted rice production systems had similar water productivity because, overall, decreased water inputs and yield decreases balanced each other out. Therefore, cultivating rice on raised beds offers promise although it is still in its early stages of development.

Another form of raised bed system is popular in Egypt, where rice is transplanted in furrows for higher productivity and water-saving. This system was developed by the Water Management Research Institute, National Water Research Center, Egypt. In this system, the top of the furrow is called 'border', and the bottom of the furrow is 'tape'. Every border and tape is named as 'strip'. The seedlings are transplanted at the bottom of the furrow with similar plant density (4-5 plants hill^{-1}, 10 cm apart) as recommended for the conventional transplanting method (25 hills m^{-2}). Before transplanting rice seedlings, enough amount of irrigation water is provided in the furrows so that transplanting can be easily done. Detailed dimensions of this system are shown in Fig. 4.5. The grain yield in this system increased by 5.86% and water-saving by 39.69 % than the conventional flatbed flooded transplanting method.

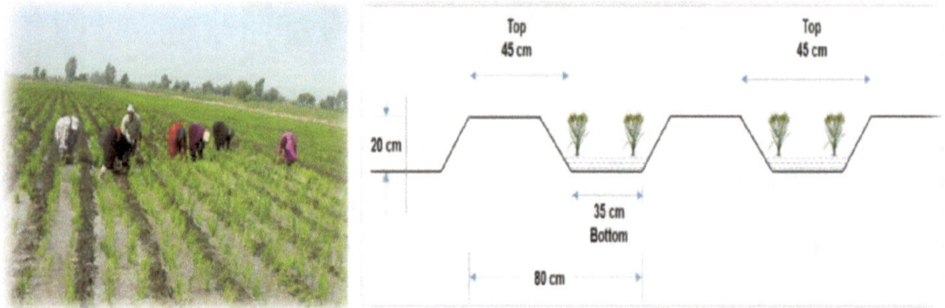

Raised bed systems (rice crop in furrow) **Fig. 4.5** Schematic diagram of strips of furrow

4.5.3.2.Saturated soil culture (SSC)

Saturated soil culture (SSC) involves keeping the rice field's soil as near to saturation as feasible with little to no ponding water, therefore, lowering the ponded water's hydraulic head, which lowers seepage and percolation losses. In actuality, SSC refers to the application of shallow irrigation to restore the depth of ponded water to around 1 cm approximately one day after its disappearance. In transplanted and direct wet-seeded rice in puddled soil, Tabbal *et al.* (2002) observed water savings of 46-48% under SSC and enhancement in water productivity (Table 4.4).

Data analysis of the 31 published field experiments revealed that water input decreased on average by 23% (5 to 50%) in SSC than the continuously flooded, with a nonsignificant yield reduction of only 6% (Bouman and Tuong, 2001).

Table 4.4 Yield, water input, and water productivity for total water input (WP_{IR}) in transplanted and wet-seeded rice under continuous flooding and SSC

Treatment	**Transplanted**			**Wet-seeded**		
	Yield ($t\ ha^{-1}$)	**Water input (mm)**	**WP_{IR} (g grain kg^{-1} water)**	**Yield ($t\ ha^{-1}$)**	**Water input (mm)**	**WP_{IR} (g grain kg^{-1} water)**
Flooded	7.4	694	1.06	7.6	631	1.20
SSC	6.7	373	1.81	7.3	324	2.27

Source: Tabbal *et al.* (2002)

A raised and furrow system is an effective way to keep the soil around saturation. Rice is grown on beds and the water in the furrows is kept close to the surface of the beds (Fig. 4.6). Several modifications have been proposed in raised bed systems for different purposes, which will be described later in this chapter in detail. To facilitate SSC practices, Borell *et al.* (1997) experimented with 120 cm broad raised beds spaced apart by 30 cm wide and 15 cm deep furrows. Under SSC, yield losses were 16-34% and water savings were 34% when compared to flooded rice. SSC is practically challenging to apply since it requires regular (daily or once every two days) applications of small volumes of irrigation water to keep furrows full just to the top in raised beds or to simply maintain a standing water depth of 1 cm in the paddy field.

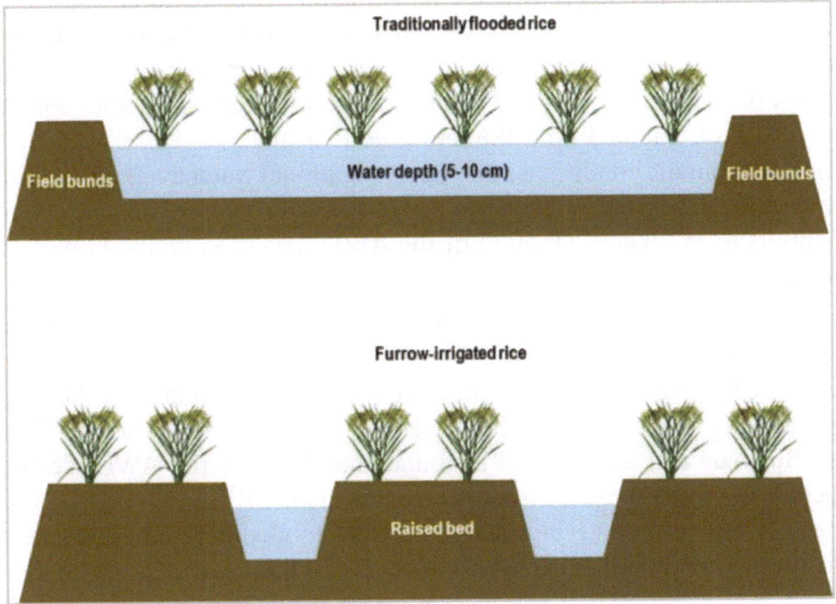

Fig. 4.6 Schematic diagram of traditionally flooded and raised-furrow system rice fields

4.5.3.3. Alternate wetting and drying (AWD)

In AWD irrigation systems, paddy fields are alternately flooded and dried, rather than continuous flooding (CF) throughout the crop season (Zhang *et al.*, 2009). The number of days of non-flooded soil (aerobic condition) in AWD before the next irrigation is applied can vary from 1 to 10 days. AWD irrigation method is a promising, water-saving, economically feasible, and eco-friendly alternative to CF.

AWD in direct seeding rice

AWD has been extensively adopted in China on heavy-textured soil as 'mid-season drainage' (Ye *et al*., 2013), and in some other Asian countries (for example, Bangladesh, India, Philippines, and Vietnam (Bouman, 2007) and the USA (Carrijo *et al*., 2017). However, except in China, in other countries, AWD has been adopted on a limited scale (Cabangon *et al*., 2016). Several factors like climatic conditions, soil type, irrigational water availability, rainfall pattern, and success in maintaining or enhancing grain yield are some major constraints in the wider adoption of the AWD irrigation method (Wassmann *et al*., 2009).

According to several researchers, employing AWD under the SRI increases yield (Thakur *et al*., 2014), while several others reported a decline in grain yield with AWD under the conventional transplanting method (Tabbal *et al*., 2002; Cabangon *et al*., 2004; Belder *et al*., 2004). After examining 31 field tests, Bouman and Tuong (2001) found that 92% of the AWD treatments reduced yields in a range of over 0% to 70% than CF. Large variations in the number of days in the drying phase (the non-flooded period), soil types, and hydrological conditions were the main causes of the AWD results.

Despite a large number of benefits of AWD, it is not practically possible to implement by the farmers, as it is difficult to identify the correct timing of irrigation after drying of the field. To address this issue, IRRI and the Institute for Agro-Environmental Science (NIAES), Japan made very easy guidelines for AWD water application and developed a device, called 'field water polyvinyl chloride (PVC) tube' to monitor water level in the field (Minamikawa *et al*., 2015) (Fig. 4.7).

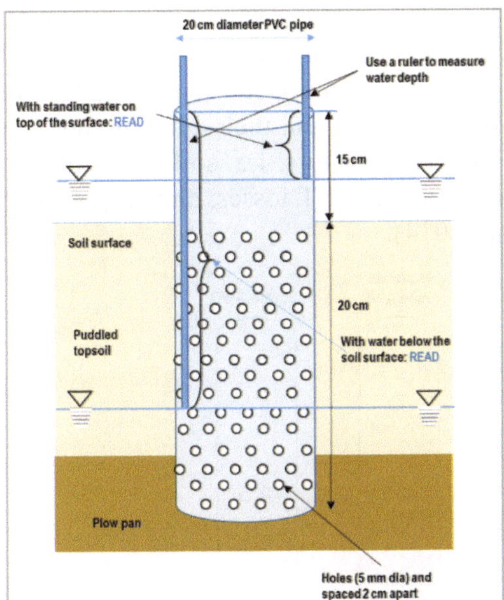

Fig. 4.7 Field water tube for monitoring water table depth in the rice field (Adapted from Bouman *et al.*, 2007)

A field water tube was installed in a rice field for AWD irrigation

The water level in the rice field under the soil's surface can be observed using the field water tube. Field water tubes with half-perforations can be made from PVC pipe, bamboo, tin cans, or plastic bottles with a diameter of 10–20 cm. Farmers can easily monitor the water table depth of rice fields using these tubes. Farmers should apply irrigation at a depth of 5 to 10 cm and wait for the next irrigation cycle until the perched water table, as a result of evapotranspiration, seepage, and percolation, drops to a specific level below the soil's surface. The fields should be re-irrigated when the field water level (FWL) in water pipes reaches 15 cm below the soil surface, and reflooding at this FWL is termed as 'safe AWD' (Bouman *et al.*, 2006). At 'safe AWD', the grain yield either increases or maintains along with water-saving of 15-30%. Also, the roots of plants can obtain sufficient water from the saturated soil and perched groundwater for crop growth and development. Farmers should implement the 'safe AWD' technique during the vegetative growth stage (tillering to panicle initiation) and then also at the grain-filling stage (Lampayan *et al.*, 2015b).

AWD is categorised as safe-AWD, moderate/ mild-AWD, and severe-AWD based on soil water potential or FWL (Fig. 4.8). Among these 'safe AWD' is now suggested for water-saving with no yield penalty, and is gradually adopted by farmers in different Asian countries, *viz.*, Bangladesh, India, Philippines, and Vietnam (Sandhu *et al.*, 2017; Kar *et al.*, 2018). Soil moisture is shifted from continuous flooding to non-flooding conditions under 'safe AWD' practice (Tan *et al.*, 2015). The drying periods can vary from 1 day to >10 days depending on soil type and soil organic content (Bouman *et al.*, 2006; Ye *et al.*, 2013). Also, the cycle of soil drying events and their duration, level of flooding and water-saving depends on several factors, *viz.* soil type, climatic conditions, availability of water, percolation and ET losses, and crop growth stage (Bouman *et al.*, 2006; Chu *et al.*, 2014).

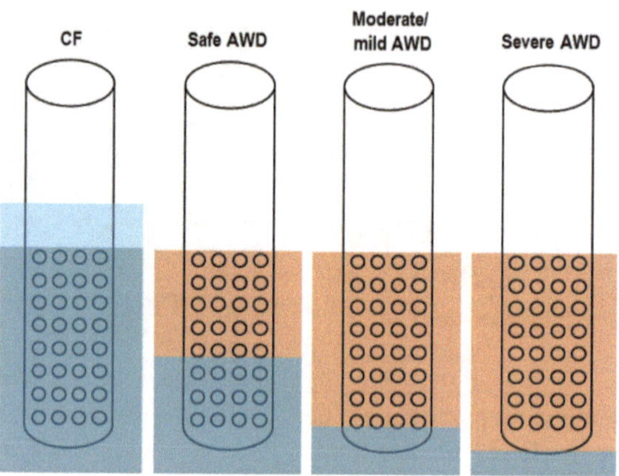

Fig. 4.8 Different types of AWD based on FWL (Field water level in water pipes below the soil surface) and SWP (Soil water potential) of the root zone compared with the conventional flooded (CF) method (Adapted from Ishfaq *et al.*, 2020)

In an AWD irrigation system, a wetting and drying cycle should be adopted during the entire growth period except the initial two weeks after seeding or transplanting. During this period shallow flooding is required for the recovery of seedlings from transplanting shock, and also to suppress weed emergence (Liang *et al.*, 2013). During panicle initiation (PI) to the end of the flowering stage, a thin layer of standing water (2-3 cm) is required because this stage of crop growth is very sensitive to water deficit (Yang *et al.*, 2017). AWD irrigation system must ensure the supply of the crop's physiological water demand and also reduce unproductive water losses from rice fields, thereby, reducing the total water input.

The AWD irrigation method is being used under conventional transplanted rice (Sandhu *et al.*, 2017; Kar *et al.*, 2018), a System of rice intensification (Thakur *et al.*, 2011; 2014; Stoop *et al.*, 2002) as well as the direct-seeded rice (Carrijo *et al.*, 2017). SRI practices with AWD irrigation significantly improve the morphology of rice plants and benefit physiological processes, which results in greater grain yield (48%) and also water-saving of 22% than conventional management practice with continuous flooding (CF) (Table 4.5) (Thakur *et al.*, 2011). Usually shifting from CF to AWD or aerobic rice does not affect the growth and yield of rice plants (Mishra, 2012; Jabran *et al.*, 2017). For example, Carrijo *et al.* (2017) found that mild AWD saves water by 23% while maintaining yield levels. He *et al.* (2019) reported grain yield was sustained under AWD and in addition to that saved 40% of water.

Table 4.5 Impact of crop management and water-saving practices on root growth, Leaf Area Index (LAI), Grain Yield, and Water Productivity

Management practices	Root dry weight (g hill^{-1})	LAI	Grain yield (t ha^{-1})	Total water consumed (x 10^4 l ha^{-1})	Water productivity (g l^{-1})
SRI with AWD	12.3	3.95	6.51	951.8	0.68
Conventional management with CF	5.8	2.60	4.40	1222.7	0.36

Adapted from Thakur *et al.* (2011)

Several potential benefits of AWD (depicted in Fig. 4.9) have been found viz., greater root growth, reduced lodging (because of a better root system), periodic soil aeration, and better control of some diseases such as golden snail. Ishfaq *et al.* (2020) reviewed and found that AWD saves water by 25-70%, reduces methane emission by 11-95 %, As (1390 %), and Hg (5-90%) in rice grains; while giving similar or better paddy yield (10-20%) than the CF. The mild-AWD also improves the quality of rice grain by reducing the kernel chalkiness by 40%, increasing the head rice recovery by 6%, and zinc concentration of the grain. However, some reports also say that under AWD rats find it easier to attack the crop during dry soil periods.

Fig. 4.9 Influence of continuous flooding (CF) and alternate wetting and drying (AWD) on soil properties, rice growth, water use, and GHGs emissions. As = Arsenic, Cd = Cadmium, Fe = iron, Hg = Mercury, GFR = Grain filling rate, Downward arrow head = Decrease, Upward arrow head = Increase (Adapted from Ishfaq *et al.*, 2020)

4.5.3.4. Micro-irrigation

Micro-irrigation mainly comprises sprinkler and drip irrigation, which is used to minimize the amount of water applied to the crops and enhance water use efficiency (WUE). A drip system may include the combined application of fertilizers with water as in drip fertigation. Other micro-irrigation system includes 'pivot irrigation' and/ or irrigation using 'rain-guns'. Micro-irrigation is mostly used for horticultural crops with several benefits, like improvement in WUE, savings in energy, fertilizers, irrigation costs, and an increase in productivity and income. Its use in rice is extremely restricted, and farmers' adoption of it is not particularly encouraging. However, further research is required to fully understand its drawbacks and find solutions before micro-irrigation for rice can be used in the future.

Rice crop with drip irrigation

Rice crop with sprinkler irrigation

In the late 1970s, many researchers began irrigating rice with sprinklers or pivots. Previous sprinkler irrigation trials were conducted in Arkansas, USA (Ferguson and Gilmore, 1977; 1978) reported a 50% saving of irrigation water in irrigated rice. Experimental results in Louisiana, USA showed rice grown under a sprinkler system produced fewer florets and filled grains/panicle than rice under flood irrigation, whereas panicle number and grain weight were not significantly affected but ultimately resulted in a 25% reduction in grain yield (Westcott and Vines, 1986). 610-1220 mm of irrigation water were applied to rice using spray systems in Arkansas (Tacker *et al.*, 2001); 460-1435 mm were applied to 33 rice fields in Arkansas (Vories *et al.*, 2005); and 382-1034 mm were applied to rice in Mississippi (Smith *et al.*, 2006). In New South Wales, Australia, Blackwell *et al.* (1985) found that rice planted under sprinkler irrigation had a higher water use efficiency than rice planted under flooded conditions; however, crop production was not as high due to shorter, more upright, and robust crop in sprinkler-irrigated than in continuously flooded plots. Contrary, sprinkler irrigation produced agronomic, environmental, and economic benefits in the central Sardinia rice region of Italy by reducing water requirements by almost 50% without compromising rice yields (Spanu *et al.*, 2009). In China, drought-resistant rice cultivars gave higher grain yield under a drip system than puddled rice cultivars. These cultivars achieved above 95% of the grain yield level under puddled circumstances with drip irrigation (Adekoya *et al.*, 2014).

Researchers from the Australia and USA revealed that although micro-irrigation had lower grain yields overall, the majority of the studies demonstrated that it had great potential for reducing water usage in rice production and boosting WUE (Westcott and Vines, 1986; Inthapan and Fukai, 1988). For instance, Blackwell *et al.* (1985) discovered that yields produced in sprinkler-irrigated rice were comparatively lower; Westcott and Vines (1986) estimated yield losses of up to 25–35% as a result of sprinkler irrigation. One of the major reasons for lower yields in a sprinkler system is the infestation of blast disease - a fungal disease in which spores are spread by air. This fungus can survive on leftover straw from the preceding year, crop rotation could be a crucial blast disease control strategy. For optimal performance in microirrigated circumstances, rice types with deep root systems and disease resistance are recommended.

Today, private organizations like Netafim and Jain Irrigation System Ltd. also researching rice by drips/ sprinkler irrigation systems in many countries *viz.*, China, India, Japan, Thailand, Australia, Brazil, Spain, USA to enhance rice production. Still, some of the constraints for micro-irrigation in rice include- high installation and maintenance costs, small and fragmented rice fields, lower market price of paddy (husked) rice, theft problems, etc. It is advised to install and design micro-irrigation for rice in keeping with the view of the supply of irrigation through the same system to the succeeding high-value crops in the same field.

Table 4.6 provides a thorough explanation of how rice crops fared in various rice-growing nations when they were microirrigated.

Table 4.6 Grain yield (GY) and water usage of rice in the sprinkler and drip irrigation compared with conventional flooded irrigation across rice-growing countries

Country, Place	Comments	Reference(s)
	Sprinkler Irrigation	
Australia, New South Wales	GY- 3.40 kg/mm of water in the sprinkler, 1.85 kg/mm in ponded treatment	Blackwell *et al.*, 1985
Australia, South-eastern Queensland	Weekly sprinkler irrigation; GY- 6.8 t ha^{-1} dry conditions was <10% of the corresponding yields in wet trial	Inthapan and Fukai, 1988; Fukai and Inthapan, 1988
USA, South-east Texas	GY- Reduced by 20-28% in sprinkler than conventional flood irrigation; Water usage 931-1171 mm through the sprinkler, and 1262-1742 mm through flood	McCauley, 1990
USA, Louisiana	GY- 4.45-5.90 and 7.14-7.85 t ha^{-1} in the sprinkler and flood irrigation, respectively	Westcott and Vines, 1986
Iran, Kooshkak	GY- 3.75-4.29 t ha^{-1} in sprinkler irrigation, 5.96 t ha^{-1} in intermittent flooding, and 6.06 t ha^{-1} in continuous flooding; water usage 836-1373, 1778, and 2262 mm in the sprinkler, intermittent and continuous flooding, respectively	Pirmoradian *et al.*, 2004
Pakistan, Lahore	GY- 3.1-3.2 t ha^{-1} in the sprinkler, and 3.1 t ha^{-1} in basin irrigation; water 5612-8417 m^3 ha^{-1} in the sprinkler (operating 39-times), and 10795 m^3 ha^{-1} in basin irrigation (operating 19-times)	Kahlown *et al.*, 2007
	Drip and/ or drip-fertigation	
Philippines, IRRI	GY- 0.8-3.2 t ha^{-1} in drip, and 2.3 t ha^{-1} in wet under upland condition	Lafitte, 2002
USA, Texas	GY- similar in drip and flood irrigation; 58% saving of irrigation water compared to flood irrigation.	Medley and Wilson, 2005
USA, Virgin Island	Higher GY under drip by 10-27% (var: 'Cybonnet' and 'Neptune'), whereas higher GY in surface flooding by 6% (var: 'Bengal' and 'Taipei')	Zimmerman, 2011
China, Xinjiang	GY- 5.78-5.90 t ha^{-1} with 1103-1121 mm water through a drip (plastic mulch), 8.32-8.58 t ha^{-1} with 3420-3552 mm water in flooding	He *et al.*, 2013
China, Xinjiang	GY- 10-12% higher in drip and plastic mulch and 60% water saving compared with flood irrigation	Zhu *et al.*, 2013
India, Ludhiana, Punjab,	GY- 7.34-8.01 and 6.63-7.60 t ha^{-1} with 860 and 1455 mm water in drip and flood irrigation, respectively; water-saving by 40-42%	Sharda *et al.*, 2017

India, Madurai, Tamil Nadu,	GY- 5.51-5.68 and 4.95-5.08 t ha^{-1} with 822-1027 and 591-757 mm water in drip-fertigation at 150 and 100%PE, respectively	Govindan and Grace, 2012; Vanitha and Mohandass, 2014
India, Coimbatore, Tamil Nadu	GY- 3.74- 4.25 t ha^{-1} in drip (aerobic) with 647.5-692.9 mm water i.e., 8-17% higher yield and 7-27% increased water productivity compared to flood irrigation	Parthasarathi *et al.*, 2013; 2014
India, Bhubaneswar, Odisha	GY- 4.27, 4.39 and 3.92 t ha^{-1} in 100, 125, and 75% of ETc in drip, 4.48 t ha^{-1} with surface irrigation	Panigrahi *et al.*, 2015
India, Morena, Madhya Pradesh	GY- 5.5 t ha^{-1} with 6316 m^3 water (irrigation + rainfall) in sub-surface drip irrigation compared to 5.5 t ha^{-1} with 8420 m^3 water with flood irrigation	AICRP-IWM, 2016
India, Bhopal, Madhya Pradesh	GY- 7.07 t ha^{-1} in SRI, 3.14 t ha^{-1} in conventional method with flood irrigation	Rao *et al.*, 2017

Source: Mandal *et al.* (2019)

4.5.4 Crop production systems

Rice crops could be grown under diverse hydrological situations ranging from aerobic conditions to completely submerged situations. Under different situations, water-saving technological interventions could be different, which has been discussed in detail in this chapter. Based on different water management strategies, rice production systems could be classified as follows (Table 4.7).

Table 4.7 Classification of rice production systems by water management strategies

Flooded lowland ↔ Dryland

Characteristics	Lowland (Anaerobic)		Lowland (Anaerobic ↔ aerobic)			Aerobic	
	Convention-al flooded	Saturated soil culture	Alternate wetting and drying			Aerobic rice	Upland rice (Direct dry seeding)
			SRI	Conventional irrigated	Rainfed		
Hydrology	Flooded	Saturation	Flooded ↔ aerobic	Flooded ↔ aerobic	Flooded ↔ aerobic	Aerobic	Aerobic
Irrigation	Irrigated	Irrigated	Irrigated	Irrigated	Rainfed	Irrigated, rainfed	Rainfed
Tillage	Puddled, non-puddled Bunded	Puddled, non-puddled Bunded	Puddled, non-puddled Bunded	Puddled, non-puddled Bunded	Puddled, non-puddled Bunded	Non-puddled Bunded	Non-puddled Non-bunded
Environment	Favorable lowland	Favorable lowland	Favorable lowland	Favorable lowland	Unfavorable lowland	Favourable upland, unfavourable lowland	Unfavorable upland
Germplasm	Lowland	Lowland	Lowland	Lowland	Lowland	Improved upland x lowland	Upland
Yield level	High	High	High	High-medium	Medium-low	Medium-low	Low
Comments	High yield with high water application	High yield with less water application, but the high cost of irrigation for application	High yield with less water	High to medium yield with less water, more seed, and chemical use	Medium to low yield with less water	Medium to low yield with less water, unavailable suitable variety, and weed problem	Low yield with less water, weed/pest problem

Adapted from Bouman, 2001 after modification

4.5.4.1. Direct seeding rice (DSR)

Irregular rainfall, water shortage, increasing transplanting costs, unavailability of labour, and declining profitability of rice production have forced many farmers to shift from transplanting to direct seeding (Peng, 2014; Sun *et al.*, 2015). The planting area of direct-seeded rice is increased rapidly. An estimate suggested that at the end of 2015, the planting area of direct-seeded rice in Jiangxi Province, China had reached 4.2 lakh hectares (Wang *et al.*, 2021). In direct seeding rice (DSR), seeds are broadcasted or sowing is done in line through the seeder directly in the main field, eliminating the need for separate seedling culture, pulling from the seedbed, and transplanting of seedlings. If dry seeds are directly broadcasted on dry-ploughed and levelled fields it is called 'Direct dry seeding (DDS)', the seeds germinate when the soil is soaked by pre-monsoon light rainfall or residual moisture or light irrigation. When pre-germinated seeds are either broadcasted or sown in line directly on the puddled and levelled field, which is free from standing water is called 'Direct wet seeding (DWS)'.

Numerous studies have shown that DSR could save about 25% on labour inputs compared with the TPR method (Bhushan *et al.*, 2007; Sanusan *et al.*, 2010; Liu *et al.*, 2015). In addition to lower water and labour requirements for crop establishment, DSR also provides better opportunities for non-rice crops such as short-duration pulse crops (e.g. Mungbean) to be grown in the wet season after the rice has been harvested because of the availability of residual moisture and shorter duration of the rice crop compared to transplanted rice. Pulse grown on DSR fields produced higher yields and income than only transplanted rice. Disadvantages of DSR include a very high seed rate and the average productivity of DSR is lower than the transplanted system. Lantican *et al.* (1999) also reported lower grain yield under dry-seeded rice than transplanted rice. They also observed farmers who grew a non-rice crop before the rice season had better weed control than those who did not. Rice fields with higher levelling produced better yields and on average, a farmer lost 0.93 t ha^{-1} yield if fields were not properly levelled. It was also found that higher yield levels in DSR in smaller plots than in large and medium farms because of improved land levelling precision and water control.

A major constraint in direct seeding is weed infestation. In DSR, weed competition is greater than in a transplanted system because the weed and rice start growing at the same time. The similarities in age and morphological features of young grassy weeds and rice seedlings make hand weeding in broadcasted direct-seeded rice more difficult than in transplanted rice. Also, lodging is another factor for yield loss in direct-seeded rice, since comparatively, roots remain at a shallow depth of the field than the transplanted rice (Yadav *et al.*, 2017). The lodging of DSR is influenced by factors like variety, planting method, natural environment, and water management (Yoshinaga *et al.*, 2003; Takayuki *et al.*, 2010).

Different direct seeding methods were evaluated and compared with the transplanting method of rice (TPR) (Thakur *et al.*, 2004). Direct seeding methods include dry-seeding by broadcast (DSB), wet-seeding by broadcast (WSB), wet-seeding in line (WSL), and wet-seeding in spots (WSS). The lowest weed population was reported under TPR, while the highest was in DSB. WSS yielded 11.8% more than TPR due to higher tillering, more panicle, and better grain-filling (Table 4.8). Total water requirements for growing rice crops under transplanting, wet seeding, and dry seeding methods of planting were 1041.2, 941.5, and 915.0 ha-mm, respectively (Table 4.9). However, water use efficiency (WUE) was lowest under dry seeding and highest under WSS. Wet seeding had 5-20% lesser labor requirements compared to TPR. In lines and spots seeding more labor was required for the sowing, but weeding was found easier and less labour-intensive in these methods. Overall, they found wet seeding on the spot, a promising technology of water and labour-saving with better yield than the conventional transplanted method. Recently, it has been reported that grain yield in mechanical dry-seeding was higher by 14.4-26.3% than in wet-seeding (Wang *et al.*, 2021).

Table 4.8. Grain yield (t ha^{-1}) and yield components of rice under different planting methods.

Planting methods	Panicle number m^{-2}	Grains number/panicle	Filled grains/panicle	Yield (t ha^{-1})
TPR	329.7b	109.2a	90.0b	3.26b
DSB	205.8e	87.0c	71.7d	2.28e
WSB	261.3d	91.7bc	76.0cd	2.65d
WSL	308.8c	94.2b	79.7c	3.22b
WSS	392.8a	108.8a	95.3a	3.64a

TPR-Transplanting, DSB-Dry seeding by broadcast, WSB-Wet-seeding by broadcast, WSL-Wet-seeding in line, and WSS-Wet-seeding in spot
Source: Thakur *et al.* (2004)

Table 4.9 Water requirements (ha-mm) under different planting methods of rice on sandy clay-loam soil during 2002 & 2003 at Bhubaneswar, India

Items	Transplanting method		Wet-seeding		Dry-seeding	
	2002	2003	2002	2003	2002	2003
Nursery preparation	19.9	19.9	--	--	--	--
Crop demand for the nursery (20 days)	13.8	14.86	--	--	--	--
Land soaking (Tillage)	150	150	150	150	20	20
Land preparation	140.0	148.6	40.0	49.0	--	--
Crop demand						
Evapotranspiration	405	430.25	396	428.05	396	428.05
Seepage & Percolation	270	320.0	310	360.0	468	498
TOTAL	998.7	1083.6	896.0	987.05	884.0	946.05
Mean	1041.15		941.53		915.03	
Relative (%)	100		90.4		87.9	

Source: Thakur *et al.* (2004)

Direct dry seeding (DDS)

Direct wet seeding (DWS)

4.5.4.2. Aerobic rice

Growing rice by direct sowing in unsaturated, non-puddled soil is known as aerobic rice production, which conserves water. Contrary to traditional flooded rice, aerobic rice receives irrigation when soil water content falls below a predetermined threshold, which helps to maintain soil moisture content between field capacity and saturation (Peng *et al.*, 2006). The amount of water applied as irrigation should match evapotranspiration and application losses. When rice is grown aerobically as an upland crop, a large amount of water could be saved due to a significant reduction in SP losses. In addition to these losses, aerobic rice farming eliminates a significant quantity of water needed for wetland preparation, which lowers the evaporation rate as there is no ponded water in the field.

Aerobic rice generally gives a yield penalty of 20–40% when compared to puddled rice, but it can reduce water use by 30–50% (Ghosh and Singh, 2010; Kumar *et al.*, 2019). Mohanty *et al.* (2021) also found that aerobic rice produces 22.5-46.0% less yield than puddled transplanted rice. Usually, upland rice varieties are drought-tolerant and low-yielding and tend to lodge under high levels of fertilizer application and supplemental irrigation. Alternatively, high-yielding lowland rice varieties grown under aerobic conditions with sufficient irrigation could save water, but the yield penalty is severe. One of the major constraints in aerobic rice is very severe weed infestation (Jabran and Chauhan, 2015).

The development of new "aerobic rice" varieties with the high-yielding traits of lowland varieties and the drought-tolerant traits of upland varieties could result in high yields under aerobic and irrigated conditions (Lafitte *et al.*, 2002, Atlin *et al.*, 2006). Improvements in weed control, nutrient application, and irrigation management are required to reduce yield loss under aerobic rice production technology (Kumar *et al.*, 2016; Ghosh and Singh, 2010).

An aerobic rice variety 'Apo' was evaluated under both aerobic irrigated and flooded conditions for yield and water-saving estimation (Bouman *et al.*, 2005; Peng *et al.*, 2006). Yields under aerobic conditions were 4.0-5.7 t ha^{-1} and 3.5-4.2 t ha^{-1} in the dry and wet seasons, respectively. The mean yield of all the investigated cultivars was 22% lower in the wet season and 32% lower in the dry season under aerobic conditions compared to flooded conditions. In aerobic fields, the total water input was 790-1,430 mm, whereas in flooded fields, it was 1,240-1,880 mm. Compared to flooded fields, aerobic fields had 250-300 mm less seepage and percolation, 80 mm less evaporation, and 25 mm less transpiration on average. They also required 190 mm less water for soil preparation. Mandal *et al.* (2013) also reported 51-53% water saving in aerobic rice compared with transplanted flooded rice (Table 4.10). A reduction in 16% yield under aerobic conditions (80-90% of field capacity) as compared to traditional flooded was observed. However, estimated water productivity, concerning rainfall and irrigation water input, increased in aerobic rice (4.71 kg grain ha^{-1} mm^{-1}) compared to traditional ones (3.04 kg grain ha^{-1} mm^{-1}) (Table 4.11).

Table 4.10. Water input through irrigation and rainfall for different aerobic soil moisture regimes and traditional flooded rice

Water input	Irrigation and rainfall	Aerobic rice	Flooded rice
Water input (Year 2008)	Pre-sowing irrigation/ irrigation for land preparation (mm)	56	362
	Irrigation during crop growth (mm)	457	829
	Rainfall (mm)	126	126
	Total (mm)	639	1317
Water input (Year 2009)	Pre-sowing irrigation/ irrigation for land preparation (mm)	61	401
	Irrigation during crop growth (mm)	556	936
	Rainfall (mm)	4	4
	Total (mm)	620	1341

Source: Mandal *et al.* (2013)

Table 4.11 Grain yield of rice and water productivity for total water input as influenced by irrigation regimes and varieties grown under aerobic and traditional flooded conditions

Treatments	Grain yield (t ha^{-1})	Water productivity (kg grain ha^{-1} mm^{-1})
Irrigation regimes		
Aerobic (I_1)	3.36b	4.60a
Aerobic (I_2)	3.15b	4.99a
Aerobic (I_3)	2.39c	4.54a
TFR (I_4)	4.04a	3.04b

I_1: aerobic soils with irrigation at 80-90% of field capacity throughout the season, I_2: aerobic soils with irrigation at 60-70% soil moisture during vegetative and 80-90% soil moisture from panicle initiation, and I_3: aerobic soils with irrigation at 60-70% of field capacity soil moisture throughout season I_4: traditional flooded rice (TFR); Mean values with different letters vary significantly according to Duncan's multiple range test ($P<0.05$).

Source: Mandal *et al.* (2013)

In addition to reducing yield, the transition from continuous floods to aerobic soil moisture regimes affects how soil carbon and nitrogen are mineralized and how available they are to crops (Mohanty *et al.*, 2013). Through nitrification-denitrification, intermittent irrigation or the AWD soil moisture regime causes the generation of N_2O (Verhoevena *et al.*, 2018). Accordingly, compared to flooded systems, aerobic rice systems typically result in much lower N-use efficiency and a higher possibility for nitrogen losses (Belder *et al.*, 2005). To maximize nitrogen uptake in situations where there is a water deficit, a suitable nitrogen management strategy that includes the nitrogen source, the timing of administration, and the dosage is necessary (Razavipour *et al.*, 2018). According to Nie *et al.* (2012), there are differences in the impacts of different types of nitrogen on aerobically grown rice plants. They recommend that the proper source of nitrogen fertilizer be used, especially ammonical-N fertilizers, to avoid yield loss in the aerobic system. In contrast to the traditional approach of N-application using prilled urea in aerobic rice, Mohanty *et al.* (2021) proposed that the application of neem-coated urea based on leaf colour chart could boost yield by 11-18%.

Crop management practices for aerobic rice: A few aerobic rice varieties for a temperate environment like northern China are HD277, HD297, and HD502, while in a tropical environment like the Philippines, the most promising variety is 'Apo'. Moreover, more research is needed to develop high-yielding aerobic rice varieties and their sustainable management systems (water, nutrients, and weed management).

To cultivate aerobic rice, the land must be dry prepared by harrowing and ploughing to create a levelled and smooth field. Dry seeding should be done with a row spacing of 20–25 cm, with a depth of 1-2 cm in heavy (clayey) soils and 2-3 cm in light-textured (loamy) soils. Seeds can be sown manually (e.g., by pushing them into slits made with a stick or a tooth harrow) or with the use of direct-seeding technology. The recommended seed rates per hectare are 60–80 kg.

Aiming for a production level of 4-6 t ha^{-1}, the overall amount of nitrogenous fertilizer application could likely adhere to local recommendations for lowland rice. It is dependent upon the native soil's N content as well as other N sources (such as farmyard manure or other organic sources). If no local recommendations are available, an amount of 90 kg N ha^{-1} could be a suitable starting point. To reduce nitrogen losses through leaching, the initial N application should be made 10-12 days after emergence. This is because emerging seedlings cannot absorb N quickly, therefore it will readily leach out. The second and third split applications of fertilizers should be given during the maximum tillering and panicle initiation stages of the crop, respectively. It has been suggested to use neem-coated urea instead of prilled urea to enhance use efficiency and reduce the yield gap in aerobic rice. Also, site-specific nutrient management (SSNM) for aerobic rice should be developed in the future.

To encourage seed emergence, a modest irrigation (3 cm) should be used after sowing if the crop is produced during the *rabi* or dry season. Maintaining the soil water content to the field's capacity should be the goal of subsequent irrigation. The same techniques used for upland crops-flash flooding, furrowing, or sprinklers-can be used to provide irrigation.

Aerobic rice with sprinkler irrigation under trial at the research farm of ICAR-Indian Institute of Water Management, Bhubaneswar, India

Weed management is one of the important activities in aerobic rice because sometimes it can cause complete crop failure. Jabran and Chauhan (2015) reviewed and found weeds under aerobic rice could cause 23-100% reductions in grain yield. A 15-307% increase in grain yield could be achieved by implementing different weed control practices in aerobic rice. Some of the practices, *viz.*, sowing of competitive crop cultivars, soil solarization, stale seedbed preparation, mulching, proper fertilization, and intercropping, were found significant for managing weeds in aerobic rice systems. A single weed control method is not sufficient under aerobic rice, therefore combining hand weeding and mechanical control with herbicide application is more effective. Effective herbicides in aerobic rice systems are 2, 4-D, bensulfuron, bispyribac-sodium, ethoxysulfuron, imazosulfuron, metsulfuron, pendimethalin, penoxsulam, pretilachlor, and triclopyr. Hence, integrated weed management is the most appropriate for weed management and the subsequent success of aerobic rice systems.

It is advised against growing aerobic rice on the same plot of land consecutively. Weed infestation, pest or disease attack, and nutritional deficiencies might occasionally cause areas grown to aerobic rice to crop failure.

4.5.4.3. The System of Rice Intensification

The System of Rice Intensification (SRI), developed in Madagascar in the 1980s based on specific modifications to conventional rice production techniques (Laulanie 1993), offers a chance to lower water consumption while simultaneously increasing rice yield (Satyanarayana *et al.*, 2007; Uphoff, 2003). It differs from traditional agronomic management practices for irrigated rice in the following ways: (a) it transplants young seedlings, ideally 8-12 days old (at the 2-3 leaf stage), quickly, carefully, and at a shallow depth of 1-2 cm; (b) it transplants single seedlings in a square pattern, with inter-row and inter-plant spacing of 20 cm or more; (c) it maintains mostly aerobic soil conditions instead of flooding fields continuously during the vegetative growth period; (d) it adds organic manures like compost or mulch depending on availability; and

(e) it uses a mechanical weeder to control weeds and simultaneously aerates the soil (Stoop *et al.*, 2002).

SRI is a methodology, not technology or a fixed set of practices (Uphoff, 2003), to be tested and optimized under a range of different agroecological environments (Stoop *et al.*, 2002). It consists of a set of interdependent agronomic practices that modify current plant, soil, water, nutrient, and weed management (Uphoff, 2003). That there are important interactions between some of the practices, e.g., between plant density and the soil moisture regime, poses considerable problems in properly assessing SRI (Stoop *et al.*, 2009). The SRI guideline to maintain somewhat moist but not always flooded paddy soil through alternate soaking and drying (AWD) or minimal daily water applications (saturated soil culture, or SSC) is of special importance in this case. It has been found that SRI practices reduce water requirements (Thakur *et al.*, 2011; Satyanarayana *et al.*, 2007) while increasing irrigated rice yields by 25-50% or even more (Sinha and Talati, 2007; Senthilkumar *et al.*, 2008).

The SRI method raises rice yields significantly with reduced production costs has been shown in >60 countries (SRI-Rice, 2021). For example, yield and other benefits of SRI methodology have been reported in many countries: Madagascar (Uphoff, 1999), Bangladesh (Husain *et al.*, 2004; Latif *et al.*, 2009), China (Wang *et al.*, 2002; Wu *et al.*, 2015; Yuan, 2002; Zhao *et al.*, 2009;), Gambia (Ceesay *et al.*, 2006), India (Satyanarayana *et al.*, 2007; Senthilkumar *et al.*, 2008; Sinha and Talati, 2007, Thakur *et al.*, 2010b, 2011, 2016b), Indonesia (Sato and Uphoff, 2007), Iraq (Hameed *et al.*, 2011), Myanmar (Kabir and Uphoff, 2007), Nepal (Dahal and Khadka, 2012), Panama (Turmel *et al.*, 2011), Sri Lanka (Namara *et al.*, 2008), and Thailand (Mishra *et al.*, 2021).

SRI rice field at ICAR-IIWM, Bhubaneswar

Two rice plants of the same variety (Surendra) and the same age; the plant on the left was raised under SRI methods and the plant on the right was grown under conventional method

The practices that constitute the SRI management strategy have been elaborated (Table 4.12). These have proved to be advantageous for irrigated rice production and, with certain modifications, also for several other rainfed crops (Abraham *et al.*, 2014). The beneficial effects of these various practices are largely based on positive feedback mechanisms between roots and shoots, which have also been documented in studies that were not connected to SRI, notably:

i. Transplanting *young seedlings* promotes early crop establishment, in part because this avoids or minimizes 'transplanting shock,' thereby enhancing the growing plants' tillering and rooting (Pasuquin *et al.*, 2008).
ii. Transplanting a *single seedling* per hill is preferable to transplanting three to five seedlings per hill, as suggested for conventional techniques since it improves tillering, root development, RuBisCO contents, and cytokinin levels (San-oh *et al.*, 2006).
iii. *Wide spacing* greatly improves each hill's performance under SRI management by reducing inter-plant competition for nutrients, water, light, and air. This permits prolific tillering, and associated root development, along with increased grain development that can more than compensate for reduced plant populations on an area basis (Thakur *et al.*, 2010a, 2014).
iv. Studies have shown that *mild AWD* can both save water and boost grain output in rice cultivation, making it an effective water-saving technique (Zhang *et al.*, 2009; Thakur *et al.*, 2018; Ishfaq *et al.*, 2020).
v. *Organic manure application* under AWD has been seen to increase significantly the uptake of N, P, and K, causing an increase in filled grains panicle^{-1}, 1000-grain weight, and grain yield (Yang *et al.*, 2004). The beneficial effect of an integrated (organic and mineral) fertilizer strategy has been significant for grain yields also under waterlogged conditions (Yang *et al.*, 2004).
vi. *Inter-cultivation/weeding* is essential because, under an AWD moisture regime, weed growth readily becomes problematic. Mechanical weed control has the advantage of aerating the soil while incorporating the weeds into the soil which enhances both root growth and health and soil populations of beneficial soil microbes (Anas *et al.*, 2011).

These respective individual practices, because of the interactions involved, e.g. between soil moisture/irrigation regime and plant spacing/density, are most effective when used in combination with other practices of SRI, and details are provided in Table 4.12.

Table 4.12 SRI practices and their effects

Practices	Effects	References
Transplanting of young seedlings	• No or reduced transplanting shock • Early and increased tillering and root growth • Earlier transplanting date into the main field extends the time for rooting and tillering	Menete *et al.*, 2008; Mishra and Salokhe, 2008; Pasuquin *et al.*, 2008
Single seedling per hill transplanted at shallow depth	• Seed requirements are greatly reduced • Reduced competition for nutrients, water, radiation, and space within a hill • Open canopy structure gives greater light interception by leaves and less shading of lower leaves, enhancing the supply of photosynthates, especially to the roots. • Early root growth enhanced: leading to increased cytokinin flux toward the shoots, delayed senescence of leaves and roots, and increased photosynthesis	San-oh *et al.*, 2004; 2006; Thakur *et al.*, 2010b
Wider spacing	• More space (below and above ground) for roots and shoots to access nutrients, water, and light • Promotes more profuse growth of roots and tillers	Thakur *et al.*, 2010a; 2013
Moist and non-flooded water management regime	• Aerobic (non-hypoxic) conditions of the soil favour root health and functioning and also support more abundant and diverse communities of beneficial aerobic soil organisms • No degeneration of roots, which under flooded soil conditions become degraded by as much as 75% by the phase of flowering • Water savings of up to 40% • Energy savings for pumped water • Reduced GHG emissions	Jagannath *et al.*, 2013; Jain *et al.*, 2014; Satyanarayana *et al.*, 2007; Suryavanshi *et al.*, 2013; Thakur *et al.*, 2011; Yang *et al.*, 2004; Yang *et al.*, 2012; Yang and Zhang, 2010; Zhao *et al.*, 2009
Inter-cultivation to control weeds	• Churning up and aerating the surface soil • Activates beneficial microbial, physical, and chemical soil dynamics • Weed biomass is incorporated into the soil as green manure • Weeding costs can be reduced	Satyanarayana *et al.*, 2007; Xu *et al.*, 2013

Increased use of organic manures	• Improves soil structure and porosity • Promotes root growth and root activity • Sustained nutrient supply over a longer period • Favours growth and activity of soil biota • Improves micronutrient content in grains	Gopalakrishnan *et al.*, 2014; Thakur *et al.*, 2020; Thies and Grossman, 2006; Yang *et al.*, 2004

Source: Thakur *et al.* (2016b)

4.6. Global status of water-saving technologies in rice

Several water-saving technologies and approaches have been experimented with, tested, and disseminated (Table 4.13) to enhance rice production in water-scarce situations including the following:

i) saturated soil culture (SSC) (Bouman and Tuong, 2001; Tabbal *et al.*, 2002),

ii) alternate wetting and drying (AWD) (Lampayan *et al.*, 2015a,b; Carrijo *et al.*, 2017),

iii) intermittent irrigation (Feng *et al.*, 2007),

iv) micro-irrigation (Govindan and Grace, 2012; Panigrahi *et al.*, 2015; Parthasarathi *et al.*, 2014),

v) raised-bed systems (Choudhury *et al.*, 2007),

vi) system of rice intensification (SRI) (Stoop *et al.*, 2002; Thakur *et al.*, 2014),

vii) aerobic rice (Rejesus *et al.*, 2013),

viii) direct-seeded rice (DSR) (Farooq *et al.*, 2011; Weerakoon *et al.*, 2011),

ix) mulching (Tao *et al.*, 2006), and

x) transplanting under non-puddled soil conditions (Malik *et al.*, 2011)

Additionally, water-saving techniques for rice farming include field channel irrigation, laser land levelling, bund height control, mechanical transplanting, soil matric potential-based irrigation with tensiometers, soil moisture-based sensor irrigation, drum seeding, and short-duration rice varieties.

Several water-saving methods are being used around the world, including aerobic rice, DSR, AWD, SRI, and aerobic rice. In addition to increasing yield, SRI and AWD also help sequester more carbon, save resources (such as labour, water, energy, and time), lower greenhouse gas emissions, and create a production system that is robust to climate change (Thakur *et al.*, 2016a; 2018).

Table 4.13 Global status of different water-saving irrigation methods/ production systems in rice

Water-saving irrigation methods/ production systems in rice	Status	Countries	Comments	References
Saturated soil culture (SSC)	Experimentation	India, Philippines	Irrigation is to be applied in 2-3 days, hence full control of irrigation water by the farmers is required.	Bouman and Tuong (2001); Tabbal *et al.* (2002)
Alternate wetting and drying (AWD)	Large-scale dissemination and adoption by farmers	Australia, Bangladesh, Brazil, China, India, Indonesia, Iran, Japan, Laos, Malawi, Myanmar, Nepal, Philippines, Senegal, Thailand, Uganda, USA, Vietnam	Full control of irrigation is required. 'Safe AWD' should be followed to avoid yield loss. Reports show many advantages with AWD apart from water saving.	Carrijo *et al.* (2017); Ishfaq *et al.* (2020); Lampayan *et al.* (2013, 2015a, 2015b); Liu *et al.* (2013); Sumana *et al.*, (2019a, 2019b); Yang *et al.* (2017)
Intermittent irrigation	Adopted by farmers	Henan Province, China	Full control of irrigation is required. Need to standardize for different soils and climatic conditions.	Feng *et al.* (2007)
Micro-irrigation	Experimentation	Australia, India, USA	More research is needed, and high investment and technical skills are required, not suitable for small and marginal farmers having small pieces of land.	Govindan and Grace (2012); Panigrahi *et al.* (2015); Parthasarathi *et al.* (2014); Rao *et al.* (2017); Sharda *et al.* (2017); Zhu *et al.* (2013)

Raised bed systems / Furrow irrigation	Experimentation	Bangladesh, Egypt, India, Indonesia, Philippines	Require mechanization for large-scale adoption. Furrow should be utilized for rice cultivation and beds for non-rice crops for diversification and higher income.	Choudhury *et al.* (2007); Singh *et al.* (2005)
System of rice intensification (SRI)	Experimentation and adoption by a large number of farmers in > 60 countries	Bangladesh, Cambodia, China, Cuba, Gambia, India, Indonesia, Iraq, Laos, Madagascar, Malaysia, Myanmar, Nepal, Pakistan, Panama, Philippines, Sierra Leone, Sri Lanka, Thailand, Tanzania, Vietnam	An agroecological method departs in several practices followed under the conventional transplanting method. Significantly improves grain yield and quality using less water, fewer seeds, and chemicals. Reduced greenhouse gas (GHG) emissions, less runoff water pollution, and more climate resilience are additional benefits.	Barrett *et al.* (2004); Ceesay *et al.* (2006); Dahal and Khadka (2012); Hameed *et al.* (2011); Husain *et al.* (2004); Kabir and Uphoff (2007); Latif *et al.* (2009); Mishra *et al.* (2021); Namara *et al.* (2008); Sato and Uphoff (2007); Satyanarayana *et al.* (2007); Senthilkumar *et al.* (2008); Sinha and Talati (2007); Thakur *et al.* (2010b, 2011, 2016a, 2016b); Turmel *et al.* (2011); Uphoff (1999); Wang *et al.* (2002); Wu *et al.* (2015); Yuan (2002); Zhao *et al.* (2009)
Aerobic rice	Experimentation and adoption by some farmers	Australia, Bangladesh, Brazil, China, India, Laos, Malaysia, Philippines	Great potential for water-saving, but requires high-yielding and disease-resistant varieties. Weed management is a challenge.	Bouman *et al.* (2005); Ding *et al.* (2010); De Borja Reis *et al.* (2018); Lampayan *et al.* (2010); Liu *et al.* (2019); Mandal *et al.* (2013); Rejesus *et al.* (2013); Silwal *et al.* (2020); Templeton and Bayot (2011); Yang *et al.* (2005); Zaman *et al.* (2018)

Direct-seeding	Large-scale dissemination and adoption by farmers	Australia, Bangladesh, India, Malaysia, Pakistan	High seed rates, poor yield, and weed infestation are major challenges.	Akbar *et al*. (2011); Farooq *et al*. (2011); Ho and Romli (1998); Weerakoon *et al*. (2011)
Mulching (non-flooded)	Experimenta-tion and adoption by some farmers	China, India, Malaysia, Philippines	Handling problem in straw mulch; In the case of plastic mulch, leftover plastic after harvest may be an environmental hazard.	Liu *et al*. (2005); Mohammed *et al*. (2016); Tao *et al*. (2006); Yan *et al*. (2018); Xu *et al*. (2007)
Transplanting in non-puddled soil	Experimenta-tion	Bangladesh, India, Philippines	More research is required; a mechanical transplanter in non-puddled soil needs to be developed.	Haque *et al*. (2006); Kar *et al*. (2018); Malik *et al*. (2011)

4.7. Technological choices under water scarcity

The technological choices and their attractiveness for the farmers depend on the land and soil type, water availability or degree of water scarcity, irrigation infrastructure and control of farmers on the water, and their socio-economic status.

When there is very little water available with absolute or physical water scarcity, farmers cannot keep their fields continuously flooded. Fig. 4.10 presents a gradient in relative water availability, suitable production system, and yield response. It also suggests technologies to reduce water inputs. On the right-hand side of the (horizontal) water availability axis, water is amply available and farmers can practice continuous flooding of lowland rice and obtain high yields. On the left-hand side, water availability is extremely low, such as in rainfed uplands, and yields are very low. Going from left to right along the water-availability axis, water gets increasingly available and yields will increase, and *vice-versa*.

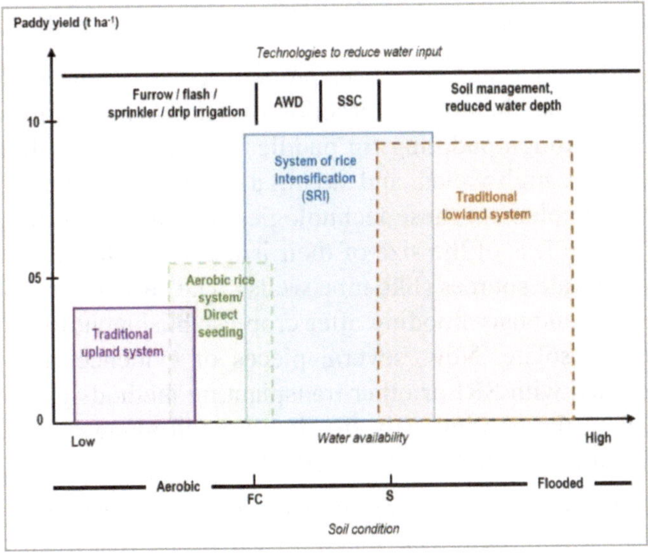

Fig. 4.10. Schematic diagram of yield responses to water availability and soil conditions in different rice production systems and their respective technologies to reduce water inputs. AWD- alternate wetting and drying; SSC- saturated soil culture; FC-field capacity; S-saturation point; *Source*: Tuong *et al.* (2005) after modification

When the water availability is moderate and an irrigation facility is available, one can adopt either an aerobic rice system or SRI, however, yield levels are usually low in aerobic rice. SRI usually produces a comparable or higher yield than flooded rice using less water by adopting AWD or SSC irrigation method (Thakur *et al.*, 2011; http://sri.ciifad.cornell.edu/). Water use and yield response in different rice production systems (Fig. 4.11) indicate that under low water availability, SRI could be one of the options to produce a high grain yield.

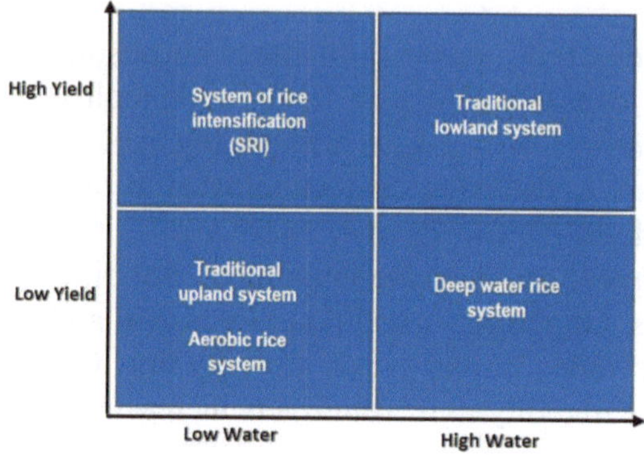

Fig. 4.11. Schematic presentation of water use and yield response in a different rice production system

Even with sufficient water available, proper land levelling, maintenance of bund with appropriate height, construction of field channels for irrigation and drainage, and thorough puddling (in puddled rice systems) will contribute to good crop growth, high yields, and saving a significant amount of water. All farmers should implement these technologies to conserve water and achieve high output, regardless of the size of their irrigation systems or whether they use shared or private sources (like tube wells). The earlier concept of keeping paddy fields continuously flooding after crop establishment to get the highest yields is now obsolete. Now, several pieces of evidence show that AWD irrigation coupled with SRI or other transplanting methods gives a similar or better yield than the continuously flooded method along with several other benefits (Thakur *et al.*, 2018; Ishfaq *et al.*, 2020).

In the deepwater rice system, seeds are sown in dry, ploughed soil and allowed to establish as young seedlings. When a field where rice is growing, floods accelerate the growth in the internodes of the stem allowing the plant to keep some of its foliage above the water level.

In a traditional lowland rice system, water levels should be around 2-3 cm after transplanting and gradually increase to 5-10 cm with increasing crop height. In direct wet seeding, the field should be kept just at saturation but not flooding from sowing to 10-12 days after emergence, and thereafter when crop height grows, the ponded water's depth should progressively rise. With direct dry seeding or aerobic rice, the soil should be moist but not saturated from sowing till emergence, or else the seeds may rot in the soil. If there is not enough rainfall to moisten the soil after sowing, flush irrigation should be used. Once a plant has three leaves, saturate the soil; as the plant grows taller, progressively raise the ponded water's depth.

In some large-scale irrigation systems, the turnaround time between land soaking and transplanting is relatively large. Under these circumstances, community nurseries or a commercial provider of seedlings could be organized, so that seedlings could get transplanted a few days after land soaking and puddling only and water could be saved. Alternative options could be both direct wet and direct dry seeding. Dry seeding should be preferred for clayey soils that don't require puddling to reduce permeability loss.

Water management strategies must be taken into account for the duration of the rice growing season due to the continued decline in water supply. During the growth season, the water's ponding depth can be lowered to 2-3 cm from the usual 5–10 cm. This will minimize seepage and percolation losses. Other options are saturated soil culture (SSC) and AWD. In SSC, soils should be kept saturated and the depth of ponded water is reduced to 0-1 cm. Maintaining a 3-5 cm water depth during the flowering stage is advised to prevent any water stress that could lead to a significant yield penalty. However, SSC would need complete control over irrigation water since it would need small volumes of irrigation water applied frequently- once every two to three days. In the 'safe AWD' practice, water losses could be reduced by a small to a considerable amount without any yield reduction. The amount of water losses reduced under SSC or AWD depends primarily on soil type and depth of the groundwater table. In heavy clay soil and shallow groundwater tables (10-40 cm deep), water losses are small and hence reductions in water losses under SSC or AWD are equally small. While in loamy or sandy soils and/or deeper groundwater tables, reductions in water losses by SSC or AWD practice would be higher, but the risk of a yield reduction also becomes higher.

Either SSC or AWD requires full water control by the farmer. With its own water sources, such as tubewells or small storage facilities (on-farm reservoirs or ponds), SSC or AWD practice is not a problem. During the rainy season also, SSC or AWD could be practiced by having a small pond in the rice field to collect extra rainwater from the field and could be used for supplemental irrigation and growing fish for a short period, as detailed in the integrated SRI section (4.8.3). Water supply to farmer groups in canal irrigation systems should be planned to achieve a specific AWD pattern.

As water becomes even more scarce, aerobic rice systems or direct dry-seeding become a viable alternative, where water required for puddling could be saved. The irrigation method should be either flash-flood or furrow or sprinkler systems. The amount of deep percolation under aerobic conditions is determined by the irrigation technique and the soil's water-holding capacity, which is reflected in the irrigation application efficiency (EA). The EA in flash-flood irrigation, furrow irrigation, and spray irrigation can be as high as 60, 70, and 80%, respectively, with careful irrigation scheduling and quantity in proportion to crop transpiration and soil water-holding capacity (Bouman *et al.*, 2007). Further, increase in water scarcity or unavailability of water, the only option left for irrigation will be through a drip-irrigation method, however, the cost of this system will be very high.

When water is available at a high cost, adopting any of the water-saving technologies becomes an economic issue. Adopting certain water-saving solutions might occasionally result in water reduction at the cost of yield loss. Adoption of technology becomes appealing if the financial savings from using less irrigation water surpasses the financial loss from lower yield. Farmers have to decide which water-saving technology should be adopted for rice cultivation, considering water availability, field situation, economic benefits, and social acceptance.

4.8. Technologies for crop diversification and enhancing income

During the rainy season, water management in rice is a challenge. Also, due to the low productivity of rainfed rice, it is mostly a non-numerative crop and farmers grow it only to fulfil food requirements. There are several technologies available through which farmers can grow diverse crops/ multi-enterprises in the rainy season apart from rice. It provides food and nutritional security and enhances income. Some of those technologies are discussed in this section.

4.8.1. Raised and sunken bed system

"Permanent beds" and "fresh beds" are the two categories of beds. Fresh beds are prepared for only one crop either for rice or non-rice crops. After harvest of the crop fields are ploughed for another crop. Permanent beds, on the other hand, are built once and then reshaped only afterwards with subsequent cropping. They are used in crop rotations (rice-nonrice crops) and have advantages in terms of cost savings, and timeliness of planting, and could be used for crop diversification. It is important to note that the raised beds are very beneficial for the non-rice crop in heavy and waterlogged soils because furrows could be used for drainage and beds remain dry.

In heavy rainfall and waterlogged areas, the cultivation of rice is the only option. However, crop diversification in the rice culture system is possible through alternate raised and sunken bed techniques. A permanent alternate raised and sunken bed is a land modification system where both rice and other crops can be grown simultaneously in the rainy season. Land modification improves the physical condition of the soil and increases the aeration status of raised beds where crops other than rice are grown successfully. Fields may be modified into alternate raised and sunken beds by digging soil of one strip (4-5 m wide) to a depth of 20-30 cm and putting the dugout soil over the adjacent strip (4-5 m side) (Fig. 4.12). The width of beds may be fixed as per convenience up to 5 m and length as per availability of land. Elevation of the raised beds thus maybe 40- 60 cm higher than the adjacent sunken beds. The top 20-30 cm soil of the raised beds remains unsaturated. By this land modification method, the aerobic soil condition of the raised beds facilitates the growing of several vegetable crops, pulses, fruits like papaya, etc. and rice may be grown simultaneously in the sunken beds. Removal of topsoil may reduce the soil fertility level and create poor soil physical condition of the sunken beds. To minimize these adverse effects, application of FYM or compost @ 10 t ha^{-1} or growing of green manuring crops during the dry season and mixing the same before rice planting is recommended for the initial 1-2 years.

Brinjal on the raised bed and rice + colocasia in sunken bed

Brinjal on the raised bed and rice in sunken bed

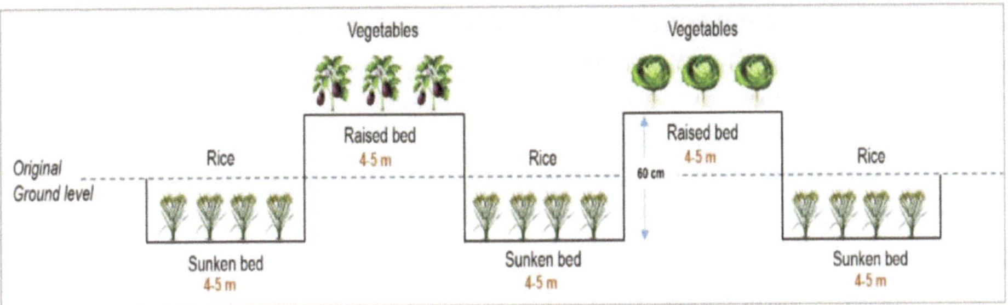

Fig. 4.12. Design and dimensions of alternate raised and sunken bed technique

4.8.2. On-dyke crop diversification

In several parts of India and other Asian countries, farmers are also going for 'bund cropping' on paddy fields to get additional income and control pests. Farmers usually grow pulses such as cowpea, black gram, vegetables, marigolds, and sunflowers on the bunds around paddy fields. This method is also being promoted as integrated pest management (IPM) to control pests like paddy leaf roller, stem borer, etc. Crops on bund also attract insects such as ladybird beetle which eat eggs of the other pests. In West Bengal of India, several farmers grow vegetables on the bund of paddy fields after raising and broadening the earthen boundary. On average, they get 5-10% of the paddy field as bund for vegetable cultivation and reported an additional income of 15,000/bund (1×400 m) per annum (Basu *et al.*, 2009).

Pulse crop on the bunds of paddy fields

4.8.3. Integrated SRI

Rainfall is the primary source of agricultural water in the majority of rainfed locations. For the most part, rice agriculture in India and other rice-growing nations is plagued by inconsistent rainfall distribution throughout the cropping season. Rainfed rice farming is linked to low productivity, inefficient fertilizer use, and environmental contamination. The use of conventional SRI techniques is restricted during the rainy/monsoon season since it is very challenging to implement any water-saving irrigation techniques. Farmers find it challenging to implement alternating wetting and drying (AWD) or intermittent irrigation. This has frequently been cited as the cause of the low adoption rates and/or SRI discontinuation in Timor Leste, Indonesia, and Cambodia.

Engineering solutions along with adjustments to soil and water management and agronomic methods can be used to address these limitations. By gathering and preserving extra rainwater, farms can produce water resources that would otherwise be in short supply. Through additional irrigation and other advantageous applications, this can raise agricultural productivity. The "5% model" is a low-cost water harvesting system that PRADAN, an Indian NGO, has been showcasing. Collecting and storing rainwater for use during the monsoon encourages farmers to turn 5% of their rainfed paddy fields into catchment ponds. This enables them to provide supplemental irrigation to their crop, raising income and household food security. Similar innovation in Cambodia has greatly increased smallholder incomes and local employment as farmers reassign some of their rainfed paddy land areas to pond construction to support diversified farming systems (http://sri.ciifad.cornell.edu/countries/cambodia/cambSidMPREng.pdf).

Integrated system of rice intensification (ISRI), combines methods of rice cultivation to enhance rice productivity within a diversified farming system that is based on constructing a farm pond in the paddy field. This supports a multi-enterprise system in place of rice monoculture. The pond conserves water for fish production while at the same time providing supplemental irrigation for the rice crop and for fruits and vegetables that are planted on the pond dykes. With ISRI technology, about 10% of the total rice field is converted into a small dugout pond at the lowest part of the field to harvest rainwater that either falls on or runs through the field. The water stored in the pond is used for short-duration fish culture during the monsoon season, and its embankment is planted with horticultural crops, bananas, and papaya in these trials. The rainwater conserved in the pond is also used for supplemental irrigation to support *kharif* paddy and tree crops. The design of this ISRI model technology (in a land area of 350 m²) is shown in Fig. 4.13.

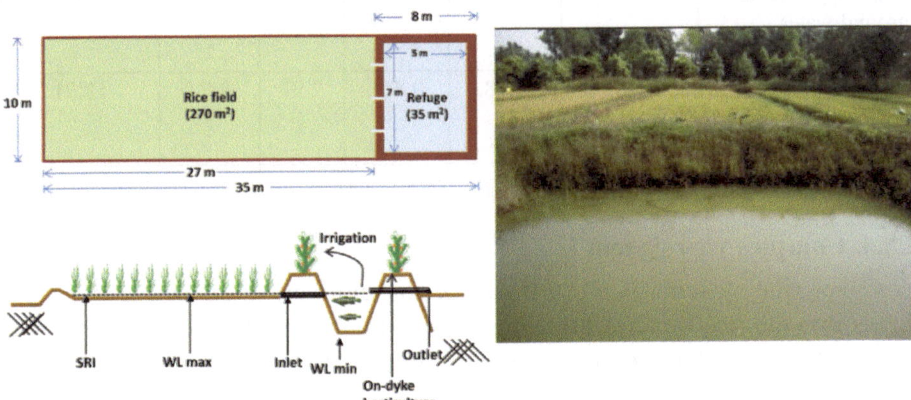

Fig. 4.13. Layout design of an integrated system of rice intensification

An integrated system of rice intensification

Integrated SRI significantly improved rice productivity (Table 4.14), income, and water productivity (Table 4.15) than rice grown by following conventional methods and SRI in rainfed conditions (Thakur *et al.*, 2015). It also helps in improving the food security and livelihoods of smallholding households.

Table 4.14 Effect of rice production systems on grain yield during the 2009 and 2010 *Kharif* (wet) Seasons

Rice production system	**Rice grain yield (tonnes per ha)**		
	2009	2010	Mean
Conventional rainfed rice (CRR)	2.36	3.41	2.89 d
SRI rainfed rice (SRR)	4.21	4.61	4.41 c
SRI rainfed rice with supplemental groundwater irrigation (SIRR)	5.96	5.43	5.70 b
SRI rainfed rice with supplemental pond irrigation (ISRI)	6.22	6.09	6.16 a

Source: Thakur *et al.* (2015)

Table 4.15 Economic analysis of different rice production systems and their water productivity, each from the same land area (350 m^2) (combined values from 2 years of experimental results)

Rice production system	**Cost of cultivation (₹)**	**Benefits (₹)**	**Net profit (₹)**	**B: C ratio**	**Water used (m^3)**	**Net water productivity (₹ per m^3)**
Conventional rainfed rice (CRR)	1,183	1,618	435	1.37	487.6	0.31
SRI rainfed rice (SRR)	1,155	2,468	1,313	2.14	487.6	2.69
SRI rainfed rice with supplemental ground-water irrigation (SIRR)	1,355	3,192	1,837	2.36	661.5	2.77
SRI rainfed rice with supplemental pond irrigation (ISRI)	4,782	14,183	9,401	2.97	497.2	18.91

Source: Thakur *et al.* (2015)

4.8.4. Complex rice system

Complex rice systems (CRSs) integrate the cultivation of azolla, fish, and ducks into the rice system. It is a promising approach to ecologically address pest problems and increase rice yields organically in East Java, Indonesia (Khumairoh *et al.*, 2012; 2018). As system complexity rises, they report increased yields (Fig. 4.14) and income (Fig. 4.15) as well as decreased vulnerability to extreme weather.

In complex rice systems, rice was grown using some practices from the SRI, *i.e.* young seedlings transplanted at wider spacing using compost, but in contrast to SRI recommendations, the fields were flooded for fish culture and ducks in the system. Ducks and fish are important components in flooded rice production systems in Asia (Lu and Li, 2006; Cheng-fang *et al.*, 2008), and in CRS *Azolla* and compost were added. *Azolla* - a floating fern that symbiotically lives with the blue-green algae (*Anabaena azollae)*, which fixes atmospheric N and increases soil fertility. Compost provides nutrients and organic matter to enhance crop nutritional status and soil structure in this system. Fish and ducks can consume *Azolla*, weeds, and pest species, which enhances the system's nitrogen cycling and reduces the number of weeds and pests (Cagauan *et al.*, 2000; Jana and Jana, 2003).

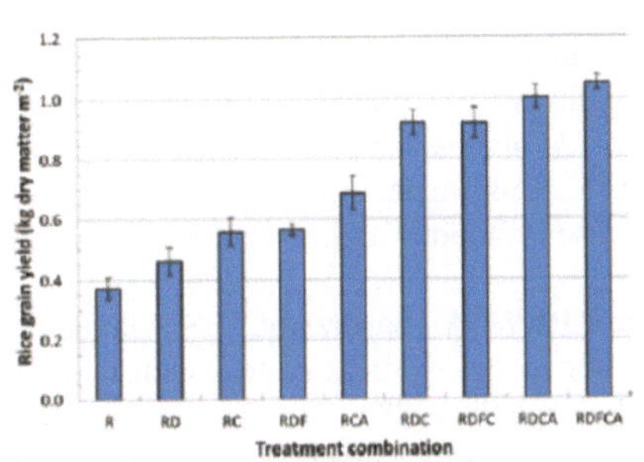

Fig. 4.14 Grain yield increases with an increase in complexity in the rice cultivation system. R = rice; D = with ducks; C = with compost; F = with fish; A = with Azolla. *Source:* Khumairoh *et al.* (2012)

Result (USD ha-1)
6000
5000
4000
3000
2000
1000
0
Costs
Revenues
Gross margin
RC
RDFCA

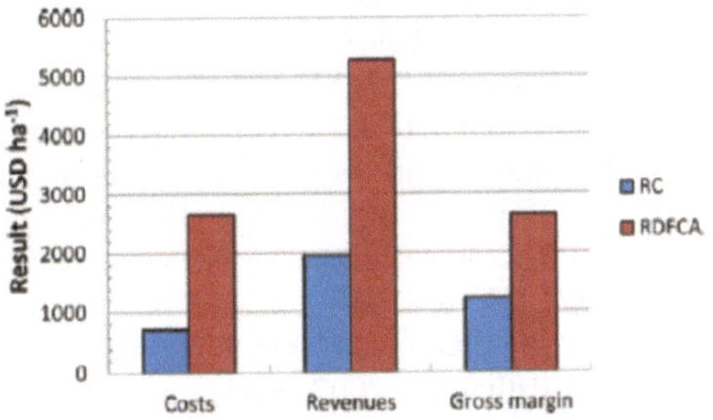

Fig. 4.15 Economic results (USD ha-1) in terms of costs, revenues, and gross margin in rice cultivation with compost (RC) and a complex system integrating rice, ducks, fish, compost, and Azolla (RDFCA).
Source: Khumairoh *et al.* (2012)

References

Abraham, B., Araya, H., Berhe, T., Edwards, S., Gujja, B., Khadka, R.B., Koma, Y.S., Sen, D., Sharif, A., Styger, E., Uphoff, N. and Verma, A. 2014. The system of crop intensification: reports from the field on improving agricultural production, food security, and resilience to climate change for multiple crops. *Agriculture & Food Security,* 2014, 3:4.

Adekoya, M. A., Liu, Z., Eli, V.d, Zhou, L., Deyan, K., Qin, J., Ma, R., Xinqiao, Y., Liu, G., Lin, C. and Lijun, L. 2014. Agronomic and ecological evaluation on growing water-saving and drought-resistant rice (*Oryza sativa* L.) through drip irrigation. *Journal of Agricultural Sciences*, Canada, 6 (5):110-119.

AICRP-IWM 2016. All India Coordinated Research Project on Irrigation Water Management (ICAR), Rajmata Vijayaraje Scindiya Krishi Vishwa Vidyalaya, Zonal Agricultural Research Station, Morena, Madhya Pradesh, India, pp. 47-52.

Akbar, N., Jabran, K. and Ali, M.A. 2011. Weed management improves yield and quality of direct seeded rice. *Australian Journal of Crop Science,* 5:688.

Anas, I., Rupela, O.P., Thiyagarajan, T.M. and Uphoff, N. 2011. A review of studies on SRI effects on beneficial organisms in rice soil rhizospheres. *Paddy and Water Environment,* 9:53-64.

Atlin, G.N., Lafitte, H.R., Tao, D., Laza, M., Amante, M. and Courtois, B. 2006. Developing rice cultivars for high-fertility upland systems in the Asian tropics. *Field Crops Research,* 97:43-52.

Balasubramanian, V., Ladha, J.K., Gupta, R., Naresh, R.K., Mehla, R.K., Singh, B. and Singh, Y. 2003. Technology options for rice in the rice-wheat system in South Asia. In: Ladha, J.K., Hill, J.E., Duxbury, J.M., Gupta, R.K., Buresh, R.J. (Editors) Improving the productivity and sustainability of rice-wheat systems: issues and impacts. ASA Special Publication 65, ASA Inc., CSSA Inc., SSSA Inc., Madison, Wis., USA. pp. 115-147.

Barrett, C.B., Moser, C.M., McHugh, O.V. and Barison, J. 2004. Better technology, better plots, or better farmers? Identifying changes in productivity and risk among Malagasy rice farmers. *American Journal of Agricultural Economics,* 86:869-888.

Basu, D., Banerjee, S. and Goswami, R. 2009. Vegetable cultivation on ail (Bund): A successful farmer-led technology transfer. *Acta Horticulturae,* 809:155-160. Doi: 10.17660/ActaHortic.2009.809.13

Belder, P., Bouman, B.A.M., Cabangon, R., Lu, G., Quilang, E.J.P., Li, Y., Spiertz, J.H.J. and Tuong, T.P. 2004. Effect of water-saving irrigation on rice yield and water use in typical lowland conditions in Asia. *Agricultural Water Management,* 65:193-210.

Belder, P., Bouman, B.A.M., Spiertz, J.H.J., Peng, S., Castaneda, A.R. and Visperas, R.M. 2005. Crop performance, nitrogen and water use in flooded and aerobic rice. *Plant and Soil,* 273:167-182.

Bhushan, L., Ladha, J.K., Gupta, R.K., Singh, S., Tirol-Padre, A., Saharawat, Y.S., Gathala, M. and Pathak, H. 2007. Saving of water and labor in a rice-wheat system with no-tillage and direct seeding technologies. *Agronomy Journal,* 99:1288.

Bird, J., Eagle, A., Horwath, W., Hair, M., Zilbert, E. and van Kessel, C. 2002. Long-term studies find benefits, challenges in alternative rice straw management. *California Agriculture,* 56(2):69-75.

Blackwell, J., Meyer, W.S. and Smith, R.C.G. 1985. Growth and yield of rice under sprinkler irrigation on a free-draining soil. *Australian Journal of Experimental Agriculture,* 25(3): 636-641.

Borell, A., Garside, A. and Shu, F.K. 1997. Improving efficiency of water for irrigated rice in a semi-arid tropical environment. *Field Crops Research,* 52:231-248.

Bouman, B.A.M., Lampayan, R.M. and Tuong, T.P., 2007. Water Management in Irrigated Rice: Coping With Water Scarcity. International Rice Research Institute, Los Baños, Philippines p. 54.

Bouman, B.A.M. 2001. Water-efficient management strategies in rice production. *International Rice Research Notes,* 16(2):17-22.

Bouman, B.A.M. 2007. Water Management in Irrigated Rice: Coping With Water Scarcity. International Rice Research Institute, Los Baños, Philippines pp. 54.

Bouman, B.A.M, Yang, X., Wang, H.Q., Wang, Z., Zhao. J. and Chen, B. 2006. Performance of aerobic rice varieties under irrigated conditions in North China. *Field Crops Research,* 97: 53-65.

Bouman, B.A.M., Peng, S., Castaneda, A.R. and Visperas, R.M. 2005. Yield and water use of irrigated tropical aerobic rice systems. *Agricultural Water Management,* 74: 87-105.

Bouman, B.A.M. and Tuong, T.P. 2001. Field water management and increase its productivity in irrigated rice. *Agricultural Water Management,* 49:11-30.

Cabangon, R.J. and Tuong, T.P. 2000. Management of cracked soils for water saving during land preparation for rice cultivation. *Soil and Tillage Research,* 56:105-116.

Cabangon, R.J., Tuong, T.P., Castillo, E.G., Bao, L.X., Lu, G., Wang, G.H., Cui, L., Bouman, B.A.M., Li, Y., Chongde, C. and Jianzhang, W. 2004. Effect of irrigation method and N-fertilizer management on rice yield, water productivity and nutrient-use efficiencies in typical lowland rice conditions in China. *Paddy and Water Environment,* 2:195-206.

Cabangon, R., Lampayan, R., Bouman, B. and Tuong, T.P. 2016. Water Saving Technologies for Rice Production in the Asia Regions. http://www.fftc.agnet.org/library.php?func=view&style=&typeid=4&id=20140303145242&print=1#pict1.

Cagauan, A.G., Branckaert, R.D. and Van Hove, C. 2000. Integrating fish and azolla into rice-duck farming in Asia. *The ICLARM Quaterly*, 23:4-10.

Carrijo, D.R., Lundy, M.E. and Linquist, B.A. 2017. Rice yields and water use under alternate wetting and drying irrigation: a meta-analysis. *Field Crops Research*, 203:173-180.

Ceesay, M., Reid, W.S., Fernandes, E.C.M. and Uphoff, N. 2006. The effects of repeated soil wetting and drying on lowland rice yield with system of rice intensification (SRI) methods. *International Journal of Agricultural Sustainability,* 4:5-14.

Cheng-fang L., C. Cou-gui,W. Jin-ping, Z.Ming, Y.Wei-ling, and S. Ahmad. 2008. Nitrogen losses from integrated rice-duck and rice-fish ecosystems in southern China. *Plant and Soil*, 307:207-217.

Choudhury, B.U., Bouman, B.A.M. and Singh, A.K. 2007. Yield and water productivity of rice-wheat on raised beds at New Delhi. *Field Crops Research*, 100: 229-239.

Chu, G., Chen, T., Wang, Z., Yang, J. and Zhang, J. 2014. Morphological and physiological traits of roots and their relationships with water productivity in water-saving and drought-resistant rice. *Field Crops Research*, 162:108-119.

Chung, I., Kim, K., Ahn, J., Lee, S., Kim, S. and Hahn, S. 2003. Comparison of allelopathic potential of rice leaves, straw, and hull extracts on barnyard grass. *Agronomy Journal,* 95(4):1063-1070.

Dahal, K.R. and Khadka, R.B. 2012. Performance of rice with varied age of seedlings and planting geometry under System of Rice Intensification (SRI) in western terai, Nepal. *Nepal Journal of Science and Technology,* 13:1-6.

De, Borja, Reis, A.F., de Almeida, R.E.M., Lago, B.C., Trivelin, P.C., Linquist, B. and Favarin, J.L., 2018. Aerobic rice system improves water productivity, nitrogen recovery and crop performance in Brazilian weathered lowland soil. *Field Crops Research*, 218:59-68.

Devasinghe, D., Premarathne, K. and Sangakkara, U. 2011. Weed management by rice straw mulching in direct seeded lowland rice (*Oryza sativa* L.). *Tropical Agricultural Research,* 22(3):263-272.

Ding, S., Wu, H., Bouman, B.A.M., Wang, H., Peng, S. and Chen, Y. 2010. Adoption and impact of aerobic rice in North Anhui. Research to Impact: Case Studies for Natural Resource Management for Irrigated Rice in Asia. International Rice Research Institute, Los Baños, Philippines, pp. 335-360.

Dong, B., Molden, D., Loeve, R., Li, Y.H., Chen, C.D. and Wang, J.Z. 2004. Farm level practices and water productivity in Zanghe Irrigation System. *Paddy and Water Environment,* 2:217-226.

El-Shahawy, T.A., El-Rokiek, K., Sharara, F. and Khalaf, K. 2006. New approach to use rice straw waste for weed control. I. efficacy of rice straw extract against broad and narrow leaved weeds in cucumber (*Cucumis sativa* L.). *International Journal of Agriculture and Biology,* 8(2):262-268.

El-Shahawy, T.A. and Zydenbos, S. 2010. Rice straw as an allelopathic agent for controlling weeds. In: 17th Australasian weeds conference. New Frontiers in New Zealand: together we can beat the weeds. Christchurch, New Zealand, pp 143-146, 26-30 Sept 2010.

FAO. 2021. Food and Agriculture Organization. Available at: http://www.fao.org/faostat/en/#data

FAO. 2023 as on 29.08.2023 FAOSTAT_data_en_8-29-2023 https://www.fao.org/faostat/en/#data/QCL

Farooq, M., Siddique, K.H.M., Rehman, H., Aziz, T., Lee, D. and Wahid, A. 2011. Rice direct seeding: experiences, challenges and opportunities. *Soil and Tillage Research,* 111: 87-98.

Feng, L., Bouman, B.A.M., Tuong, T.P., Cabangon, R.J., Li, Y., Lu, G. and Feng, Y. 2007. Exploring options to grow rice using less water in northern China using a modelling approach-I: field experiments and model evaluation. *Agricultural Water Management,* 88:1-13.

Ferguson, J.A. and Gilmore, J.T. 1977. Centre pivot sprinkler irrigation of rice. Arkansas Farm Research, March-April, pp.12-20.

Ferguson, J.A. and Gilmore, J.T. 1978. Water and nitrogen relations of sprinkler irrigated rice. Arkansas Farm Research, May-June, pp.22-30.

Fukai, S. and Inthapan, P. 1988. Growth and yield of rice cultivars under sprinkler irrigation in south-eastern Queensland. 3. Water extraction and plant water relations dash comparison with maize and grain sorghum. *Australian Journal of Experimental Agriculture,* 28(2): 249- 252.

Ghosh, A. and Singh, O.N. 2010. Determination of threshold regime of soil moisture tension for scheduling irrigation in tropical aerobic rice for optimum crop and water productivity. *Experimental Agriculture,* 46: 489-499.

Gopalakrishnan, S., Vadlamudi, S., Bandikinda, P., Sathya, A., Vijayabharathi, R., Rupela, O., Kudapa, H., Katta, K. and Varshney, R.K. 2014. Evaluation of Streptomyces strains isolated from herbal vermicompost for their plant growth-promotion traits in rice. *Microbiological Research,* 169:40-48.

Govindan, R. and Grace, T. M. 2012. Influence of drip fertigation on growth and yield of rice varieties (*Oryza sativa* L.). *Madras Agricultural Journal,* 99(4/6):244-247.

Hameed, K.A., Mosa, A.K.J. and Jaber, F.A. 2011. Irrigation water reduction using system of rice intensification compared with conventional cultivation methods in Iraq. *Paddy and Water Environment,* 9:121-127.

Hanafi, E.M., El Khadrawy, H., Ahmed, W. and Zaabal, M. 2012. Some observations on rice straw with emphasis on updates of its management. *World Applied Sciences Journal,* 16(3):354-361.

Haque, M.E., Bell, R.W., Islam, M.A. and Rahman, M.A. 2006. Minimum tillage unpuddled transplanting: An alternative crop establishment strategy for rice in conservation agriculture cropping systems. *Field Crops Research,* 185: 31-39.

He, G., Wang, Z. and Cui, Z. 2019. Managing irrigation water for sustainable rice production in China. *Journal of Cleaner Production,* 118928

He, H., Ma, F., Yang, R., Chen, L., Jia, B., Cui, J., Fan, H., Wang, X. and Li, L. 2013. Rice performance and water use efficiency under plastic mulching with drip irrigation. *PLoS ONE,* 8(12): e83103.doi:10.1371/journal.pone.0083103.

Ho, N.K. and Romli, Z. 1998. Impact of direct seeding on rice cultivation: lessons from the Muda area of Malaysia. Direct Seeding: Research Strategies and Opportunities. International Rice Research Institute, Los Ban˜os, Philippines.

Hobbs, P. and Gupta, R.K. 2003. Rice-wheat cropping systems in the Indo-Gangetic Plains: issues of water productivity. I. Relation to new resource-conserving technologies. In: Kijne, J.W., Barker, R. and Molden, D. (editors) Water productivity in agriculture: limits and opportunities for improvement. CABI Publishing, Wallingford, UK. pp. 239-253.

Husain, A.M., Chowhan, P.B., Uddin, A.F.M. and Rahman, A.B.M. 2004. Final Evaluation Report on Verification and Refinement of the System of Rice Intensification (SRI) Project in Selected Areas of Bangladesh. PETRRA-Project, IRRI, Dhaka, Bangladesh.

Inthapan, P. and Fukai, S. 1988. Growth and yield of rice cultivars under sprinkler irrigation in south-eastern Queensland. 2. Comparison with maize and grain sorghum under wet and dry conditions. *Australian Journal of Experimental Agriculture,* 28(2):243-248.

Ishfaq, M., Farooq, M., Zulfiqar, U., Hussain, S., Akbar, N., Nawaz, A. and Anjum, S.A. 2020. Alternate wetting and drying: A water-saving and ecofriendly rice production system. *Agricultural Water Management,* 241:106363.

Jabran, K., Riaz, M., Hussain, M., Nasim, W., Zaman, U., Fahad, S. and Chauhan, B.S. 2017. Water-saving technologies affect the grain characteristics and recovery of fine-grain rice cultivars in semi-arid environment. *Environmental Science and Pollution Research,* 24(14):12971-12981.

Jabran, K. and Chauhan, B.S. 2015. Weed management in aerobic rice systems. *Crop Protection,* 78:151-163.

Jagannath, P., Pullabhotla, H. and Uphoff, N. 2013. Meta-analysis evaluating water use, water saving, and water productivity in irrigated production of rice with SRI vs. standard management methods. *Taiwan Water Conservancy,* 61(4):14-49.

Jain, N., Dubey, R., Dubey, D.S., Singh, J., Khanna, M., Pathak, H. and Bhatia, A. 2014. Mitigation of greenhouse gas emission with system of rice intensification in the Indo-Gangetic Plains. *Paddy and Water Environment,* 11:593-601.

Jana, B. B. and Jana, S. 2003. The potential and sustainability of aquaculture in India. *Journal of Applied Aquaculture,* 13:283-316.

Jat, M.L., Chandna, P., Gupta, R., Sharma, S.K. and Gill, M.A. 2006. Laser Land Leveling: A Precursor Technology for Resource Conservation. Rice-Wheat Consortium Technical Bulletin Series 7. New Delhi, India: Rice-Wheat Consortium for the Indo-Gangetic Plains. pp 48.

Kabir, H. and Uphoff, N. 2007. Results of disseminating the system of rice intensification with farmer field school methods in Northern Myanmar. *Experimental Agriculture,* 43:463-476.

Kahlown, M.A., Raoof, A., Zubair, M. and Kemper, W.D. 2007. Water use efficiency and economic feasibility of growing rice and wheat with sprinkler irrigation in the Indus Basin of Pakistan. *Agricultural Water Management,* 87(3):292-298.

Kar, I., Sudhir-Yadav, Mishra, A., Behera, B., Khanda, C., Kumar, V. and Kumar, A. 2018. Productivity trade-off with different water regimes and genotypes of rice under nonpuddled conditions in Eastern India. *Field Crops Research,* 222:218-229.

Kato-Noguchi, H. and Ino, T. 2005. Concentration and release level of momilactone B in the seedlings of eight rice cultivars. *Journal of Plant Physiology,* 162(9):965-969.

Khumairoh, U., Groot, J.C.J. and Lantinga, E.A. 2012. Complex agro-ecosystems for food security in a changing climate. *Ecology and Evolution,* 2(7):1696-1704.

Khumairoh, U., Lantinga, E.A., Rogier P. O. Schulte, R.P.O., Suprayogo, D. and Groot, J.C.J. 2018. Complex rice systems to improve rice yield and yield stability in the face of variable weather conditions. *Scientific Reports,* 8:14746 doi: 10.1038/s41598-018-32915-z

Kumar, A., Nayak, A.K., Mohanty, S. and Das, B.S. 2016. Greenhouse gas emission from direct seeded paddy fields under different soil water potentials in Eastern India. *Agriculture, Ecosystems & Environment,* 228:111-123.

Kumar, A., Nayak, A.K., Das, B.S., Panigrahi, N., Dasgupta, P., Mohanty, S., Kumar, U., Panneerselvam, P. and Pathak, H. 2019. Effects of water deficit stress on agronomic and physiological responses of rice and greenhouse gas emission from rice soil under elevated atmospheric CO_2. *Science of Total Environment,* 650: 2032-2050.

Lafitte, R.H., Courtois, B. and Arraudeau, M. 2002. Genetic improvement of rice in aerobic systems: progress from yield to genes. *Field Crops Research,* 75:171-190.

Lafitte, R.H. 2002. Relationship between leaf relative water content during reproductive stage water deficit and grain formation in rice. *Field Crops Research,* 76:165-174.

Lampayan, R.M., Bouman, B.A.M., Flor, R.J. and Palis, F.G. 2013. Developing and disseminating alternate wetting and drying water saving technology in the Philippines. In: Kumar, A. (Ed.), Mitigating Water-Shortage Challenges in Rice Cultivation: Aerobic and Alternate Wetting and Drying Rice Water-Saving Technologies IRRI. Asian Development Bank, Manila, Philippines.

Lampayan, R.M., Rejesus, R.M., Singleton, G.R. and Bouman, B.A.M., 2015a. Adoption and economics of alternate wetting and drying water management for irrigated lowland rice. *Field Crops Research,* 170:95-108.

Lampayan, R.M., Samoy-Pascual, K.C., Sibayan, E.B., Ella, V.B., Jayag, O.P., Cabangon, R.J. and Bouman, B.A.M., 2015b. Effects of alternate wetting and drying (AWD) threshold level and plant seedling age on crop performance, water input, and water productivity of transplanted rice in Central Luzon. Philippines. *Paddy and Water Environment,* 13:215-227.

Lantican, M.A., Lampayan, R.M., Bhuiyan, S.I. and Yadav, M.K. 1999. Determinants of improving productivity of dry-seeded rice in rainfed lowlands. *Experimental Agriculture,* 35:127-140.

Latif, M.A., Ali, M.Y., Islam, M.R., Badshah, M.A. and Hasan, M.S. 2009. Evaluation of management principles and performance of the System of Rice Intensification (SRI) in Bangladesh. *Field Crops Research,* 114:255-262.

Laulanié, H. 1993. Le système de riziculture intensive malgache. *Tropicultura,* 11:110-114.

Liang, X.Q., Chen, Y.X., Nie, Z.Y., Ye, Y.S., Liu, J., Tian, G.M., Wang, G.H. and Tuong, T.P. 2013. Mitigation of nutrient losses via surface runoff from rice cropping systems with alternate wetting and drying irrigation and site-specific nutrient management practices. *Environmental Science and Pollution Research,* 20:6980-6991.

Liu, T.Q., Fan, D.J., Zhang, X.X., Chen, J., Li, C.F. and Cao, C.G. 2015. Deep placement of nitrogen fertilizer reduces ammonia volatilization and increases nitrogen utilization efficiency in no-tillage paddy fields in central China. *Field Crops Research,* 184:80-90.

Liu, H., Zhan, J., Hussain, S. and Nie, L. 2019. Grain yield and resource use efficiencies of upland and lowland rice cultivars under aerobic cultivation. *Agronomy,* 9(10):591.

Liu, L., Chen, T., Wang, Z., Zhang, H., Yang, J. and Zhang, J. 2013. Combination of site-specific nitrogen management and alternate wetting and drying irrigation increases grain yield and nitrogen and water use efficiency in super rice. *Field Crops Research,* 154:226-235.

Liu, X., Ai, Y., Zhang, F., Lu, S., Zeng, X. and Fan, M. 2005. Crop production, nitrogen recovery and water use efficiency in rice-wheat rotation as affected by non-flooded mulching cultivation (NFMC). *Nutrient Cycling in Agroecosystems,* 71:289-299.

Lu, J. and Li, X. 2006. Review of rice-fish-farming systems in China-one of the Globally Important Ingenious Agricultural Heritage Systems (GIAHS). *Aquaculture,* 260:106-113.

Maclean, J.L., Dawe, D., Hardy, B. and Hettel, G.P. (Eds.). 2002. Rice almanac. International Rice Research Institute, Los Baños, Philippines, 253 pp.

Malik, R., Kamboj, B., Jat, M., Sidhu, H., Bana, A., Singh, V., Sharawat, Y., Pundir, A., Sahnawaz, R., Anuradha, T., Kumaran, N. and Gupta, R. 2011. No- till and un-puddled mechanical transplanting of rice. Cereal Systems Initiative for South Asia. New Delhi, India.

Mandal, K.G., Kundu, D.K., Thakur, A.K., Kannan, K., Brahmanand, P.S. and Kumar, A. 2013. Aerobic rice response to irrigation regimes and fertilizer nitrogen rates. *Journal of Food, Agriculture & Environment,* 11(3&4):1148-1153.

Mandal, K.G., Thakur, A.K. and Ambast, S.K. (2019). Current rice farming, water resources and micro-irrigation. *Current Science*, 116 (4): 568-576.

McCauley, G.N. 1990. Sprinkler vs. flood irrigation in traditional rice production regions of southeast Texas. *Agronomy Journal,* 82:677-683.

Medley, J. and Wilson, L. T. 2005. The use of sub-surface drip irrigation for rice. Proceeding of 10th National Convention System on Cotton and Rice Conference, Texas, USA, pp. 33-34.

Menete, M.Z.L., van Es, H.M., Brito, R.M.L., DeGloria, S.D. and Famba, S. 2008. Evaluation of system of rice intensification (SRI) component practices and their synergies on salt-affected soils. *Field Crops Research*, 109:34-44.

Minamikawa, K., Tokida, T., Sudo, S., Padre, A. and Yagi, K. 2015. Guidelines for Measuring CH_4 and N_2O Emissions from Rice Paddies by a Manually Operated Closed Chamber Method. National Institute for Agro-Environmental Science, Tsukuba, Japan.

Mishra, A. 2012. Intermittent irrigation enhances morphological and physiological efficiency of rice plants. *Agriculture (Polnohospodárstvo)*, 58:121-130.

Mishra, A. and Salokhe, V.M. 2008. Seedling characteristics and the early growth of transplanted rice under different water regimes. *Experimental Agriculture,* 44:365-383.

Mishra, A., Ketelaar, J. W. H., Uphoff, N. and Whitten, M. 2021. Food security and climate-smart agriculture in the Lower Mekong Basin of Southeast Asia: Evaluating impacts of system of rice intensification with special reference to rainfed agriculture. *International Journal of Agricultural Sustainability*, 19(2):152-174.

Mohammed, U., Aimrun, W., Amin, M.S.M., Khalina, A. and Zubairu, U.B. 2016. Influence of soil cover on moisture content and weed suppression under system of rice intensification (SRI). *Paddy and Water Environment,* 14:159-167.

Mohanty, S., Nayak, A.K., Bhaduri, D., Swain, C.K., Kumar, A., Tripathi, R., Shahid, M., Behera, K.K. and Pathak, H. 2021. Real-time application of neem-coated urea for enhancing N-use efficiency and minimizing the yield gap between aerobic direct-seeded and puddled transplanted rice. *Field Crops Research*, 264:108072

Nader, G. and Robinsons, P. 2010. Rice producers' guide to marketing rice straw. UCANR Publication, Oakland.

Namara, R., Bossio, D., Weligamage, P. and Herath, I. 2008. The practice and effects of the System of Rice Intensification (SRI) in Sri Lanka. *Quarterly Journal of International Agriculture,* 47:5-23.

Nie, L., Pang, S., Chen, M., Shah, F., Huang, J., Cui, K. and Xiang, J. 2012. Aerobic rice for water-saving agriculture: A review. *Agronomy for Sustainable Development,* 32: 411-418.

Panigrahi, P., Rautaray, S. K., Panda, R. K., Thakur, A. K. and Raichaudhuri S. 2015. Response of rain-fed rice to supplemental irrigation with drip and surface irrigation methods in eastern India. *International Journal of Tropical Agriculture,* 33(2): 971-975.

Parthasarathi, T., Mohandass, S., Senthilvel, S. and Vered, E. 2013. Effect of drip irrigation systems on yield of aerobic rice. *Environmental Ecology,* 31 (4A): 1826-1829.

Parthasarathi, T., Vanitha, K., Mohandass, S. and Vered, E. 2014. Importance of large root system on increasing yield in drip irrigated aerobic rice. *The Bioscan*, 9(4):1541-1544.

Pasuquin, E., Lafarge, T. and Tubana, B. 2008. Transplanting young seedlings in irrigated rice fields: Early and high tiller production enhanced grain yield. *Field Crops Research,* 105:141-155.

Peng, S., Bouman, B.A.M., Visperas, R.M., Castañeda, A., Nie, L. and Park, H-K. 2006. Comparison between aerobic and flooded rice: agronomic performance in a long-term (8-season) experiment. *Field Crops Research*, 96:252-259.

Peng, S. 2014. Reflection on China's rice production strategies during the transition period. *Scientia Sinica Vitae*, 44:845-850.

Pirmoradian, N., Sepaskhan, A.R. and Maftoun, M. 2004. Effects of water-saving irrigation and nitrogen fertilization on yield and yield components of rice (*Oryza sativa* L.). *Plant Production Science,* 7(3):337-346.

Rao, K.V.R., Gangwar, S., Keshri, R., Chourasia, L., Bajpai, A. and Soni, K. 2017. Effects of drip irrigation system for enhancing rice (*Oryza sativa* L.) yield under system of rice intensification management. *Applied Ecology and Environmental Research,* 15(4):487-495.

Razavipour, T., Moghaddam, S.S., Doaei, S., Noorhosseini, S.A. and Damalas, C.A. 2018. Azolla (*Azolla filiculoides*) compost improves grain yield of rice (*Oryza sativa* L.) under different irrigation regimes. *Agricultural Water Management,* 209: 1-10.

Rejesus, R.M., Martin, A.M. and Gypmantasiri, P. 2013. Meta-impact assessment of the irrigated rice research consortium. Special IRRI Report. Int. Rice Res. Inst., Los Baños (Philippines) p. 174.

San-oh, Y., Mano, Y., Ookawa, T. and Hirasawa, T., 2004. Comparison of dry matter production and associated characteristics between direct-sown and transplanted rice plants in a submerged paddy field and relationships to planting patterns. *Field Crops Research,* 87:43-58.

San-oh, Y., Sugiyama, T., Yoshita, D., Ookawa, T. and Hirasawa, T., 2006. The effect of planting pattern on the rate of photosynthesis and related processes during ripening in rice plants. *Field Crops Research*, 96:113-124.

Sandhu, N., Subedi, S.R., Yadaw, R.B., Chaudhary, B., Prasai, H., Iftekharuddaula, K., Thanak, T., Thun, V., Battan, K.R., Ram, M., Venkateshwarlu, C., Lopena, V., Pablico, P., Maturan, P.C., Cruz, M.T.S., Raman, K.A., Collard, B. and Kumar, A. 2017. Root traits enhancing rice grain yield under alternate wetting and drying condition. *Frontiers in Plant Science,* 8:1879-1891.

Sanusan, S., Polthanee, A., Audebert, A., Seripong, S. and Mouret, J.C. 2010. Growth and yield of rice (*Oryza sativa*) as affected by varieties, seeding depth and water deficits at vegetative stage. *Asian Journal of Plant Sciences*, 9:36-43.

Sato, S. and Uphoff, N. 2007. A review of on-farm evaluation of system of rice intensification (SRI) methods in eastern Indonesia: CAB Reviews. Commonwealth Agricultural Bureau International, Wallingford, pp. 67-87.

Satyanarayana, A., Thiyagarajan, T.M. and Uphoff, N. 2007. Opportunities for water saving with higher yield from the system of rice intensification. *Irrigation Science,* 25:99-115.

Senthilkumar, K., Bindraban, P.S., Thiyagarajan, T.M., de Ridder, N. and Giller, K.E., 2008. Modified rice cultivation in Tamil Nadu, India: yield gains and farmers' (lack of) acceptance. *Agricultural Systems,* 98:82-94.

Sharda, R., Mahajan, G., Siag, M., Singh, A. and Chauhan, B.S. 2017. Performance of drip irrigated dry seeded rice (*Oryza sativa* L.) in south Asia. *Paddy and Water Environment,* 15:93-100.

Silwal, S., Bhattarai, S.P. and Midmore, D.J. 2020. Aerobic rice with or without strategic irrigation in the subtropics. *Agronomy,* 10(11):1831.

Sharma, P.K., Lav, J.K., Bhushan, Ladha, Naresh, R.K., Gupta, R.K., Balasubramanian, B.V. and Bouman, B.A.M., 2002. Crop-water relations in rice-wheat cropping under different tillage systems and water-management practices in a marginally sodic, medium-textured soil. In: Bouman, B.A.M., Hengsdijk, H., Hardy, B., Bindraban, P.S., Tuong, T.P. and Ladha, J.K. (editors).Water-Wise Rice Production. International Rice Research Institute, Los Ba˜nos, Philippines, pp. 223-235.

Singh, A.K., Choudhury, B.U. and Bouman, B.A.M. 2002. Effects of rice establishment methods on crop performance, water use, and mineral nitrogen. In: Bouman, B.A.M., Hengsdijk, H., Hardy, B., Bindraban, P.S., Tuong, T.P. and Ladha, J.K. (editors).Water-Wise Rice Production. International Rice Research Institute, Los Ba˜nos, Philippines, pp. 237-246.

Singh, R., Kundu, D.K., Mohanty, R.K., Ghosh, S., Kumar, A. and Kannan, K. 2005. Raised and sunken bed technique for improving water productivity in lowlands. Research Bulletin 28, Water Technology for Eastern Region, Bhubaneswar, Odisha, India, p. 42.

Sinha, S. K. and Talati, J., 2007. Productivity impacts of the system of rice intensification (SRI): A case study in West Bengal, India. *Agricultural Water Management,* 87:55-60.

Smith, M.C., Massey, J.H., Branson, J., Epting, J.W., Pennington, D., Tacker, P.L., Thomas, J., Vories, E.D. and Wilson, C. 2006. Water use estimates for various rice production systems in Mississippi and Arkansas. *Irrigation Scence,* 25(2):141-147.

Spanu, A., Murtas, A. and Ballone, F. 2009. Water use and crop coefficients in sprinkler irrigated rice. *Italian Journal of Agronomy*, 2:47-58.

Stoop, W.A., Adam, A. and Kassam, A. 2009. Comparing rice production systems: A challenge for agronomic research and for the dissemination of knowledge-intensive farming practices. *Agricultural Water Management,* 96:1491-1501.

Stoop, W.A., Uphoff, N. and Kassam, A. 2002. A review of agricultural research issue raised by the system of rice intensification (SRI) from Madagascar: Opportunities for improving system for resource poor farmers. *Agricultural Systems,* 71:249-274.

SRI-Rice. 2021. SRI-Rice website: Available at: http://sri.cals.cornell.edu/countries/index.html

Sumana, M., Ullah, H., Datta, A., Paothong, K., Kachenchart, B., Shrestha, S. and Shrestha, P. R. 2019a. Effect of water and rice straw management practices on yield and water productivity of irrigated lowland rice in the Central Plain of Thailand. *Agricultural Water Management,* 211:89-97.

Sumana, M., Ullah, H., Datta, A., Shrestha, P.R., Shrestha, S. and Kachenchart, B. 2019b. Effects of water and rice straw management practices on water savings and greenhouse gas emissions from a double-rice paddy field in the Central Plain of Thailand. *European Journal of Agronomy,* 107: 18-29.

Sun, L.M., Hussain, S., Liu, H.Y., Peng, S.B., Huang, J.L., Cui, K.H. and Nie, L.X. 2015. Implications of low sowing rate for hybrid rice varieties under dry direct-seeded rice system in Central China. *Field Crops Research*, 175:87-95.

Surendran, U., Raja, P., Jayakumar, M. and Subramoniam, S.R. 2021. Use of efficient water saving techniques for production of rice in India under climate change scenario: A critical review. *Journal of Cleaner Production*, 309:127272.

Suryavanshi, P., Singh, Y.V., Prasanna, R., Bhatia, A. and Shivay, Y.S. 2013. Pattern of methane emission and water productivity under different methods of rice crop establishment. *Paddy and Water Environment,* 11:321-329.

Tabbal, D.F., Bouman, B.A.M., Bhuiyan, S.I., Sibayan, E.B. and Sattar, M.A. 2002. On-farm strategies for reducing water input in irrigated rice, case studies in the Philippines. *Agricultural Water Management,* 56:93-112.

Tacker, P., Vories, E., Wilson, C. Jr., and Slaton, N. 2001. Water management, In Rice Production Handbook, Chapter 9 (ed.), N. A. Slaton, University of Arkansas, Arkansas, Little Rock, Cooperative Extension Service, MP192-10M-1-01RV, pp. 75-86.

Takayuki, K., Naoki, H., Kazuhiro, U. and Ken, I. 2010. Lodging resistance locus *prl5* improves physical strength of the lower plant part under different conditions of fertilization in rice (*Oryza sativa* L.). *Field Crops Research,* 115:107-115.

Tan, X., Shao, D., Gu, W. and Liu, H. 2015. Field analysis of water and nitrogen fate in low land paddy fields under different water managements using HYDRUS-1D. *Agricultural Water Management,* 150:67-80.

Tao, H., Brueck, H., Dittert, K., Kreye, C., Lin, S. and Sattelmacher, B. 2006. Growth and yield formation of rice (*Oryza sativa* L.) in the water-saving ground cover rice production system (GCRPS). *Field Crops Research,* 95:1-12.

Templeton, D. and Bayot, R. 2011. An impact assessment of the 'Developing a system of temperate and tropical aerobic rice (STAR) in Asia' project. Impact Assessment Series IA-05. Challenge Program for Water and Food.

Thakur, A. K., Uphoff, N., and Stoop, W. A. 2016b. Scientific underpinnings of the system of rice intensification (SRI): What is known so far? *Advances in Agronomy,* 135:147-179.

Thakur, A.K., Kassam, A., Stoop, W.A. and Uphoff, N. 2016a. Modifying rice crop management to ease water constraints with increased productivity, environmental benefits, and climate-resilience. *Agriculture, Ecosystems and Environment,* 235:101-104.

Thakur, A.K., Mandal, K.G., Mohanty, R.K. and Ambast, S.K. 2018. Rice root growth, photosynthesis, yield and water productivity improvements through modifying cultivation practices and water management. *Agricultural Water Management,* 206:67-77.

Thakur, A.K., Mandal, K.G. and Raychaudhuri, S. 2020. Impact of crop and nutrient management on crop growth and yield, nutrient uptake and content in rice. *Paddy and Water Environment,* 18:139-151.

Thakur, A.K., Mohanty, R.K., Patil, D.U. and Kumar, A., 2014. Impact of water management on yield and water productivity with system of rice intensification (SRI) and conventional transplanting system in rice. *Paddy and Water Environment,* 12:413-424.

Thakur, A.K., Mohanty, R.K., Singh, R. and Patil, D.U. 2015. Enhancing water and cropping productivity through Integrated System of Rice Intensification (ISRI) with aquaculture and horticulture under rainfed conditions. *Agricultural Water Management,* 161:65-76.

Thakur, A.K., Rath, S. and Mandal, K.G. 2013. Differential responses of system of rice intensification (SRI) and conventional flooded rice management methods to applications of nitrogen fertilizer. *Plant and Soil,* 370:59-71.

Thakur, A.K., Rath, S., Patil, D.U. and Kumar, A. 2011. Effects on rice plant morphology and physiology of water and associated management practices of the system of rice intensification and their implications for crop performance. *Paddy and Water Environment,* 9:13-24.

Thakur, A.K., Rath, S., Roychowdhury, S. and Uphoff, N., 2010a. Comparative performance of rice with system of rice intensification (SRI) and conventional management using different plant spacings. *Journal of Agronomy and Crop Science,* 196:146-159.

Thakur, A.K., Roy Chowdhury, S., Kundu, D.K. and Singh R. 2004. Evaluation of planting methods in irrigated rice. *Archives of Agronomy & Soil Science,* 50: 631-640.

Thakur, A.K., Uphoff, N. and Antony, E., 2010b. An assessment of physiological effects of system of rice intensification (SRI) practices compared with recommended rice cultivation practices in India. *Experimental Agriculture,* 46:77-98.

Thies, J.E. and Grossman, J.M., 2006. The soil habitat and soil ecology, in: Uphoff, N., *et al.* (Eds.), Biological Approaches to Sustainable Soil Systems. CRC Press, Boca Raton, FL, pp. 59-78.

Tuong, T. P. and Bouman, B. A. M. 2003. Rice production in water scarce environments. In Water Productivity in Agriculture: Limits and Opportunities for Improvement (eds.), Kijne, J.W., Barker, R., and Molden, D., CABI Publishing, Wallingford, UK. pp. 53-67.

Tuong, T. P., Bouman, B. A. M., and Mortimer, M. 2005. More rice, less water- Integrated approaches for increasing water productivity in irrigated rice-based systems in Asia. *Plant Production Science,* 8:231-241.

Turmel, M., Turner, B.L. and Whalen, J.K. 2011. Soil fertility and the yield response to the System of Rice Intensification. Renew. *Agricultural Food Systems,* 26:185-192.

Uphoff, N. 2003. Higher yields with fewer external inputs? The system of rice intensification and potential contributions to agricultural sustainability. *International Joutnal of Agricultural Sustainability,* 1:38-50.

Uphoff, N. 1999. Agroecological implications of the system of rice intensification (SRI) in Madagascar. *Environment, Development and Sustainability,* 1:297-313.

Vanitha, K. and Mohandass, S. 2014. Effect of humic acid on plant growth characters and grain yield of drip fertigated aerobic rice (*Oryza sativa* L.). *The Bioscan*, 9(1):45-50.

Verhoevena, E., Decock, C., barthel, M., Bertora, C., Sacco, D., Romani, M., Sleutel, S. and Six, J. 2018. Nitrification and coupled nitrification-denitrification at shallow depths are responsible for early season N_2O emissions under alternate wetting and drying management in an Italian rice paddy system. *Soil Biology and Biochemistry,* 120: 58-69.

Vories, E. D., Tacker, P. L. and Hogan, R. 2005. Multiple inlet approach to reduce water requirements for rice production. *Applied Engineering Agriculture,* 21(4):611-616.

Wang, S.H., Cao, W.X., Jiang, D., Tai, T.B. and Zhu, Y. 2002. Physiological characteristics of high-yield techniques with SRI rice, in: Uphoff, N., Fernandes, E.C.M., Longping, Y., Jimimg, P., Rafaralahy, S., Rabenandrasana, J. (Eds.), Assessment of the System of Rice Intensification: Proceedings of an international conference, Sanya, April 1-4, 2002, Cornell Intl. Inst. for Food Agric & Dev, Ithaca, NY, pp. 116-124. (http://ciifad.cornell.edu/sri/proc1/sri_27.pdf)

Wang, W.X., Du, J., Zhou, Y.Z., Zeng, Y.J., Tan, X.M., Pan, X.H., Shi, Q.H., Wu, Z.M. and Zeng, Y.H. 2021. Effects of different mechanical direct seeding methods on grain yield and lodging resistance of early *indica* rice in South China. *Journal of Integrative Agriculture,* 20(5):1204-1215.

Wassmann, R., Jagadish, S.V.K., Sumfleth, K., Pathak, H., Howell, G., Ismail, A., Serraj, R., Redona, E., Singh, R.K. and Heuer, S. 2009. Regional vulnerability of climate change impacts on Asian rice production and scope for adaptation. *Advances in Agronomy,* 102:91-133.

Weerakoon, W.M.W., Mutunayake, M.M.P., Bandara, C., Rao, A.N., Bhandari, D.C. and Ladha, J.K. 2011. Direct-seeded rice culture in Sri Lanka: lessons from farmers. *Field Crops Research,* 121: 53-63.

Westcott, M.P. and Vines, K.W. 1986. A comparison of sprinkler and flood irrigation for rice. *Agronomy Journal,* 78: 637-640.

Wu, W., Ma, B. and Uphoff, N. 2015. A review of the system of rice intensification in China. *Plant and Soil*, 393(1-2):361-381.

Xu, C., Wang, D., Chen, S., Chen, L. and Zhang, X. 2013. Effects of aeration on root physiology and nitrogen metabolism in rice. *Rice Science,* 20:148-153.

Xu, G.W., Zhang, Z.C., Zhang, J.H. and Yang, J.C. 2007. Much improved water use efficiency of rice under non-flooded mulching cultivation. *Journal of Integrative Plant Biology,* 49:1527-1534.

Yadav, S., Singh, U.M., Naik, S.M., Venkateshwarlu, C., Ramayya, P.J., Raman, K.A., Sandhu, N. and Kumar, A. 2017. Molecular mapping of OTLs associated with lodging resistance in dry direct-seeded rice (*Oryza sativa*). *Frontiers in Plant Science*, 8: 1431.

Yan, F., Sun, Y., Hui, X., Jiang, M., Xiang, K., Wu, Y., Zhang, Q., Tang, Y., Yang, Z., Sun, Y. and Ma, J. 2019. The effect of straw mulch on nitrogen, phosphorus and potassium uptake and use in hybrid rice. *Paddy and Water Environment*, 17:23-33.

Yan, F., Sun, Y., Hui, X., Yin, Y., Wang, H., Wang, C., Guo, C., Yang, Z., Sun, Y. and Ma, J. 2018. Effects of wheat straw mulch application and nitrogen management on rice root growth, dry matter accumulation and rice quality in soils of different fertility. *Paddy and Water Environment*, 16:507-518.

Yang, J. and Zhang, J. 2010. Crop management techniques to enhance harvest index in rice. *Journal of Experimental Botany,* 61:3177-3189.

Yang, C., Yang, L., Yang, Y. and Ouyang, Z. 2004. Rice root growth and nutrient uptake as influenced by organic manure in continuously and alternately flooded paddy soils. *Agricultural Water Management,* 70:67-81.

Yang, J., Zhang, H. and Zhang, J. 2012. Root morphology and physiology in relation to the yield formation of rice. *Journal of Integrative Agriculture,* 11:920-926.

Yang, J., Zhou, Q. and Zhang, J. 2017. Moderate wetting and drying increases rice yield and reduces water use, grain arsenic level, and methane emission. *Crop Journal,* 5:151-158.

Yang, X., Bouman, B.A.M., Wang, H., Wang, Z., Zhao, J. and Chen, B. 2005. Performance of temperature aerobic rice under different water regimes in north China. *Agricultural Water Management,* 74: 107-122.

Ye, Y., Liang, X., Chen, Y., Liu, J., Gu, J., Guo, R. and Li, L. 2013. Alternate wetting and drying irrigation and controlled-release nitrogen fertilizer in late-season rice. Effects on dry matter accumulation, yield, water and nitrogen use. *Field Crops Research,* 144:212-224.

Yoshinaga, S., Toriyama, K., Heong, K.L. and Hardy, B. 2003. Direct-seeding cultivation of rice in Japan: Stabilization of seedling establishment and improvement of lodging resistance. *Japanese Journal of Crop Science*, 70:194-201.

Yuan, L.P. 2002. A scientist's perspective on experience with SRI in China for raising the yields of super hybrid rice, in: Uphoff, N., Fernandes, E.C.M., Longping, Y., Jimimg, P., Rafaralahy, S., Rabenandrasana, J. (Eds.), Assessment of the System of Rice Intensification: Proceedings of an international conference, Sanya, April 1-4, 2000, Cornell Intl. Inst. for Food Agriculture & Dev, Ithaca, NY, pp. 23-25. (http://ciifad.cornell.edu/sri/proc1/sri_06.pdf)

Zaman, N.K., Abdullah, M.Y., Othman, S. and Zaman, N.K. 2018. Growth and physiological performance of aerobic and lowland rice as affected by water stress at selected growth stages. *Rice Science,* 25(2):82-93.

Zhang, H., Xue, Y., Wang, Z., Yang, J. and Zhang, J. 2009. An alternate wetting and moderate soil drying regime improves root and shoot growth in rice. *Crop Science,* 49:2246-2260.

Zhao, L.M., Wu, L.H., Li, Y.S., Lu, X.H., Zhu, D.F. and Uphoff, N. 2009. Influence of the system of rice intensification on rice yield and nitrogen and water use efficiency with different application rates. *Experimental Agriculture,* 45:275-286.

Zhu, Q. C., Wei, C. Z., Li, M. N., Zhu, J.L. and Wang, J. 2013. Nutrient availability in the rhizosphere of rice grown with plastic film mulch and drip irrigation. *Journal of Soil Science and Plant Nutrition,* 13 (4):943-953.

Zimmerman, T. 2011. Evaluation of irrigation systems for rice production on St. Croix, USVI, Virgin Islands Water Resource Research Institute, University of Virgin Island, St. Thomas, pp.8.

5

Water Management Strategies for Dry-season Crops Under Rice-based System

5.1. Introduction

Rice-fallows are a rainfed rice mono-crop production system, where only a single crop (rice) is grown and these lands remain without any crop for the rest of the period of the year (fallow). An estimate shows that rice-fallow areas extended over 22.3 M ha in South Asia, maximum in India (88.3%) followed by Bangladesh (8.7%), Nepal (1.4%), Sri Lanka (1.1%), Pakistan (0.5%), and Bhutan (0.02%) (Gumma *et al.*, 2016). The eastern part of India accounts for ~ 84% (9.7 million ha) of the country's total rice-fallow area (i.e. 11.7 million ha) (Ali *et al.*, 2014; Singh *et al.*, 2017), despite high rainfall during rainy / monsoon season. The reasons are that after the rainy season, irrigation facilities are lacking, very fast depletion of the soil residual moisture due to temperature rise, the poor economic condition of the farmers, and a lack of awareness or knowledge of available technologies. Additionally, delayed harvesting of long-duration rice (150-160 days) results in late sowing of winter crops, which, in turn, exacerbates the temperature and moisture stresses in winter crops (Kumar *et al.*, 2018a; 2018b). Overall, the challenges of growing another crop in rice-fallow areas are the lack of irrigation facilities including declining groundwater levels, the dominance of the long-duration rice varieties cultivation, higher evaporative loss of the residual soil moisture, degradation of soil quality, and poor financial conditions of the farming community with diminishing profit. Recently, climate change with erratic rainfall and delayed monsoon arrival added more challenges for these areas in several places. Deficit soil moisture imposes stress on the crops and negatively affects the winter crops productivity in the rice-fallow system, and several times it results in complete crop failure (Singh *et al.*, 2012; Ghosh *et al.*, 2016).

Despite several challenges in rice-fallow areas, there is ample scope to enhance the winter-cropped area through strategic land, crop, and irrigation management practices (Mishra *et al.*, 2018; Kumar *et al.*, 2018a). Proper water management in combination with better crop/ land management would improve water productivity and could help to achieve 'more crop per drop'. Therefore, a multidisciplinary and integrated approach is required to address this important issue.

In this chapter, we are discussing water management strategies (either through land modification conserving water, or adopting water-saving methods) and technologies (including crop selection) available for enhancing productivity together with income from rice-fallow areas focusing on post-rainy season strategies.

5.1. Technological options based on moisture availability during the harvest of *Kharif* crops

Usually, moisture availability during the post-rainy season depends on many factors, but the most important are the land conditions. For example, if it is low land, then moisture at the time of *Kharif* rice harvest remains high, contrary, if it is upland, then the land becomes dry at the time of rice harvest. Therefore, we need to select technological options for the growing post-rice season crops based on moisture availability. A relative option for the technologies has been presented in Table 5.1. A schematic diagram of moisture availability conditions during the harvest of *Kharif* rice and respective technological options for the winter (*rabi*) crop has been presented in Fig. 5.1.

Table 5.1. Relative technological options based on the field moisture condition during the *Kharif* rice harvest

Field moisture conditions during *Kharif* rice harvest	Technological options	Reference
Lowland with high moisture	Draining excess water to create a favourable soil condition for winter pulses or wheat or oilseed crops	Layek *et al.* (2014)
Land with moderate moisture	Paira / relay /utera cropping with standing stubbles/mulching	Rautaray (2008)
Land with dry soil	No-tillage along with standing stubbles/ mulching along with life-saving irrigations	Das *et al.* (2013)

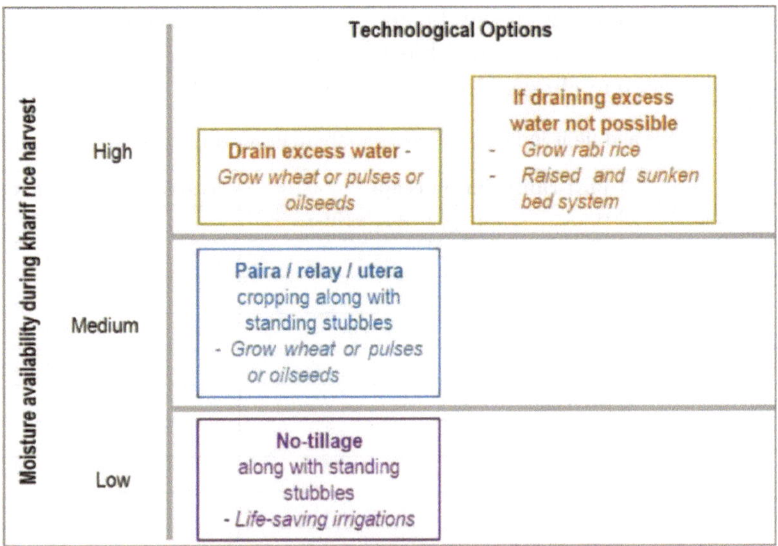

Fig. 5.1. Schematic diagram of moisture availability conditions during the harvest of *Kharif* rice and respective technological options for the winter (*rabi*) crop

The details of these technological possibilities have been discussed in detail in the subsequent sections of this chapter.

5.2. Crop selection based on irrigation water availability after *Kharif* rice harvest

The choice of an appropriate winter crop is very crucial and the crop should be low water-requiring and stress-tolerant to withstand the biotic and abiotic stresses (Bandyopadhyay *et al.,* 2015; Singh *et al.,* 2017). Fast growth for early maturity and ground cover with crops with deep roots have been suggested as some of the important desirable plant characteristics for water-limited rice-fallow conditions (Bandyopadhyay *et al.*, 2016; Hazra and Bohra, 2020). Pulses like chickpeas, Lathyrus, lentils, etc. are short-duration and low-water-requiring crops and have a deep root system, thereby, the ability to absorb moisture from deeper soil (Kumar *et al.*, 2019a; 2019b). Similarly, oilseed crops like groundnut, mustard, safflower, sunflower, linseed, etc. could also be integrated with the rice-fallow system. Timely sowing of winter crops and the adoption of suitable soil moisture conservation practices play a key role in the success (Layek *et al.*, 2014). Now, several improved agro-tools (e.g. Happy or Super Seeder) are readily available in the market for the sowing of *rabi* crops (wheat or pulses or oilseeds) under the leftover crop residues of *Kharif* rice (Sidhu *et al.*, 2008).

A list of crops that can be grown during the post-rainy season and their water requirements has been shown in Table 5.2, and based on water availability for irrigation, farmers can choose different crops (Table 5.3).

Table 5.2 Crop types and their indicative water requirements during the post-rainy season

Crop types	Post-rainy season crops	Water requirements (mm)
Cereals / Millets	Wheat / Sorghum / other Millets	450-650
	Maize	500-800
Pulses	Pea	350-500
	Chickpea / Lentil / *Lathyrus*	250-300
Oilseeds	Mustard	310-400
	Soybean	450-700
	Groundnut	500-700
	Safflower	250-300
	Linseed	450-500
	Sunflower	600-1000
Vegetables	Beans	300-500
	Onion	350-550
	Cabbage	380-500
	Chillies	400-600
	Potato	500-700
	Tomato	600-800
Other crops	Tobacco	400-600
	Sugar beet	550-750

Source: FAO

Table 5.3. An indicative picture to show water available for irrigation and options to select crops

Water availability (mm)	Water availability (10^6 x litre) for a one-hectare area	Crops option
300-500	3-5	Wheat/sorghum / other millets, pea, chickpea/lentil / *lathyrus,* mustard, soybean, safflower, linseed, beans, onion, cabbage, chillies, tobacco, etc.
500-700	5-7	Wheat/sorghum / other millets, maize, soybean, groundnut, sunflower, chillies, potato, tomato, tobacco, sugar beet, etc.
700-900	7-9	Maize, sunflower, tomato, sugar beet, etc.
900-1100	9-11	Sunflower, cotton, rice, etc.
>1100	>11	Rice, sugarcane, banana, cotton, citrus, etc.

The irrigation water availability during the post-rainy season and respective technological options for the winter (*rabi*) crop have been depicted in Fig. 5.2.

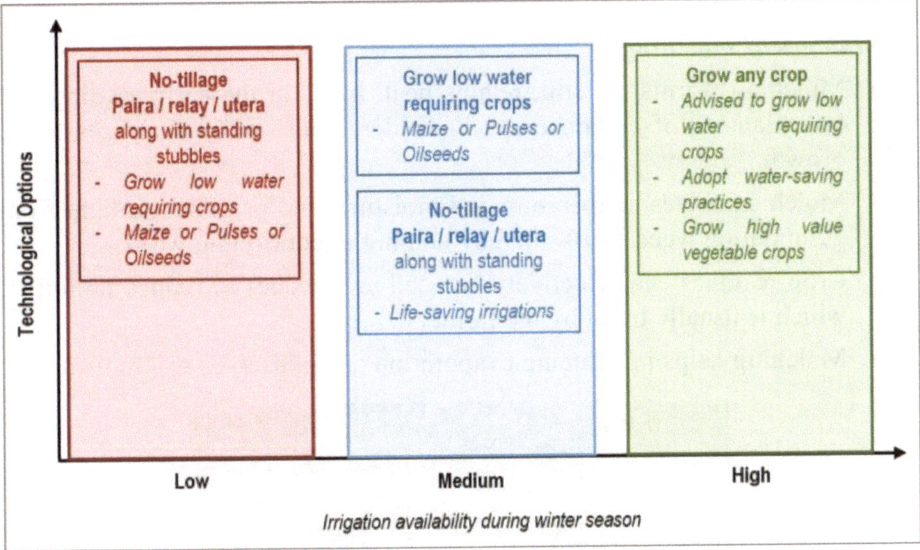

Fig. 5.2. Schematic diagram of irrigation availability during *rabi* season and respective technological options for winter (*rabi*) crop

5.3. Soil moisture conservation & utilization techniques

5.3.1. Resource conservation techniques

Resource conservation techniques (RCTs) prevent rapid soil moisture losses (Kar *et al.,* 2004), improve soil organic carbon (SOC) as well as bio-physical properties, and hence could be strategic approaches for post-rice crops (Ali *et al.,* 2014). Pulses/oilseed crops should be sown just after the harvest of rice crops for better utilization of the residual soil moisture. Soil moisture loss is prevented by the zero tillage (ZT) approach and the planting of post-rice crops could be advanced by seven days (Mishra *et al.,* 2016). Several researchers have reported moisture conservation as well as yield enhancement of pulse crops under ZT or reduced tillage (RT) than conventional tillage (CT) operations. For example, Kar and Kumar (2009) reported that RT increases pulse yield by ~33-44% over CT in rice-fallow. Similarly, Ghosh *et al.* (2010) also reported a higher yield of pulses with RT.

5.3.2. Mulching

Any material, like crop straw, crop residues, leaves, green manuring leaves or loose soil, etc. which is spread over the soil surface for reducing evaporation loss, and soil erosion and to protect roots from extremely high or low temperatures, is called a mulch; and the practice of applying these mulches is called mulching. Different types of plastic/polythene mulches are also used (Figs. 5.3 & 5.4).

Advantages of mulching

- Mulching warms the soil, retains heat, helps in the early seeding and transplanting of certain crops in cold regions, and encourages faster growth.
- Mulch stabilizes temperature and moisture, and prevents sunlight from germinating weed seeds, thereby helping in controlling weeds.
- Crop residues can effectively be used as mulches to reduce pollution, which is usually burnt by the farmers.
- Mulching helps in reducing evaporation and conserves soil moisture.

Fig. 5.3. Rice straw and plastic mulching in capsicum to conserve soil moisture

Fig. 5.4. Plastic mulching in vegetable crops

In a rice-fallow system, effective moisture conservation practices, like, mulching, tillage operations, etc. could help in mitigating moisture-related stress. Mulching is the process of applying materials to the soil surface, such as crop straw/residues/leaves, loose soil, green manuring leaves, plastic, etc., to prevent soil erosion and reduce evaporation, as well as to shield plant roots from extreme heat or cold. Mulch could be applied either during the start of the crop growing season or after crop germination or it may be reapplied as and when required. It has several benefits, viz., warms the soil, allows early and faster crop growth, especially in cold regions, minimizes growth of weeds, conserves soil moisture, etc. Rice crop residues, which are often burnt by the farmers, cause severe air pollution and it is a matter of concern for all of us, could be effectively used as mulches to conserve soil moisture. Certainly, a tradeoff between availability, use as fodder or mulching, or some other purposes should be worked out clearly. Some of the mulch materials, like *Gliricidia,* could be grown on the bunds of field plots, which ensures the availability of organic mulches, which are rich in N also. *Gliricidia* leaves as mulching is most effective in cutting down evaporation during a dry spell to supply moisture to the crop.

Multi-location trials in different states in India revealed that retention of rice stubbles/mulching and no-tillage (NT) maintains higher soil moisture levels, improves soil properties in rice-fallow areas, and greatly enhances the productivity of pulses (chickpea, Lathyrus, lentil). Paddy straw/ water hyacinth mulch increases the groundnut productivity which is sown after rice harvest (Chaudhary *et al.,* 2014).

5.3.3. Utera/relay/para cropping

Utera/ relay/ para cropping is the sowing of seeds of the next crop before the harvest of a standing *kharif* rice crop to utilize residual moisture of the field under a rainfed ecosystem. Usually, seeds are broadcasted 10-12 days before rice harvest. To facilitate uniform germination of utera crops, rice fields should be properly levelled to maintain uniform soil moisture. Line or mechanical transplanting of rice gives a higher yield of utera crops and also saves seeds (Mishra and Kumar, 2018). In utera, fundamental principles of conservation agriculture, zero-tillage, and residue retentions on the surface have been followed (Ali *et al.*, 2014). Apart from utilizing available moisture in rainfed areas, another advantage of utera cropping is by growing two or three pulses or oilseeds or cereal crops, the risks of crop failure are reduced, thereby improving income for the farmers. Rautaray (2008) conducted on-farm farmers' participatory trials in 10 villages of Assam and found 73% greater net returns under the improved rice-utera pea cropping sequence in comparison to the farmer's practice. Better performance in terms of yield under utera cropping than zero tillage (with or without mulch) happens because of early sowing and better utilization of residual soil moisture. Amongst different crops tried, Lathyrus gave the maximum yields and profits followed by linseed and lentil. Wheat as a utera crop in the rainfed rice ecosystem in Bihar is shown in Fig. 5.5.

Fig. 5.5. Wheat as utera crop in rainfed rice ecosystem

5.3.4. Rice ratooning

In low-land areas or where double-season rice systems are prevalent, total rice production from unit land area is higher compared to single-season rice systems, which significantly contributes to the total global rice supply. In the traditional system of double-season rice, seeds are sown in a seedbed, seedlings are uprooted from the nursery and transplanted in clumps in the main field in both seasons. However, the areas under double rice cropping have been decreasing continuously in Asian countries due to water scarcity, irregularities in rainfall, labour shortage, and a low degree of mechanization (Santos *et al.*, 2003).

Ratoon rice is one way to boost food production at a reduced cost by obtaining an additional rice yield with few agricultural inputs (Harrell *et al.*, 2009). From the hills (stubble) left behind after the main crop is harvested, ratoon rice is grown as the second crop of rice (Fig. 5.6). In general, ratooning rice yields lower than the main crop, therefore, it has been not adopted for large-scale. However, Xu *et al.* (2015) report the grain yields in the ratoon season range from 3.0 to 3.8 t ha^{-1}. Adoption of suitable management approaches for both the main crop and the ratoon crop could also result in a higher yield from the ratoon crop (Liu *et al.*, 2012). These practices are land preparation, appropriate sowing time, maintaining adequate plant density, use of cultivars having high ratoon yield, proper water management, application of adequate rates of fertilizers, and control of weeds and diseases.

Fig. 5.6. Rice ratoon

Dong *et al.* (2017) report that direct seeding on the puddled soil (wet-seeded rice: WSR) reduces the labour required for the rice-ratoon system more than the traditional transplanting system. They compared the performance of a wet-seeded rice-ratoon rice system (WSR-RR) with a transplanted rice-ratoon rice system (TR-RR) in central China using two cultivars. The ratoon season yields (4.05-5.83 t ha^{-1}) and annual grain yields (12.4-15.7 t ha^{-1}) of WSR-RR were similar to those of TR-RR (Table 5.4). Therefore, WSR-RR could be an alternative rice planting system to TR-RR.

Table 5.4 The main season yield, ratoon yield, and annual yield of two cultivars in the wet-seeded rice-ratoon rice system (WSR-RR) compared with a transplanted rice-ratoon rice system (TR-RR)

Cultivar	Planting pattern	Main season yield (t ha^{-1})	Ratoon season yield (t ha^{-1})	Annual yield (t ha^{-1})
Huanghuazhan	TR-RR	7.88	4.67	12.6
	WSR-RR	8.34	4.05	12.4
Tianyouhuazhan	TR-RR	9.54	5.72	15.3
	WSR-RR	9.90	5.83	15.7

Source: Dong *et al.* (2017)

Compared with traditional double-season rice systems, ratoon rice has several advantages, viz., saving water, time, labour, seed, and nursery supplies (Munda *et al.* 2009). The economy lies, however, in land preparation (not required for ratoon crops), seed saving, reduced crop growth period, and better utilization of the growing season. Ratoon crops typically mature in around half the time of the main crop.

5.3.5. Conservation agriculture

Conservation agriculture (CA) has been widely promoted as an alternative to tillage-based conventional agriculture and an approach to crop husbandry that can reconcile these often competing objectives (Jat *et al.*, 2020). The primary tenets of CA are three interconnected concepts that must coexist in space and time and be consistently followed to foster synergies (FAO, 2011; Pittelkow *et al.*, 2014; FAO, 2015). These principles are:

a) *Continuous minimum mechanical soil disturbance:* This translates into low-disturbance direct planting and low-disturbance no-tillage practices. Soil disturbance must be minimized in all operations, permitting disturbance of no more than 25% of the soil surface in bands no broader than 15 cm under very specified circumstances.

b) *Permanent organic soil cover:* This can be made of living crops, such as cover crops, mulch made from crop wastes, or any other organic mulch materials. The amount of soil cover should ideally cover 100% of the soil surface, but it should never be less than 30% in order to maintain and increase the levels of soil organic matter. It should also always provide enough organic carbon.

c) Diversification of crop species grown in associations and/or sequences: This includes plantation crops, mixed-, inter-, or relay-cropping, cover crops in perennial orchards, and annual crop rotations and sequences. Legumes are included because of their nitrogen-fixing properties and because they flower to aid pollinator populations.

Individual CA practices have been practised by farmers for a long time (Montgomery, 2007). The integration of these three principles ought to be implemented concurrently through locally developed and tested procedures-makes the current CA approach unique. It has been suggested that additional best management practices should be added to reinforce fundamental CA procedures (Kassam *et al.*, 2011):

i) use of quality seeds;
ii) balanced or enhanced crop nutrition to support healthy soils;
iii) integrated pests, diseases and weeds management; and
iv) efficient water management.

Therefore, the combination of all enhanced techniques performed promptly and effectively constitutes sustainable crop production intensification based on CA. Farmers can choose from a range of CA-based practice combinations with this technique and customize it to fit their specific local production variables and limits.

The CA practices leading to an increase in crop yields through a set of agroecological functions (Ranaivoson *et al.*, 2017; 2019). Research suggests that CA has multiple benefits, including the saving of labour, energy, and mineral nitrogen in farming and reduction in greenhouse gas emissions (Alam *et al.*, 2019), enhanced biological activity in soils (Naresh *et al.*, 2016), and the long-term yield and productivity increase as a consequence (Sharma *et al.*, 2012). In the short-term and irrigated environments, few studies have suggested that CA has no yield advantages if not optimally implemented (Pittelkow *et al.*, 2014). It is anticipated that production increases from the application of CA will be mostly dependent on its effects on soil water, soil mineral nitrogen (N), and weed dynamics in smallholders' low-input rainfed cropping systems. Productivity under CA varies considerably among and across the environments, depending on the crop, location, climate, management, and how many and how long the three principles of CA have been applied (Zhang *et al.*, 2022). Some reviews have revealed that no-till practices can reduce crop productivity, but no-till in combination with residue retention and crop rotation may increase yield (Jat *et al.*, 2020; Kumara *et al.*, 2020).

Generally, CA increases soil water penetration and decreases evaporation and runoff to maintain soil moisture for higher crop water transpiration (Hobbs, 2007; Scopel *et al.*, 2004). Crop yields may rise as a result of these impacts, particularly in arid locations or during dry spells. Mulching with crop residues during periods of significant rainfall can cause waterlogging in soils with poor drainage (Sissoko *et al.*, 2013). In CA systems, keeping crop wastes on the soil's surface improves nutrient availability and cycling (Iqbal *et al.*, 2011; Turmel *et al.*, 2014). Growing levels of surface residues are anticipated to enhance soil nutrient content, at least over time, as residue breakdown releases nutrients into the soil. However, using cereal leftovers as mulch material short-term immobilizes soil mineral N, which can lead to crop N deficit, particularly in low-input cropping systems, resulting in poorer yields (Govaerts *et al.*, 2006).

One of the biggest obstacles to the application of CA on smallholder farms with minimal inputs is still weed management (Lee and Thierfelder, 2017). It has been suggested that the removal of soil tillage is a weed management strategy (Giller *et al.*, 2009; Chauhan *et al.*, 2012). According to Nakamoto *et al.* (2006), tillage eliminates weeds physically and may bury certain weed seeds in deeper soil layers, reducing their exposure to favourable germination conditions. Furthermore, the effectiveness of chemical and physical weed management techniques may be diminished by the mulch layer in CA systems (Bajwa, 2014). Nonetheless, some research shows that CA may be able to lessen weed infestation by using mulch, which acts as a physical barrier to the growth of weeds (Teasdale and Mohler, 2000; Bilalis *et al.*, 2003; Campiglia *et al.*, 2012; Ranaivoson *et al.*, 2018). It is acknowledged that the amount of mulch that effectively controls weeds is very high, however it varies greatly based on the agroecological circumstances.

The area under CA in South Asia (SA) is also relatively small compared to temperate nations and South America (Somasundaram *et al.*, 2020). The coverage of a partial CA-based system (at least one crop with no-till, with or without residue retention) is estimated at around 2.5 million ha in South Asia (Jat *et al.*, 2020), equivalent to 1% of total arable land, which is much lower than the proportion in America, Europe, Australia, and China (Kassam *et al.*, 2018). The direct sowing of wheat in the Punjab states of Pakistan and India marked the beginning of CA in South Asia (Hafeez-Ur-Rehman *et al.*, 2015).

Several agronomic challenges, including the occurrence of weeds, insect pests and diseases, the lack of widespread and suitable systems for crop residue management, and the non-availability of proper seeding or planting equipment, affect the adoption rate of CA in SA (Jat *et al.*, 2021).

5.4. Adoption of water-saving irrigation methods

5.4.1. Micro-irrigation

Micro-irrigation (MI) can save a lot of water in rice-based farming systems. It can increase crop yields while utilizing less water in various crops.

5.4.1.1. Drip irrigation

Drip-irrigation (DI) saves 35-40% of water resulting in a 40% improvement in water productivity in rice cultivation (Panigrahi *et al.*, 2021), however, its use in rice or rice-rice cropping sequence was not economical. Therefore, the inclusion of capsicum and baby corn after rice makes the system economically viable (Fig. 5.7). They standardized the use of drip-irrigation (DI) in rice-based cropping sequence and found rice-capsicum-baby corn in DI could give very high economic benefits (B: C ratio: 2.87) in eastern India.

Fig. 5.7. Drip irrigation in **(a)** rice, **(b)** capsicum, and **(c)** baby corn

Drip fertigation is the application of soluble fertilizers through irrigation by using a drip system to the crops. The technique applies dissolved fertilizers with irrigation water at a low rate to the vicinity of the plant root zone. Usually, it has been demonstrated that fertigation allows the soil to absorb up to 90% of the applied nutrients in comparison to granular or dry fertilizer applications with typical absorption rates of barely up to 40%. A higher abruption of the fertilizers through fertigation leads to enhanced use efficiency. A drip irrigation system is ideal for fertigation because of the high water application efficiency associated with negligible nutrient loss due to reduced surface runoff and leaching of nitrogen and potassium, thereby minimizing groundwater and water bodies' contamination. In a drip fertigation system, the amount of nutrients, timing of application, concentrations, etc. can be easily controlled. Some of the advantages of fertigation through drip irrigation are precise nutrient application, reduced application cost, greater nutrient use efficiencies, minimal operational hazards, no soil compaction, etc.

In India, the results of drip fertigation studies in different crops conducted at different centres of AICRP on Irrigation Water Management (IWM) have been compiled and presented in Table 5.5 and crops under drip irrigation/fertigation have been shown in Fig. 5.8. Overall, results show crop yield increase by 4 to 155%, fertilizer saving by 25% (range: 20-50%), water-saving ranged between 6 to 66%, and enhancement in income ranged between 4 to 262%.

Table 5.5 Effect of drip fertigation on crop yield, irrigation water saving, fertilizer saving, and increase in income

Location	Soil texture [#]	Crop	Yield (t/ha)	Yield increase (%)	Water saving (%)	Fertilizer saving (%) (NPK)	Income increased (%)
Palampur	SiCL	Strawberry	20.4	19.3	54.2	25	32.7
Palampur	SiCL	Pea	9.5	9.9	40.0	50	25.5
Palampur	SiCL	Potato	22.3	41.1	17.3	25	34.6
Parbhani	C	Hybrid Maize	8.8	12.8	16.8	25	10.4
Parbhani	C	Cabbage	37.8	17.2	17.8	-	17.0
Parbhani	C	Turmeric	31.2	26.5	26.1	20	33.1
Faizabad	SiL	Rajmash bean	12.7	39.6	61.8	25	47.2
Faizabad	SiL	Tomato	32.9	23.8	65.0	25	41.1
Faizabad	SiL	Okra	6.8	40.9	50.1	25	58.6
Faizabad	SiL	Sugarcane	100.4	60.7	59.5	25	67.4
Faizabad	SiL	Marigold	21.2	32.3	54.7	25	40.5
Faizabad	SiL	Aonla	13.3	75.0	33.0	25	113.5
Coimbatore	SaL	Cotton+ Blackgram (CEY)	3.86	83.8	24.3	-	131.5
Coimbatore	SaL	Maize+ Blackgram (MEY)	7.74	36.3	23.6	-	55.3
Coimbatore	SaL	Chewing cane	88.1	15.0	27.0	-	26.2
Coimbatore	SaL	Small onion (using MS)	15.7	31.6	28.4	25	55.2
Coimbatore	SaL	Chilli	19.1	87.1	51.2	25	215.2
Coimbatore	SaL	Brinjal	70.7	154.8	47.3	25	261.7
Bhavanisagar	SaL	Turmeric	40.4	22.1	33.2	20	21.7

Bhavanisagar	SaL	Sugarcane	148.0	28.0	27.0	18	41.6
Sriganganagar	SaL	Summer squash	34.0	62.6	52.2	20	62.6
Sriganganagar	SaL	Brinjal	66.5	12.2	34.0	20	12.3
Sriganganagar	SaL	Bittergourd	40.8	10.3	52.9	20	11.5
Sriganganagar	SaL	Chilli	32.3	55.3	31.1	-	60.5
Sriganganagar	SaL	Sponge gourd	28.3	9.3	44.2	20	20
Kota	CL	Garlic	10.4	50.1	22.0	25	135.5
Kota	CL	Garlic	15.4	17.4	6.1	25	25.9
Kota	CL	Bittergourd	28.8	35.8	11.1	25	65.1
Dapoli	SaCL	Cashew	1.99	55.3	33.0	20	76.1
Dapoli	SaCL	Sweet corn	18.0	14.1	15.8	20	69.6
Dapoli	SaCL	Brinjal	32.3	22.8	41.7	-	17.0
Rahuri	C	Papaya	226.4	112.7	66.5	25	139.0
Gayeshpur	SaL	Banana	43.3	26.5	50.5	20	36.7
Jorhat	SaL	Assam lemon	14.9	47.3	35	20	14.9
Navsari	C	Banana	108.8	41.6	59.4	20	37.5
Navsari	C	Onion	45.7	59.0	30.0	20	38.9
Navsari	C	Turmeric	20.7	28.2	32.0	20	39.0
Navsari	C	Brinjal	36.8	21.0	40.0	20	12.0
Navsari	C	Chilli	6.1	17.3	48.0	20	11.0
Navsari	C	Papaya	61.0	20.0	27.0	20	8.0

Navsari	C	Sugarcane	182.0	68.0	36.3	20	79.0
Navsari	C	Rabi sorghum-Cowpea	10.7	38.2	17.4	20	66.6
Pantnagar	SaL	Spring-maize	15.0	3.9	27.0	-	3.8
Pantnagar	SaL	Pea	11.8	9.1	50.0	25	6.7
Pantnagar	SaL	Rice	4.7	8.7	44.7	-	10.9
Pantnagar	SaL	Mentha	0.28	36.3	26.9	-	48.4
Chiplima	SaL	Onion	21.0	32.6	28.7	-	56.1
Chalakudy	SaL	Banana	21.3	4.0	33.3	50	4.0
Chalakudy	SaL	Okra	14.0	36.6	40	50	36.6

[#]Soil textures: SaL-Sandy loam, LS- Loamy sand, SiL-Silty loam, SaCL- Sandy clay loam, SiCL- Silty clay loam, CL- Clay loam, C- Clay. CEY- Cotton equivalent yield; MEY- Maize equivalent yield. MS- Micro-sprinkler.

Source: Nanda *et al.* (2021)

Fig. 5.8. Drip irrigation/fertigation in **(a)** groundnut, **(b)** cauliflower, **(c)** cabbage, and **(d)** gourd

Despite substantial water saving through drip irrigation, there are a few constraints in its adoption, especially in India:

i) Higher initial investment,
ii) Frequent irrigation (almost daily) is required in the summer season,
iii) Drip-irrigated fields have more weed infestation,
iv) Shifting of drip laterals during intercultural operation,
v) Need for power either through oil or electricity or solar,
vi) Damage of lateral pipes due to rodents leads to higher maintenance costs, and
vii) The operation and maintenance of drip irrigation systems requires technical knowledge.

Nowadays, researchers are using soil moisture-based sensors coupled with drip irrigation for precise and more efficient irrigation management in several crops, for example, in bananas (Fig. 5.9), the sensor-based drip irrigation saves 20% water and produces 16.6% more fruit with better quality (higher total soluble solids and lower acidity) compared with manual drip irrigation in banana. Further, under sensor-based drip irrigation, water used was almost half and fruit yield was 1.6 times more than the surface irrigation in bananas. The water use efficiency (WUE) improved by 45% and 216% under sensor-based drip irrigation compared with manual drip irrigation and surface irrigation, respectively (Panigrahi *et al.*, 2019).

Fig. 5.9. Sensor-based drip irrigation in banana

5.4.1.2. Sprinkler irrigation

During the post-rainy season, the sprinkler irrigation system is another method of water-saving, which is commonly used for closely spaced crops, like, wheat, pulse, mustard, groundnut, etc. (Fig. 5.10). A sprinkler irrigation system uses a pump to apply high-pressure water. It uses a small diameter nozzle in the pipes to release water in the same way that rain does. Because of the wide variety of discharge capacity, water is disseminated via a pipe network, sprayed into the air, and irrigated in most soil types.

Fig. 5.10. Sprinkler irrigation in groundnut

Advantages of sprinkler system

- Suitable for most of the soil types except heavy clay.
- Nil conveyance loss.
- Water-saving up to 30-50%.
- Suitable for irrigation for closely spaced crops.
- Suitable for undulating land.
- Increase yield and nutrient use efficiency.

- Soluble fertilizers/chemicals use are possible.
- Reduces soil compaction.
- Mobility of the system helps system operation easy.
- Saves land.
- Provides frost protection & helps in the alteration of microclimate.
- Reduces labour costs.

The layout of the sprinkler system

A sprinkler irrigation system typically consists of different components, *viz.*, pump, mainline and sub-mainlines, laterals, and sprinkler (Fig. 5.11):

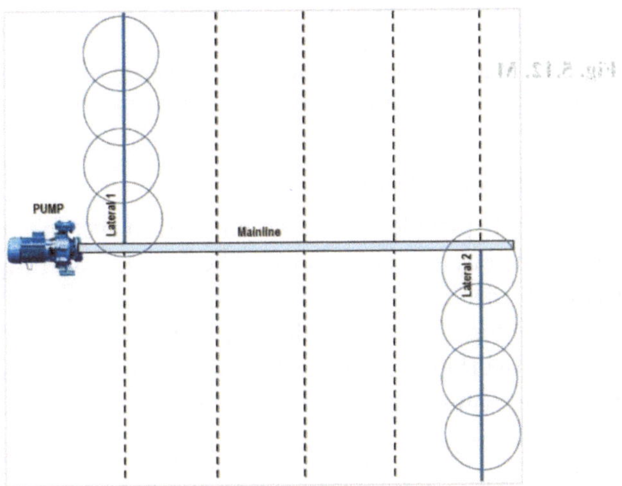

Fig. 5.11. Layout of sprinkler irrigation system

Major parts of sprinkler irrigation system (Fig. 5.12):

i) Pumping station or header Assembly
ii) By-pass valve
iii) Fertilizer tank
iv) Filtration system
v) Pressure gauges
vi) Control valves
vii) HDPE / PVC Pipes
viii) QRC pump connector
ix) Sprinkler nozzles
x) Service saddle

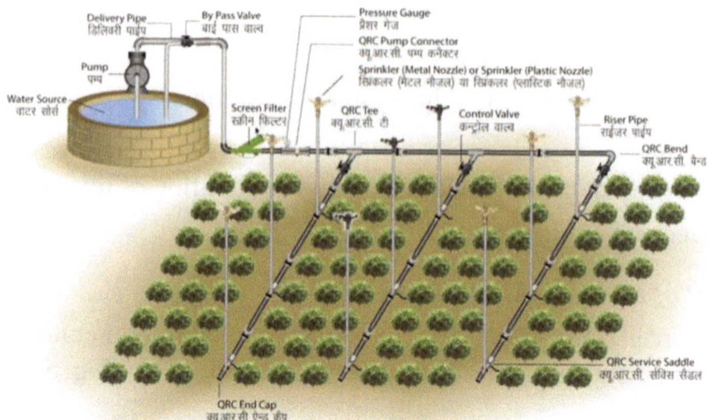

Fig. 5.12. Major components and layout of sprinkler irrigation system (*source*: https://vikaspedia.in/agriculture/agri-inputs/farm-machinary/sprinkler-irrigation-system)

A wetting pattern in the sprinkler system

A single rotational sprinkler produces an uneven wetting pattern. Normally, the area wetted is circular (Fig. 5.13). The heaviest wetting occurs near the sprinkler (Fig. 5.14). To achieve better consistency, many sprinklers should be operated close together so that their patterns overlap. To achieve adequate consistency, the overlap should be at least 65% of the wetted diameter. This determines the maximum space between sprinklers.

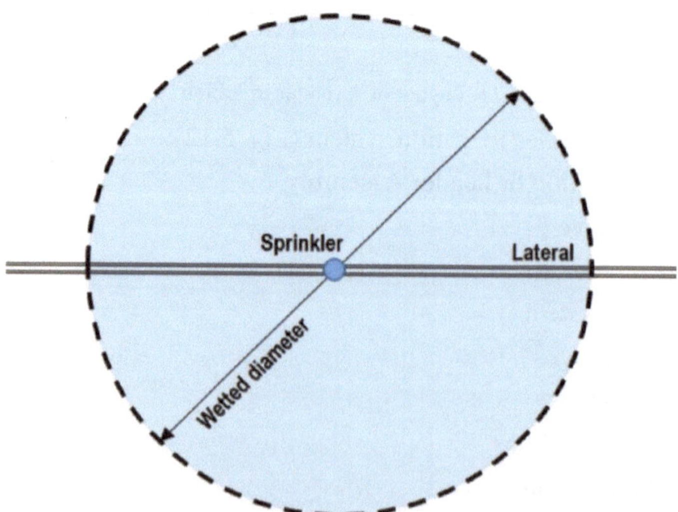

Fig. 5.13. Top view of the wetting pattern of a single sprinkler

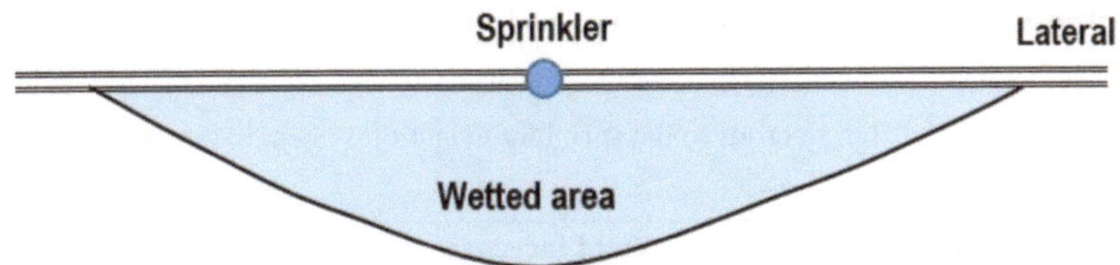

Fig. 5.14. Side view of the wetting pattern of a single sprinkler

Wind and water pressure affect the uniformity of sprinkler applications. To minimise the wind effects the sprinklers can be placed more closely. Also, sprinklers will only perform well at the appropriate operating pressure recommended by the manufacturer. The pressure above or below the recommended level affects the distribution. Low pressure is the most common problem and this happens because of increased friction due to the wearing of pumps and pipes. As a result, the water jet does not break up and most of the water falls in one area towards the outside of the wetted circle. When the operating pressure is too high then the water distribution will be poor and developed fine spray will fall close to the sprinkler.

Application rate: The average rate of water sprayed onto the crops is called the application rate and is measured in mm h^{-1} (Table 5.6). It depends on the operating pressure, the size of sprinkler nozzles and the distance between each sprinkler. While selecting any sprinkler system it is most important that the average application rate should be less than the infiltration rate of the soil. So that all the water applied will be absorbed by the soil with minimal or no runoff.

Table 5.6 Types of sprinklers based on precipitation

Sprinkler type	Precipitation rate (mm/h)
Low Volume Sprinkler	< 13
Medium Volume Sprinkler	13 - 25
Large Volume Sprinkler (Raingun)	> 25

Drop sizes in sprinkler: Water sprays from a sprinkler break up into small drop sizes between 0.5 to 4.0 mm. The large drops fall close to the periphery of the wetted circle, while the smaller ones fall close to the sprinkler. Large drops of sprinklers can cause soil erosion and also damage delicate crops so in such conditions, it is advised to use small sprinklers. The drop size of the sprinkler can be controlled by nozzle size and pressure. At low pressure, the drop size remains to be larger because the water jet does not break up easily. Therefore, to avoid soil and crop damage small-diameter nozzles should be operated above/or the normal recommended pressure.

Water-saving and yield increases in some crops have been presented in Table 5.7.

Table 5.7 Response of different crops to the sprinkler irrigation system in terms of water-saving and yield increase.

Crops	Water Saving (%)	Yield increase (%)
Chilli	33	24
Cotton	36	50
Fenugreek	29	35
Gram	69	57
Jowar	55	34
Maize	41	36
Onion	33	23
Sunflower	33	20
Wheat	35	24

Source: https://www.fao.org/3/s8684e/s8684e06.htm **(accessed on July 05, 2022)**

5.4.2. Furrow irrigation

Furrows are parallel small channels to carry irrigation water for the crop. Usually, the crops are grown on the ridges between the two furrows (Fig. 5.15). Furrow irrigation is a water-saving method and is suitable for various crops (especially row crops), soil types, and land slopes. Some crops are damaged due to water stagnation if their stem or crown is covered with water and those crops should be furrow irrigated.

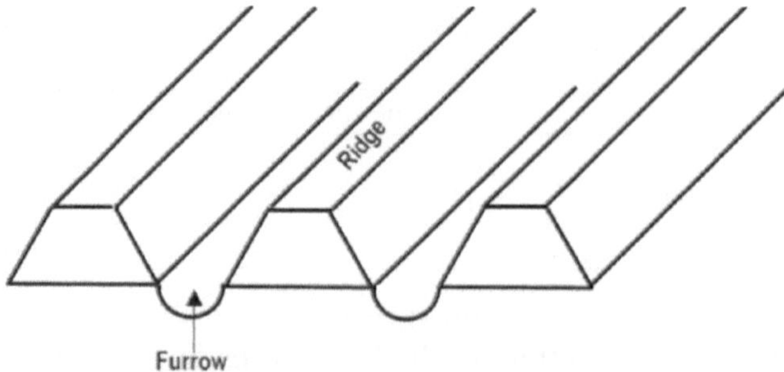

Fig. 5.15. Ridges and Furrows

Furrow irrigation is also suitable for growing fruit trees. During the initial growth of a tree, a single furrow alongside the row of trees may be sufficient. But as the tree grows, two or more furrows would be required to sufficiently provide water. The furrow irrigation is suitable for the following crops to irrigate:

- Row crops, *viz.*, wheat, maize, sugarcane, sunflower, soybean, etc.
- Crops susceptible to flooding, *viz.*, tomatoes, potatoes, beans, vegetables, etc.
- Fruit trees, *viz.*, citrus, grape, etc.

A field experiment photograph of furrow irrigation in a cabbage crop is shown in Fig. 5.16.

Fig. 5.16. Furrow irrigation in cabbage crop

Usually, flat land or land with gentle slopes (less than 0.5%) is suitable for furrow irrigation to avoid soil erosion. A gentle slope (up to 0.05%) is provided in furrows to facilitate drainage after irrigation or in case of high rainfall. Furrow irrigation is not preferred on undulating land, but it can be done using the land contours. Furrow irrigation can be used in most of the soil types. But, for very coarse sandy soils it is not recommended because of high percolation losses. For furrow irrigation, soils that easily crust are most suitable because while irrigation water does not flow over the ridge and the soil on the ridge remains friable where plants grow. In sandy soils, furrows should not be long, otherwise water will not reach the downstream end because of excessive percolation losses. However, in clay soils, furrows can longer due to the low infiltration rate of water.

Mostly, the shape, length, and spacing under furrow irrigation are determined by the slope of land, soil type, and available stream size. However, other factors that should be considered while designing a furrow system are field length, depth of irrigation, and farming practice.

When the furrows are short, stream sizes up to 0.5 litre s^{-1} will provide adequate irrigation. If stream sizes are larger then furrows should be longer because water will flow rapidly down the furrows. The stream size depends on the furrow slope so that it should not cause erosion, however, it is recommended that stream sizes greater than 3.0 litre s^{-1} should not be used (Table 5.8).

Table 5.8 Maximum furrow lengths (m) depends on furrow slope, type of soil, stream size, and net irrigation depth.

Furrow slope (%)	Max. stream size (liter s^{-1}) per furrow	Clayey soil		Loamy soil		Sandy soil	
		Net irrigation depth (mm)					
		50	75	50	75	50	75
0.0	3.0	100	150	60	90	30	45
0.1	3.0	120	170	90	125	45	60
0.2	2.5	130	180	110	150	60	95
0.3	2.0	150	200	130	170	75	110
0.5	1.2	150	200	130	170	75	110

Source: FAO (https://www.fao.org/3/s8684e/s8684e04.htm)

When large depths of irrigation are applied, furrows should be longer so that water flow gets more time in the furrows to infiltrate. In the case of mechanized farming, furrows should be as long as possible to expedite the work. However, shorter furrows can be irrigated more efficiently than longer ones because of lower percolation losses. It is always advised to construct the furrow length equal to the field length, instead of the ideal length.

The furrow shape depends on the type of soil as well as on the stream size. For example, in sandy soils, a deep V-shaped and narrow furrow is suggested because water flows faster vertically than sideways which is required to reduce water percolating downwards (Fig. 5.17a). However, in clay soils, wide and shallow is suggested to get a larger wetted area to increase infiltration rate (Fig. 5.17b). Also, the furrow should be larger for the larger stream size.

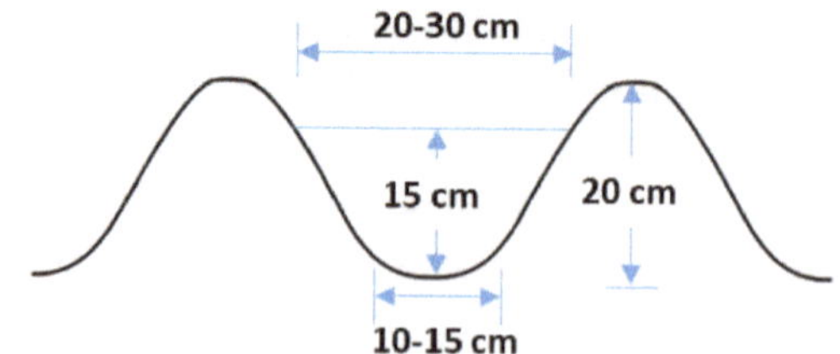

Fig. 5.17a. A deep V-shaped and narrow furrow (for sandy soil)

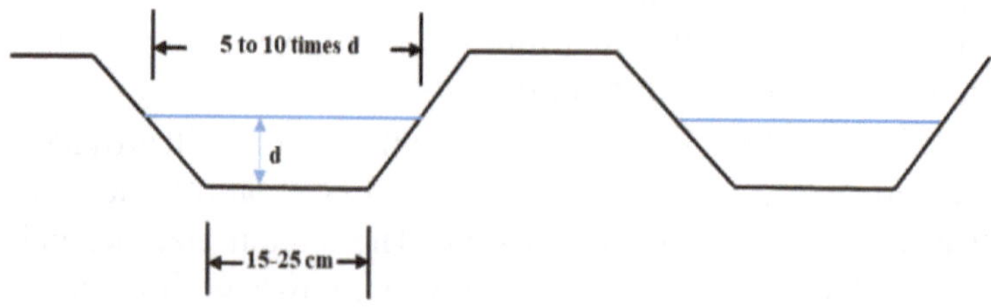

Fig. 5.17b. A wide and shallow furrow (for clay soil)

The furrow spacing depends on the type of soil and the cultivation practices. For example, in sandy soils, the ideal furrow spacing is 30-60 cm (30 cm for coarse sand and 60 cm for fine sand). Conversely, two consecutive furrows on clay soils should be spaced 75-150 cm apart. Double-ridged furrows, sometimes referred to as beds & furrows, are another option for clay soils (Fig. 5.18). Their benefit is that intercultural operations are made easier and more rows of the crop can be grown on each ridge.

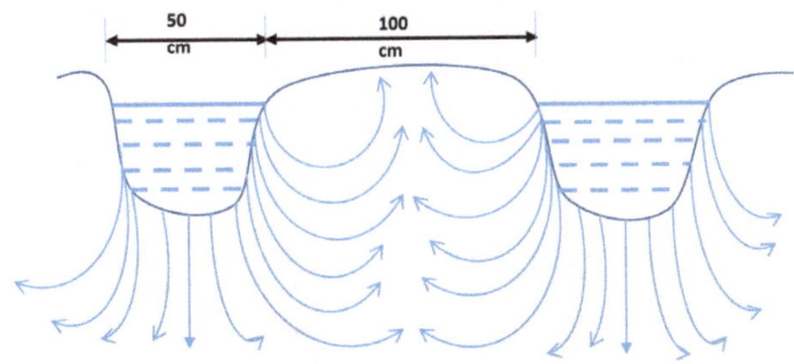

Fig. 5.18. A double-ridged furrow (Beds & Furrows)

In a mechanized farming system, a compromise between the available machinery to make furrows and the ideal spacings for the crops is required. Even though varying plant spacing is needed for different crops, it is preferable to keep the furrow widths uniform. It is important to take precautions to guarantee that there is enough lateral wetness at the standard spacing. Because of the small field sizes in India, ridgers-either animal or hand-drawn-are the most often used tool for creating furrows.

Irrigation is supplied to each furrow from the field canal using siphons or spiles. The field canal with siphons or spiles is occasionally replaced with a gated pipe. The amount of water that may be used to irrigate multiple furrows at once depends on the available flow. 'Alternate furrow irrigation' is a useful technique for conserving irrigation water during water shortage. Instead of watering every furrow at the 10-day interval, this irrigation method irrigates alternating furrows: furrows 1, 3, 5, etc. are irrigated on the fifth day, and furrows 2, 4, and 6 etc. are irrigated on the tenth day. As a result, rather than receiving a substantial amount of water every 10 days, the crop receives some every 5 days. Generally speaking, crops benefit more from smaller, more frequent applications of water than from larger applications at longer intervals.

In the event of furrow irrigation, a shallow drain ought to be constructed at the end of the field to eliminate surplus runoff. Even in ideal circumstances, runoff may account for 30% of the inflow. Waterlogging could harm crops on the field's lower side if no drains are installed. To stop soil erosion, one can also plant light plants in the drain. Reducing the water inflow can stop excessive runoff after the irrigation water reaches the end of the furrows. We refer to this as "cut-back irrigation." Additionally, runoff water may occasionally be utilized again for the farm's crops.

Furrows should be correctly spaced, have a consistent slope, and apply irrigation water quickly to provide the crop with a uniformly wetted root zone. The lateral (or sideways) water movement in the furrow is more crucial in getting the crop's root zone wetted in the ridge than the descending water movement. As was previously mentioned, the type of soil has a major influence on the lateral and downward circulation of water in furrows (Fig. 5.19a-c). Adjacent wetting patterns from the furrows should overlap each other in the best scenario for the crop, and water should travel upward through the capillary rise to wet the entire ridge and deliver water to the crop's root zone (Fig. 5.20).

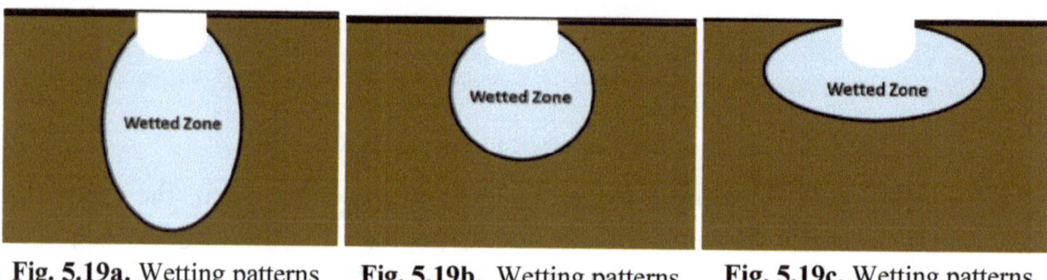

Fig. 5.19a. Wetting patterns in furrows in sandy soil

Fig. 5.19b. Wetting patterns in furrows in loam soil

Fig. 5.19c. Wetting patterns in furrows in clay soil

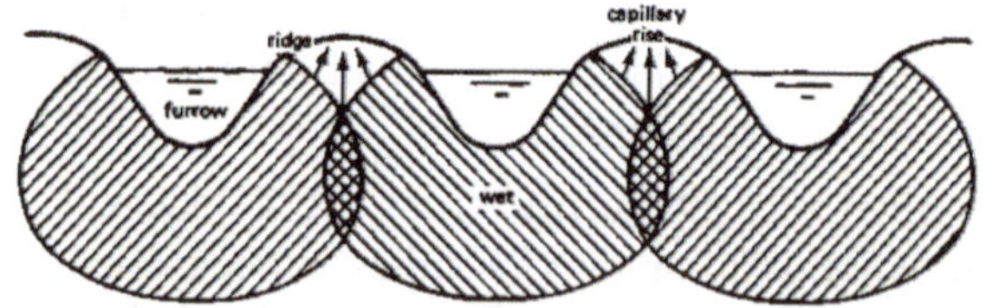

Fig. 5.20. An ideal wetting pattern for crop growth

It is crucial to have a consistent slope and a sizable stream size so that water moves quickly down the furrow to distribute water evenly along its length. Large percolation losses at the furrow's head end can be prevented in this method. Unfavourable natural conditions (such as a compacted layer, different soil types, uneven slope), poor furrow layout (such as too wide furrow spacing), and poor management (such as stream size that is too large or too small and stopping the inflow too soon) can all contribute to poor wetting patterns in the furrows. Wetting along the furrow may be uneven because to an uneven slope. The root zone won't be sufficiently wetted if the furrow spacing is too large (Fig. 5.21a). Similarly, insufficient wetting of the ridges will occur from a too-small stream (Fig. 5.21b). Furthermore, inadequate water distribution along the furrow's length will come from a little stream. Conversely, overtopping of the ridge may happen on flat slopes if the stream size is too high (Fig. 5.21c), and erosion of the bed and sides of the furrow may occur on steeper slopes if the stream size is too large.

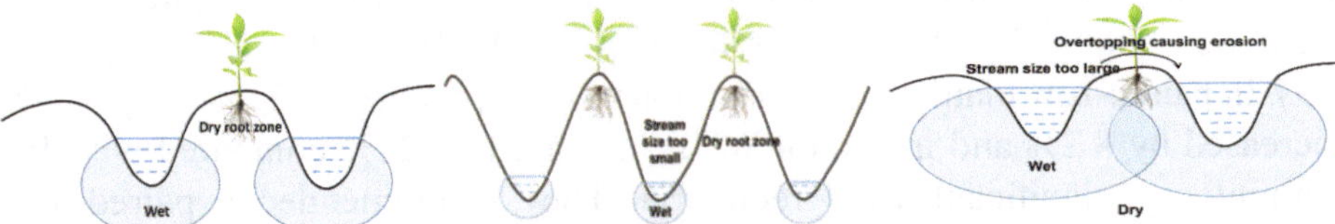

Fig. 5.21a. 'Too wide' spacing between two adjacent furrows

Fig. 5.21b. 'Too small' stream size to wet the ridge

Fig. 5.21c. 'Too large' stream size causing overtopping or erosion

Plant placement in a furrow system is flexible and dependent on the environment. For example, in heavy rainfall areas, plants should be on the top of the ridge to avoid waterlogging (Fig. 5.22a); in water-scarce areas, plants should be in the furrows to protect them from water scarcity (Fig. 5.22b); in a salinity problem area, plants should be planted away from the ridge top to avoid salt accumulation (Fig. 5.22c); in winter and early spring crops in colder areas, the seed should be placed on the sunny side of the ridge (Fig. 5.23a); and in hotter areas, seeds should be placed on the shady side of the ridge, to protect them from the sunlight (Fig. 5.23b).

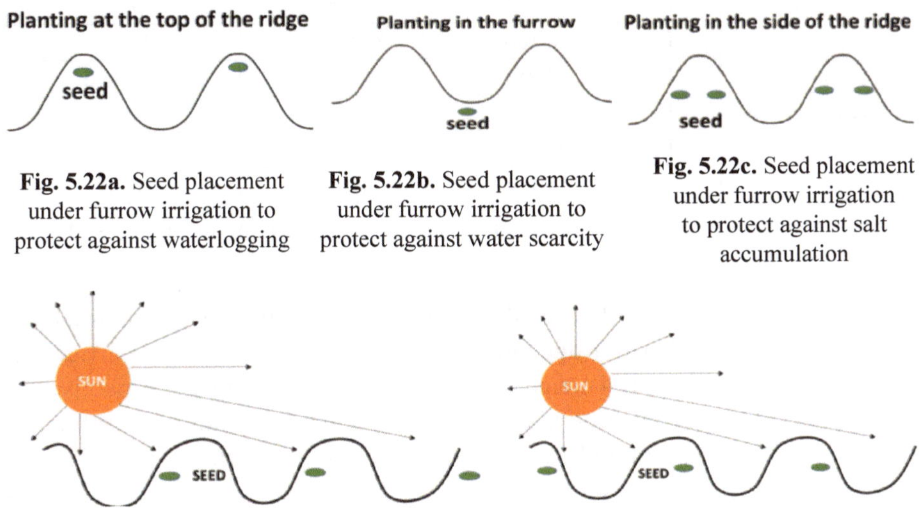

Fig. 5.22a. Seed placement under furrow irrigation to protect against waterlogging

Fig. 5.22b. Seed placement under furrow irrigation to protect against water scarcity

Fig. 5.22c. Seed placement under furrow irrigation to protect against salt accumulation

Fig. 5.23a. Winter and early spring crops- seeds are planted on the sunny side of the ridge

Fig. 5.23b. In hotter areas- seeds are planted on the shady side of the ridge

Mandal *et al.* (2018) experimented with four irrigation regimes, i.e., 2, 3, 4, and 5 numbers of irrigation (named I_1, I_2, I_3 and I_4, respectively), and three planting methods: normal single row planting with 50-cm row-to-row and 15-cm plant-to-plant spacing with 50-cm furrow spacing (P_1); paired-row planting at 25-cm spacing, 75-cm furrow spacing (P_2); and paired-row planting at 50-cm spacing with 100-cm furrow spacing (P_3) in potato. The highest tuber yield was obtained in I_4, whereas the highest WUE was found in I_3 (Table 5.9). In paired-row planting (P_2) irrigation volume reduced by 21%; crop WUE increased by 8.2% and irrigation WUE increased by 21% compared with P_1, without any significant yield reduction. They recommended a paired-row method of planting potatoes at 75 × 20 cm spacing and 4 irrigations for higher tuber yield and increased WUE in tropical sub-humid conditions.

Table 5.9 Potato tuber yield, crop evapotranspiration, and crop water use efficiency as influenced by irrigation and planting methods

Irrigation treatments	Planting methods	Tuber yield (t ha^{-1})	Evapotranspiration (mm)	Water use efficiency (WUE, kg ha-mm^{-1})
I_1: two irrigations	P_1	12.02d	268.7g	44.9abc
	P_2	11.67d	261.4gh	44.8abc
	P_3	10.36e	248.4h	41.9cde
I_2: three irrigations	P_1	14.60c	328.7e	44.4abc
	P_2	14.49c	308.6f	47.0a
	P_3	12.41d	280.1g	44.5abc
I_3: four irrigations	P_1	16.65ab	418.5bc	39.8de
	P_2	16.16b	352.6d	45.9ab
	P_3	13.95c	326.2ef	42.8bcd
I_4: five irrigations	P_1	17.09a	504.7a	33.9f
	P_2	16.80ab	432.9b	38.8e
	P_3	14.13c	403.2c	35.1f

Values in a column with different letters differ significantly according to Duncan's Multiple Range Test (DMRT) at $p<0.05$; *Adapted from: Mandal et al (2018)*

Similarly, Mandal *et al.* (2019) evaluated the effects of three planting methods of groundnut viz., flat-bed, ridge and furrow, and paired-row at four irrigation regimes viz., 1, 2, 3, and 4-irrigations (named I_1, I_2, I_3, and I_4, respectively). A schematic diagram of three planting methods of groundnut is presented in Fig. 5.24 and experimental photos of flatbed planting and paired-row planting with furrow irrigation are shown in Figs. 5.25a and 5.25b. As compared with the flat-bed method of planting, ridge and furrow, and paired-row methods resulted in an increase in pod yield by 13 and 17% along with irrigation water saving by 26.7 and 41.7%, respectively (Table 5.10).

Pod yield level was similar in ridge and furrow and paired-row methods with four irrigations, but irrigation water used was 28.4% less in the latter than in the former. In the treatment of three irrigations under the paired-row method, pod yield was reduced by only 3.8%, whereas water saving was 26.9% compared to the treatments of four irrigations x paired-row method. This study confirmed that paired-row planting and furrow irrigation had increased pod yield, saved water, and enhanced the WUE of groundnut under hot sub-humid conditions.

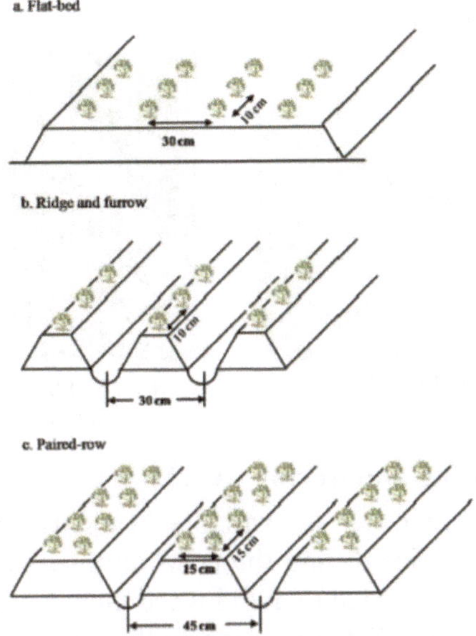

Fig. 5.24. Schematic diagram of three planting methods of groundnut used in the experiment, namely (a) flat-bed, P_{fb} (b) ridge and furrow, $P_{rf,}$ and (c) paired-row method, $P_{pr.}$

Table 5.10 Combined effect of irrigations and planting methods on groundnut pod yield, irrigation requirements, and irrigation water use efficiency (IWUE)

Irrigation treatments	Planting methods	Pod yield (kg ha^{-1})	Irrigation water depth (mm)	IWUE (kg ha-mm^{-1})
I_1: one irrigation	P_{fb}	1194i	79.3g	15.9d
	P_{rf}	1381h	57.8h	26.0b
	P_{pr}	1429gh	44.9h	33.5a
I_2: two irrigation	P_{fb}	1447g	149.4d	10.0f
	P_{rf}	1681e	109.4f	16.3d
	P_{pr}	1677ef	86.1g	20.3c
I_3: three irrigation	P_{fb}	1615f	206.1b	8.0fg
	P_{rf}	1907c	147.9d	13.4e
	P_{pr}	1977b	123.4e	16.4d

I_4: four irrigation	P_{fb}	1830d	291.4a	6.3g
	P_{rf}	2051a	217.0b	9.5f
	P_{pr}	2056a	168.9c	12.3e

P_{fb}, flat-bed method at 30 x 10 cm spacing; P_{rf}, ridge, and furrow planting at 30 x 10 cm spacing; P_{pr}, paired-row planting at 45 x 15 cm spacing; Means in the same column with the same letters are not significantly different by Duncan's multiple range test (DMRT) at p =0.05.

Adapted from: Mandal *et al.* (2019)

Fig. 5.25a. Flatbed flooding irrigation of groundnut

Fig. 5.25b. Paired-row planting of groundnut and furrows irrigation

5.4.3. Alternate wetting and drying (AWD)

In several areas, where water is available in plenty (mostly groundwater), farmers prefer to grow rice during the post-rainy season and they keep their rice fields continuously flooded to avoid weed menace. In this season rice can be grown by adopting the AWD irrigation method, where not only we can save lots of water but grain yield is also increased along with several other benefits. In AWD irrigation systems, fields are alternately flooded and dried, rather than conventional continuous flooding (CF) throughout the crop season. Usually, the number of days of non-flooded soil (aerobic condition) in AWD before the next irrigation is applied can vary from 1 day to more than 10 days. During the post-rainy season, the number of drying days should not be more than the rainy season, to avoid water stress on the rice crop. This irrigation system is a promising, water-saving, economically viable, and eco-friendly alternative to continuous flooding.

Details of this methodology and its implications on grain yield, water-saving, and other l factors have been discussed in section 4.5.3.3 of this book.

5.4.4. Modified crop production system with water-saving irrigation methods

A wide range of other post-rainy season crops, including wheat, sugarcane, maize, finger millet, tef, mustard, legumes, vegetables, and even spices, is now being extended to or adapted from the principles and practices of the SRI, which is improving irrigated rice production (Stoop *et al.* 2002; Uphoff, 2015). The system of crop intensification (SCI) refers to a range of innovations that farmers and civil society organizations have developed for these crops. Versions of SRI for wheat, sugarcane, finger millet, tef, etc. are included in the SCI rubric; each has an abbreviation (e.g., SWI, SSI, SFMI, STI...) (Adhikari *et al.* 2018).

Based on SCI experiences, it is possible to increase crop plant yields while using less water, fertilizer, seeds, and agrochemicals, and increasing climate resilience by improving root growth and soil fertility through the use of organic materials. Numerous farmer-centred technologies for agroecological crop management have a track record of supporting agricultural sustainability in the future. Countries in Asia, Africa, and Latin America are practising this new SCI methodology more frequently for various crops (Table 5.11). Currently, additional study is required to confirm the effectiveness and significance of these advances as well as to elucidate their limitations and conditions. Since there have been no harmful consequences on the health of humans or the environment to date, increasing public awareness of these agronomic possibilities should encourage further research and, where the results warrant it, the application of these approaches.

Table 5.11 An overview of SCI experimentation and experience in different crops and countries

Crops	Countries
Finger millet	Ethiopia, India, Nepal, Malawi
Wheat	Afghanistan, Ethiopia, India, Mali, Nepal, Netherlands, Pakistan, USA,
Maize	India
Sugarcane	Cuba, India, Kenya, Tanzania
Tef	Ethiopia
Mustard	India
Pulses: cowpea/black-eyed pea, chickpea/ garbanzo beans; mung bean/green gram; lentil/ black gram; pigeon pea/red gram; common/ haricot/ kidney bean; soybean; groundnut/peanut	Ethiopia, India
Vegetables: carrots; brinjal; onions; potatoes; mallow	Ethiopia, India, Leone, Pakistan, Sierra, USA
Spices: coriander; cumin; turmeric	India
Fruit trees	India, USA

Source: Adhikari *et al.* (2018)

Adhikari *et al.* (2018) have published a thorough analysis of the effects of SCI methods on yield improvement of eight crops in seven different nations.

Overall, it shows that SCI practices significantly improve the yield, and income while using lesser seeds and water than conventional practices. To show the effects on root and crop growth by adopting SCI practices, some of the pictures have been presented in Figs. 5.28a-d.

Fig. 5.26a. Root systems comparison of the finger millet plants grown by SFMI (Left) and conventional method (Right) in Jharkhand, India.

Fig. 5.26b. Finger millet plants are grown with different methods in Jharkhand, India. Left- Improved cultivar (A404) grown with SFMI method, Centre- Improved variety growth with the traditional method of broadcasting, Right- Local variety growth with the traditional method of broadcasting.

Fig. 5.26c. Two wheat fields were planted at the same time with the same verity in Chandrapura village, Khagaria district, Bihar, India in 2012. Left- Grown with SWI having spikes, Right- Grown with standard method still at vegetative stage.

Fig. 5.26d. Comparison of tef panicles of a single plant. Top-grown with TIRR methods, Bottom-grown with a standard broadcasting method. (TIRR: Tef with Improved seed, Reduced seed rate, and row planting using similar principles as the system of tef intensification, STI).

Source: Adhikari et al. 2018

The Ethiopian government needed to increase tef production quickly and extensively. As a result, to develop a less labour-intensive version of the tef intensification (STI) system, experimenting with drilling seeds in rows with wide spacing was conducted rather than raising seedlings and then transplanting them. TIRR approach increased grain yields by around 70% compared to standard production levels while using 90% fewer seeds and labour savings.

5.1 Land modification/ shaping and planting methods

Since the amount of water available for agriculture is decreasing day by day, land and water management is essential to the development of sustainable agriculture in both rainfed and irrigated regions. In India, only 36% of the cultivated area has having assured irrigation system, the rest is rainfed and farmers depend only on rainfall. Nowadays, due to climate change rainfall is also erratic and severely affects crop production and farmers' income. Apart from converting some parts of land into water harvesting structures or ponds, there are many other options available for modifying land, which greatly improves crop production, helping in diversifying crop selection, reducing water use, and enhancing benefits to farmers while minimizing crop failures.

In this section, we discuss how land modifications save irrigation water and at the same time improve farmers' income through crop diversification. Thereby, it addresses the challenges of water scarcity, climate resilience, and economic benefits.

5.5.1. Water harvesting structures (WHS) and water storage

In several parts of the country, heavy rain occurs during the *kharif* season, and because of the lack of storage facilities, plenty of water is lost through runoff to the river or sea. This results in no water availability to raise the second crops in *rabi*. Therefore, it is essential to construct farm ponds or community water reservoirs to harvest and conserve excess rainwater during the *kharif* season to provide supplemental irrigation to the *Kharif* crops and life-saving irrigation to succeeding pulse/oilseed crops (Fig. 5.27). This could be constructed either with Government funding support by the community, or through individual investment.

The area to be devoted to the farm pond depends on the rainfall amount, its intensity, and the purpose for which water is to be stored. An experiment conducted in the eastern part of India, reports that 10% of rice fields for the farm pond and 13% area for the bund sufficiently harvest rainwater for supplemental or life-saving irrigation for *kharif* rice, irrigating fruit crops grown on the bund, and short-term fish culture. This stored water remains available for irrigation up to March month (Thakur *et al.* 2015), detailed in section 4.8.3.

PRADAN, an NGO, demonstrated 'the 5% model', where 5% of rainfed paddy fields are converted into catchment ponds to trap and store rainwater during the monsoon. This enables to provide supplemental irrigation to the rice crop, which enhances production, raises income, and improves household food security. A similar type of innovation has been demonstrated in several countries for multi-enterprise agriculture, where farmers can add fisheries, duckeries, etc. It greatly increases smallholder incomes and local employment as farmers reassign some of their rainfed paddy land areas to pond construction to support diversified farming systems.

Fig. 5.27. Water harvesting pond

In areas having sandy soil or high temperatures after winter, stored water in the pond is either lost through high percolation or evaporation and becomes unavailable for *rabi* crops. Under such a situation, pond lining with Silpaulin or HDPE tarpaulins checks the percolation losses and helps in storing water for a longer period for irrigating crops during critical growth stages (Fig. 5.28). This could help in increasing both crop and land productivity for higher income for the farmers.

Fig. 5.28. Pond or ditch lining with Silpaulin to conserve rainwater

5.5.2. Bed planting methods and furrow irrigation

With the crops grown during the post-rainy season, it is a challenge to find efficient water-saving agricultural practices for the irrigated system. In this context, saving irrigation water is extremely important for stability in the production of this crop, as water scarcity is increasing day by day. An efficient planting method could save irrigation and increase water use efficiency (WUE).

To save irrigation water and enhance WUE for crops under different planting methods have been tested by researchers, *viz.*, in cotton, groundnut (Mandal *et al.* 2019), moong bean, soybean, oilseed rape, pearl millet, potato (Mandal *et al.* 2018), soybean and sugarcane.

In the bed planting system, crops are raised on the beds, and irrigation is applied through furrows. It is more appropriate for growing temporary water-logging stress crops, *viz*., cotton, maize, soybean, vegetables, etc. Due to changes in crop geometry and land configuration, this system has effective control over irrigation /drainage which has positive effects on the transport of nutrients. Water in the bed-furrow system flows from the furrows into the beds horizontally. In the bed, it is drawn downward by gravity and upward by capillary action, transpiration, and evaporation toward the soil surface.

Advantages of bed planting systems

i) Better and more effective irrigation water management: It uses one-third less water than flatbed methods and increases more than 20% crop yields.
ii) Seed saving, ex. in wheat by 20 to 30% compared to flatbed planting.
iii) Higher plant stands and better crop growth and yield.
iv) In wheat, this system makes it easier to precisely apply N and water during the grain-filling stage to increase protein content without causing lodging. Since many farmers choose not to irrigate after heading, this can have a large, positive impact on yield.
v) Weed control becomes easier and lesser pest/disease spread.
vi) Crop diversification and intercropping are possible, which is more climate-resilient and provides assured income to the farmers.

The design of beds depends on several factors, viz. crop grown, soil type, rainfall pattern, groundwater status, and the mechanical implements used. A permanent bed planting system needs to consider several factors, including crop residue management, irrigation and fertilizer management, inter-tillage operation, and weed control. In sandy-loam to loam soils for different crops (having various inter-row crop spacing), a model of bed planting is presented in Fig. 5.29 and the sunflower crop on the bed is shown in Fig. 5.30.

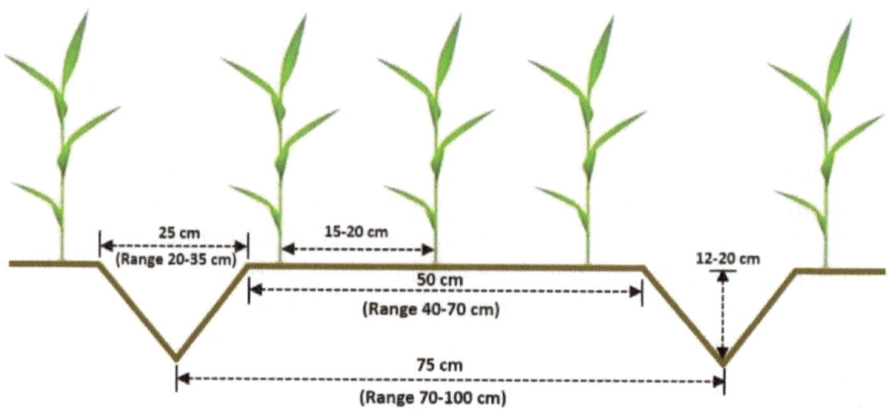

Fig. 5.29. A schematic design/ model of bed planting technique

Fig. 5.30. Bed planting of sunflower

Power tiller-operated bed planters have been developed by scientists and have been demonstrated in the farmers' fields with encouraging results. Using a bed planter, sowing on a permanent bed is also possible without further tillage operation. For bed planting, farm machinery should be developed locally.

The bed dimension depends on several factors, viz., types of crop grown, soil types, rainfall, irrigation facilities, farm implements available for making bunds, etc. It is advisable to consider factors like management of irrigation, fertilizer, crop residue, weeds, and inter-tillage operations for the development of a permanent bed system. The width of each bed including furrow could be 70-100 cm, of which the bed width may be 40-70 cm. The width and depth of the furrow may be 25-35 cm and 12-20 cm, respectively. The row spacing depends on the crops to be grown. Typically, two rows of paddy, wheat, maize, groundnuts, or chickpeas can be planted on raised beds. It has been found that in wheat crop yields with two or three rows on bed are not much different, while crop lodging has been observed to be higher with three rows per bed. It is beneficial to intercrop wheat/mustard with furrow-planted sugarcane or planting of one row of pigeon peas on beds. For effective weed control, crops that cover the surface quickly should be selected.

Crop diversification and intercropping, such as pigeon pea with green gram or sorghum, wheat with chickpea or mint, maize with potato, Indian mustard with sugarcane, rice with soybean, etc., are made easier by this technique. In 1994, bed-planted wheat-soybean rotation covered over 75% of Mexico's wheat land; yet, South Asian farmers are still experimenting with rice-wheat crop rotation. Nonetheless, a variety of crops provide advantages in terms of reduced water usage and increased output (Table 5.12).

Table 5.12. Yield enhancement and water savings in bed planting over the flat planting method

Crops	Yield under bed planting method (t ha^{-1})	Yield under flat planting method (t ha^{-1})	% Water savings over the flat planting method	%Yield increase over the flat planting method
Maize	3.27	2.38	35.5	37.4
Urd bean	1.83	1.37	26.9	33.6
Mung bean	1.62	1.33	27.9	21.8
Green peas	11.91	10.40	32.4	14.5
Wheat	5.12	4.81	26.3	6.4
Rice	5.62	5.29	42.0	6.2
Okra	34.4	29.1	33.3	18.2
Carrot	36.3	28.6	31.8	26.9
Radish	34.7	26.7	29.4	30.0
Cabbage	33.0	27.8	26.8	18.7
Pigeonpea	2.2	1.5	30.0	46.7
Chickpea	1.85	1.58	27.3	17.1
Cauliflower	25.9	18.9	36.4	37.0
Average	—	—	31.2	24.2

Source: Connor *et al.* (2002)

5.6. Integrated nutrient management to improve WUE of dry-season crops

The WUE can be improved by reducing the evaporation, particularly in dry or hot areas where the evaporation value is quite high. This can be achieved by appropriate nutrient management techniques. Nutrient application to crops increases water-use efficiency by increasing the evapotranspiration (ET), particularly the transpiration (T) component of ET and the transpiration efficiency of the crop. Thus the T/ET ratio, which is conducive to higher WUE, is higher for the fertilized than for the unfertilized crop. The field experiments on wheat showed that judicious application of limited irrigation water and enhancing water use efficiency (WUE) through integrated nutrient management could boost the productivity of wheat and enhance the WUE of the crop (Mandal *et al.*, 2005). Water and nutrients, and their positive interaction favourably influenced the yield, ET, and WUE of wheat. In central Indian vertisols, as grain yields in three post-sowing irrigations with 100% NPK were similar to two post-sowing irrigations with 100% NPK + FYM, one pre-sowing irrigation plus two post-sowing irrigation along with 100% NPK and 10 t ha^{-1} farmyard manure could be a feasible option in the event of limited irrigation water. Similarly, for Indian mustard in central Indian vertisols, Mandal et al (2010) reported that the application of organic manure in combination with NPK fertilizers could reduce the need for one post-sowing irrigation without compromising the yield of this crop, particularly under the limited water availability situation.

References

Adhikari, P. *et al.* 2018. System of crop intensification for more productive, resource-conserving, climate-resilient, and sustainable agriculture: experience with diverse crops in varying agroecologies. *International Journal of Agricultural Sustainability*, 16 (1): 1-28.

Alam, M.K., Bell, R.W. and Biswas, W.K. 2019. Decreasing the carbon footprint of an intensive rice-based cropping system using conservation agriculture on Eastern Gangetic Plains. *Journal of Cleaner Production,* 218:259-272.

Ali, M., Ghosh, P.K. and Hazra, K.K. 2014. Resource conservation technologies in rice fallow. In: Resource Conservation Technology in Pulses, pp. 83-88. Ghosh, P.K., Kumar, N., Venkatesh, M.S., Hazra, K.K. and Nadarajan. N. (Eds). Scientific Publishers, Jodhpur, Rajasthan, India.

Bajwa, A.A. 2014. Sustainable weed management in conservation agriculture. *Crop Protection,* 65:105-113.

Bandyopadhyay, K. K. *et al.* 2015. Characterization and crop planning of rabi-fallows using remote sensing and GIS. *Current Science,* 108:2051-2062.

Bandyopadhyay, P. K. *et al.* 2016. Effects of stubble length of rice in mitigating soil moisture stress and on yield of lentil (*Lens culinaris* Medik) in rice-lentil relay crop. *Agricultural Water Management,* 173:91-102.

Bilalis, D., Sidiras, N., Economou, G. and Vakali, C. 2003. Effect of different levels of wheat straw soil surface coverage on weed flora in Vicia faba crops. *Journal of Agronomy and Crop Science,* 89:233-241.

Campiglia, E., Radicetti, E. and Mancinelli, R. 2012. Weed control strategies and yield response in a pepper crop (*Capsicum annuum* L.) mulched with hairy vetch (*Vicia villosa* Roth.) and oat (*Avena sativa* L.) residues. *Crop Protection,* 33:65-73.

Canfield, D., Glazer, A. and Falkowski, P. 2010. The evolution and future of earth's nitrogen cycle. *Science*, 330:192-196.

Choudhary, M., Ghasal1, P.C., Choudhary, G.L., Prajapat, K. and Choudhary, H.R. 2014. Enhancing Productivity and Resource Use Efficiency of Rice-Fallow through Introduction of Pulses. *Popular Kheti*, 2(4):19-21.

Chauhan, B.S., Singh, R.G., Mahajan, G., *et al.* 2012. Ecology and management of weeds under conservation agriculture: a review. *Crop Protection,* 38:57-65.

Connor, D.J., Gupta, Raj K., Hobbs, Peter R. and Sayre, K.D. 2002. Bed Planting in Rice-Wheat System. Rice-Wheat Consortium for Indo-Gangetic Plains.

Damodaran, H. and Biswas, P. 2019. Top agricultural scientists body rejects zero budget natural farming. The Indian Express daily newspaper, Sept 10, 2019.

Das, A., Ramkrushna, G.I., Nagchan, S.V. and Munda, G.C. 2013. Resource conservation technologies in pulse-based cropping system in NEH region. In: Resource Conservation Technology in Pulses, pp 43-57. Ghosh, P.K., Kumar, N., Venkatesh, M.S., Hazra, K.K. and Nadarajan, N. (Eds). Scientific Publishers, Jodhpur, Rajasthan, India.

Dong, H. *et al.* 2017. The growth and yield of a wet-seeded rice-ratoon rice system in central China. *Field Crops Research,* 208:55-59.

Economic Survey. 2019. Economic Survey 2018-19. Department of Economic Affairs, Ministry of Finance, Government of India.

EPW. 2019. Mirage of Zero Budget Farming (Editorials). *Economic & Political Weekly*, 54(30), July 27.

Evenson, R.E. and Gollin, D. 2003. Assessing the impact of the green revolution, 1960-2000. *Science*, 300:758-762.

FAO. 2011. What is Conservation Agriculture? FAO Conservation Agriculture Website at: http://www.fao.org/ag/ca/1a.html.

FAO. 2015. Conservation Agriculture. http://www.fao.org/ag/ca/1a.html.

Ghosh, P.K., Das, A., Saha, R., Enboklang, K., Tripathi, A.K., Munda, G.C. and Ngachan, S.V. 2010. Conservation agriculture towards achieving food security in North East India. *Current Science,* 99:915-921.

Ghosh, P.K., Hazra, K.K., Nath, C.P., Das, A. and Acharya, C.L. 2016. Scope, constraints and challenges of intensifying rice (*Oryza sativa* L.) fallows through pulses. *Indian Journal of Agronomy,* 61(4): S122-S128.

Gilbert, S.F., Sapp, J. and Tauber, A.I. 2012. A symbiotic view of life: we have never been individuals. *Quarterly Review of Biology*, 87(4): 325-341.

Giller, K.E., Witter, E., Corbeels, M. and Tittonell, P. 2009. Conservation agriculture and smallholder farming in Africa: the heretics view. *Field Crops Research*, 114:230-234.

Govaerts, B., Sayre, K.D., Ceballos-Ramirez, J.M., *et al.* 2006. Conventionally tilled and permanent raised beds with different crop residue management: effects on soil C and N dynamics. *Plant and Soil*, 280:143-155.

Gumma, M. K. *et al.* 2016. Mapping of rice–fallow crop land areas for short-season grain legumes intensification in South Asia using MODIS 250 m time-series data. *International Journal of Digital Earth,* 9:981-1003.

Hafeez-Ur-Rehman, N.A., Wakeel, A., Saharawat, Y.S. and Farooq, M. (2015). Conservation agriculture in South Asia. In: Farroq, M., Siddique, K.H.M. (Eds.), Conservation Agriculture. Springer Publishing House: Cham. ISBN 978-3-319-11619-8.

Harrell, D.L., Bond, J.A. and Blanche, S. 2009. Evaluation of main-crop stubble height on ratoon rice growth and development. *Field Crops Research*, 114:396-403.

Hazell, P. and Rahman, A. 2014. New Directions for Smallholder Agriculture. Oxford University Press.

Hazra, K. K. and Bohra, A. 2020. Increasing relevance of pulse crops to sustainable intensification of Indian agriculture. *National Academy of Science Letters,* https ://doi.org/10.1007/ s4000 9-020-00948 -6.

Hobbs, P.R. 2007. Conservation agriculture: What is it and why is it important for future sustainable food production? *Journal of Agricultural Sciences,* 145:127-137.

IAASTD. 2009. Agriculture at a Crossroads: The Global Report. Island Press, Washington DC, USA.

Iqbal, M., Ul-Hassan, A. and van Es, H.M. 2011. Influence of residue management and tillage systems on carbon sequestration and nitrogen, phosphorus, and potassium dynamics of soil and plant and wheat production in semi-arid region. *Soil Science and Plant Analysis*, 42:528-547.

Jat, H.S., Datta, A., Choudhary, M., Sharma, P.C. and Jat, M.L. 2021. Conservation agriculture: factors and drivers of adoption and scalable innovative practices in Indo- Gangetic Plains of India – a review. *International Journal of Agricultural Sustainability*, 19:40-55.

Jat, M.L., Chakraborty, D., Ladha, J.K., Rana, D.S., Gathala, M.K., McDonald, A. and Gerard, B. 2020. Conservation agriculture for sustainable intensification in South Asia. *Nature Sustainability*, 3: 336-343.

Kar, G. and Kumar. A. 2009. Evaluation of post-rainy season crops with residual soil moisture and different tillage methods in rice-fallow of eastern India. *Agricultural Water Management*, 96: 931-938.

Kar, G., Singh, R. and Verma, H.N 2004. Productive and profitable management of rainfed lowland rice through intensive cropping and efficient water use. Research Bulletin No.17. Water Technology Centre for Eastern Region (WTCER), Bhubaneshwar, Odisha, India, p 56.

Kassam, A., Friedrich, T. and Derpsch, R. 2018. Global spread of conservation agriculture. *International Journal of Environmental Sciences*, 76:29-51.

Kassam, A., Friedrich, T., Derpsch, R., Lahmar, R., Mrabet, R., Basch, G., González-Sánchez, E.J. and Serraj, R. 2012. Conservation agriculture in the dry Mediterranean climate. *Field Crops Research*, 132:7-17.

Kassam, A.H., Friedrich, T., Shaxson, F., Reeves, T., Pretty, J. and de Moraes Sa, J.C. 2011. Production Systems for Sustainable Intensification – Integrating Productivity with Ecosystem Services. Technology Assessment – Theory and Prexis, Special Issue on Feeding the World, July.

Khadse, A., Rosset, P.M. and Ferguson, B.G. 2017. Taking agro ecology to scale: The zero budget natural farming peasant movement in Karnataka, India. *The Journal of Peasant Studies,* 45:1-28.

Kumar, R. et al. 2018a. Evaluation of crop establishment techniques in rice-fallows of Eastern Indo-Gangetic Plains. in *National Conference on Organic Waste management for Food and Environmental Security under Theme Crop Residue Management*, 8-10 Feb., 2018. ICAR -Indian Institute of Soil Science, Bhopal, Madhya Pradesh, India.

Kumar, R. et al. 2019a. Sustainable intensification of rice fallows of Eastern India with suitable winter crop and appropriate crop establishment technique. *Environmental Science and Pollution Research,* 26:29409-29423.

Kumar, R., Mishra, J.S., Upadhyay, P.K. and Hans, H. 2019b. Rice fallows in the Eastern India: problems and prospects. *Indian Journal of Agricultural Sciences,* 89:567-577.

Kumar, R., Mishra, J.S. and Kumar, R. 2018b. Enhancing the productivity of rice fallows area of Eastern India through inclusion of pulses and oilseeds. *Indian Farming,* 68:7-10.

Kumara, T.M.K., Kandpal, A. and Pal, S. 2020. A meta-analysis of economic and environmental benefits of conservation agriculture in South Asia. *Journal of Environmental Management*, 269:110773

Lal, R. 2009. Soil degradation as a reason for inadequate human nutrition. *Food Security*, 1: 45-47.

Layek, J., Chowdhury, S., Ramkrushna, G. I. and Das, A. 2014. Evaluation of different lentil cultivars in low land rice-fallow under no-till system for enhancing cropping intensity and productivity. *Indian Journal of Hill Farming,* 27:4-9.

Lee, N. and Thierfelder, C. 2017. Weed control under conservation agriculture in dryland smallholder farming systems of southern Africa. A review. *Agronomy for Sustainable Development*, 37.

Liu, K.L., Qin, J.T., Zhang, B. and Zhao, Y.W. 2012. Physiological traits, yields and nitrogen translocation of ratoon rice in response to different cultivations and planting periods. *African Journal of Agricultural Research*, 7:2539-2545.

Lorimer, J. 2017. Probiotic environmentalities: rewilding with wolves and worms. *Theory, Culture and Society*, 34(4):27-48.

LVC. 2010. La Via Campesina. Sustainable peasant and family farm agriculture can feed the world. La Via Campesina, Jakarta, Indonesia.

Mandal, K.G., Thakur, A.K. and Mohanty, S. 2018. Planting techniques and irrigation influenced crop growth, light interception and yield–evapotranspiration relationship of potato. *International Journal of Plant Production*, 12:285-296.

Mandal, K.G., Thakur, A.K. and Mohanty, S. 2019. Paired-row planting and furrow irrigation increased light interception, pod yield and water use efficiency of groundnut in a hot sub-humid climate. *Agricultural Water Management*, 213:968-977.

Mandal, K. G., Hati, K.M., Misra, A.K., Bandyopadhyay, K.K. and Mohanty, M. 2005. Irrigation and nutrient effects on growth and water-yield relationship of wheat (*Triticum aestivum* L.) in central India. *Journal of Agronomy and Crop Science,* 191 (6):416-425.

Mandal, K.G., Hati, K.M. Misra, A.K. and Bandyopadhyay, K.K. 2010. Root biomass, crop response and water-yield relationship of mustard (*Brassica juncea* L.) grown under combinations of irrigation and nutrient application. *Irrigation Science*, 28: 271-280.

Mishra, J.S. and Kumar, R. 2018. Zero tillage options of pulses in rice-based cropping system. Farm Mechanization for Production. Pp 122-136. Khare, D., Nahatkar, S.B., Shrivastava, A.K., Jha A.K. (Eds). Scientific Publishers, India, New Delhi.

Mishra, S. 2018. Zero Budget Natural Farming: Are this and similar practices the answers? Working Paper No. 70, Nabakrushna Choudhury Centre for Development Studies, Bhubaneswar.

Mishra, J.S., Kumar, R. and Bhatt, B.P. 2018. Low-cost technologies for management of rice fallows in Eastern India. In: *XXI Biennial National Symposium of Indian Soc. Agron.*, 24–26 Oct. 2018. MAPUAT, Udaipur, Rajasthan, India. 7-9.

MoA&FW. 2019. Agriculture Census 2015-16 (Phase-I). All India Report on Number and Area of Operational Holdings. Ministry of Agriculture & Farmers Welfare, Government of India.

Montgomery, D. 2007. Dirt: The Erosion of Civilizations. University California Press, Berkeley, Los Angeles.

Munda, G.C., Das, A. and Patel, D.P. 2009. Evaluation of transplanted and ratoon crop for double cropping of rice (*Oryza sativa* L.) under organic input management in mid altitude sub-tropical Meghalaya. *Current Science*, 96:1620-1627.

Munster, D. 2018. Performing alternative agriculture: Critique and recuperation in Zero Budget Natural Farming, South India. *Journal of Political Ecology*, 25:748- 764.

Nakamoto, T., Yamagishi, J. and Miura, F. 2006. Effect of reduced tillage on weeds and soil organisms in winter wheat and summer maize cropping on Humic Andosols in Central Japan. *Soil and Tillage Research*, 85:94-106.

Nanda, P., Sandal, S.K., Mohanty, S., Dasgupta, P. and Mishra, A. 2021. Drip fertigation in crops: Experiences from AICRP on Irrigation Water Management. ICAR-Indian Institute of Water Management, Bhubaneswar, Odisha, India. 8p

Naresh, R.K., Dwivedi, A., Gupta, R.K., Rathore, R.S., Dhaliwal, S.S., Singh, S.P., Kumar, P., Kumar, R., Singh, V., Singh, V. and Singh, O. 2016. Influence of conservation agriculture practices on physical, chemical and biological properties of soil and soil organic carbon dynamics in the subtropical climatic conditions: a review. *Journal of Pure and Applied Microbiology*, 10: 1061-1080.

NITI Aayog. 2018. Strategy for New India @ 75. NITI Aayog, New Delhi.

Niyogi, D.G. 2018. Andhra farmers taste success with Zero Budget Natural Farming. *Down to Earth,* (Online). https://www.downtoearth.org.in/author/deepanwita-gita-niyogi-2399

Palekar, S. 2005. The philosophy of spiritual farming. 2nd ed. Zero Budget Natural Farming Research, Development & Extension Movement, Amravati, Maharashtra, India.

Panigrahi, P., Panda, R.K., Rautaray, S., Raychaudhury, S., Thakur, A.K. and Mishra, A. 2021. Techno-economic feasibility of year-round use of drip irrigation in a rice-based cropping system. *Irrigation and Drainage*, https://doi.org/10.1002/ird.2612

Panigrahi, P., Raychaudhuri, S., Thakur, A.K., Nayak, A.K., Sahu, P. and Ambast, S.K. 2019. Automatic drip irrigation scheduling effects on yield and water productivity of banana. *Scientia Horticulturae*, 257: https://doi.org/10.1016/j.scienta.2019.108677

Pingali, P.L. 2012. Green revolution: impacts, limits, and the path ahead. *Proceedings of the National Academy of Sciences of the United States of America*, 109(31):12302-12308. doi:10.1073/pnas.0912953109

Pittelkow, C.M., Liang, X., Linquist, B.A., van Groenigen, K.J., Lee, J., Lundy, M.E., van Gestel, N., Six, J., Venterea, R.T. and van Kessel, C. 2014. Productivity limits and potentials of the principles of conservation agriculture. *Nature*, 517:365-368.

Ranaivoson, L., Naudin, K., Ripoche, A., *et al.* 2017. Agro-ecological functions of crop residues under conservation agriculture. A review. *Agronomy for Sustainable Development,* 37.

Ranaivoson, L., Naudin, K., Ripoche, A., *et al.* 2018. Is mulching an efficient way to control weeds? Effects of type and amount of crop residue in rainfed rice-based cropping systems in Madagascar. *Field Crops Research*, 217:20-31.

Ranaivoson, L., Naudin, K., Ripoche, A., Rabeharisoa, L. and Corbeels, M. 2019. Effectiveness of conservation agriculture in increasing crop productivity in low-input rainfed rice cropping systems under humid subtropical climate. *Field Crops Research*, 239:104-113.

Ranjit, Kumar, Kumar, S., Yashavanth, B.S., Meena, P.C., Ramesh, P., Indoria, A.K., Kundu, S. and Manjunath, M. 2020. Adoption of Natural Farming and its Effect on Crop Yield and Farmers' Livelihood in India. ICAR-National Academy of Agricultural Research Management, Hyderabad, India.

Rautaray, S.K. 2008. Performance of rice (*Oryza sativa*) - utera pea (*Pisum sativum*) cropping system as influenced by management practices. *Indian Journal of Agricultural Sciences*, 78(8):699-701.

Rosset, P.M. and Martinez-Torres, M.E. 2012. Rural social movements and agroecology: Context, theory and process. *Ecology and Society*, 17(3):

Saldanha, L.F. 2018. A Review of Andhra Pradesh's Climate Resilient Zero Budget Natural Farming, Environment Support Group, Bangalore. http://esgindia.org

Santos, A.B., Fageria, N.K. and Prabhu, A.S. 2003. Rice ratooning management practices for higher yields. *Communications in Soil Science and Plant Analysis*, 34(5-6):881-918.

Scopel, E., Da Silva, Fa.M.M., Corbeels, M., *et al.* 2004. Modelling crop residue mulching effects on water use and production of maize under semi-arid and humid tropical conditions. *Agronomie*, 24:383-395.

Sharma, A.R., Jat, M.L., Saharawat, Y.S., Singh, V.P. and Singh, R. 2012. Conservation agriculture for improving productivity and resource-use efficiency: prospects and research needs in Indian context. *Indian Journal of Agronomy*, 57:131-140.

Shotwell, A. 2016. Against Purity: Living Ethically in Compromised Times. Minnepolis, MN: University of Minnesota Press.

Singh, R., Kundu, D.K., Mohanty, R.K., Ghosh, S., Kumar A. and Kannan, K. 2005. Raised and sunken bed technique for improving water productivity in lowlands. Research Bulletin 28, WTCER (ICAR), Bhubaneswar, 42p.

Singh, R.G. *et al.* 2012. Opportunities for managing rice fallow systems with conservation agriculture technologies. *Indian Farmig,* 62:31-34.

Singh, R.N. *et al.* 2017. Influence of rice (*Oryza sativa* L.) habit groups and moisture conservation practices on soil physical and microbial properties in rice + lathyrus relay cropping system under rice fallows in Eastern Plateau of India. *Indian Journal of Agricultural Sciences,* 87:633-639.

Sissoko, F., Affholder, F., Autfray, P., *et al.* 013). Wet years and farmers' practices may offset the benefits of residue retention on runoff and yield in cotton fields in the Sudan-Sahelian zone. *Agricultural Water Management,* 119:89-99.

Smith, J., Jagadeesh, Y., Smith, P. and Nayak, D.R. 2020. Potential yield challenges to scale-up of zero budget natural farming. *Nature Sustainability,* doi.org/10.1038/ s41893-019-0469-x.

Smith, P. *et al.* 2013. How much land-based greenhouse gas mitigation can be achieved without compromising food security and environmental goals? *Global Change Biology*, 19(8). https://doi.org/10.1111/gcb.12160

Somasundaram, J., Sinha, N.K., Dalal, R.C., Lal, R.L., Mohanty, M., Naorem, A.K., Hati, K.M., Chaudhary, R.S., Biswas, A.K., Patra, A.K. and Chaudhari, S.K. 2020. No-till farming and conservation agriculture in South Asia-Issues, challenges, prospects and benefits. *Critical Reviews in Plant Sciences,* 39:236-279.

Stoop, W.A., Uphoff, N. and Kassam, A. 2002. A review of agricultural research issue raised by the system of rice intensification (SRI) from Madagascar: Opportunities for improving system for resource poor farmers. *Agricultural Systems*, 71:249-274.

Teasdale, J.R. and Mohler, C.L. 2000. The quantitative relationship between weed emergence and the physical properties of mulches. *Weed Science*, 48:385-392.

Thakur, A.K., Mohanty, R.K., Singh, Rajbir and Patil, D.U. 2015. Enhancing water and cropping productivity through Integrated System of Rice Intensification (ISRI) with aquaculture and horticulture under rainfed conditions. *Agricultural Water Management*, 161:65-76.

Turmel, M.-S., Speratti, A., Baudron, F., *et al.* 2014. Crop residue management and soil health: a systems analysis. *Agricultural Systems*, 134:6-16.

Uphoff, N. 2015. The system of rice intensification: Responses to frequently-asked questions. CreateSpace/Amazon. Retrieved from http://sri.cals.cornell.edu/aboutsri/SRI_FAQs_Uphoff_2016.pdf

Wallenstein, M. 2017. To restore our soils, feed the microbes. The Conversation. Online article, visited on 22 Dec 2019. https://source.colostate.edu/restore-soils-feed-microbes/

Xu, F.X., Xiong, H., Zhang, L., Zhu, Y.C., Jiang, P., Guo, X.Y. and Liu, M. 2015. Progress in research of yield formation of ratooning rice and its high-yielding key regulation technologies. *Scientia Agricultura Sinica*, 48:1702-1717.

Zhang, T., Wei Xiong, W., Sapkota, T.B., Jat, M.L., Montes, C., Krupnik, T.J., Jat, R.K., Karki, S., Nayak, H., Faisal, A.A. and Jat, H.S. 2022.The optimization of conservation agriculture practices requires attention to location-specific performance: Evidence from large scale gridded simulations across South Asia. *Field Crops Research*, 282:108508

6

Integrated Rice-Fish Farming Systems

6.1. Introduction

The growing human population presents a number of issues, including food scarcity, malnutrition, inadequate natural resources (land and water) for agricultural production, and environmental deterioration (MacDonald, 2010). Today, viable intensification of agri-production systems can fulfill rising food, nutrition and environmental demands (Pretty and Bharucha, 2014). Production of rice (*Oryza sativa* L.), a vital element of food security, which feeds nearly half of the world's population, occupies 143 million hectares of land (FAOSTAT, 2023), and uses up to 90% of all irrigation water in Asia (Liu *et al.*, 2018). For instance, it has been determined that systems for producing rice are one of the main industries that wastes water resources and greatly increases greenhouse gas emissions (FAO, 2013). Therefore, it is necessary to look for ways to manage the rice production system more effectively. In paddy rice systems, co-cultivating aquatic animals (fish, shrimp, crab, shellfish, and ducks) and rice has been suggested as a way to reduce the risks of environmental pollution/ contamination linked with rice production while simultaneously optimizing the use of land and water resources to provide humans with both grains and meat (Hu *et al.*, 2016; Ahmed and Garnett, 2011). Co-culture systems have drawn more interest recently due to their potential to reduce the stresses associated with environmental and food insecurity.

Rice a major crop, primarily grown in flooded fields that meets the food demand of approximately three billion consumer on a daily basis (Akinbile *et al.*, 2011) and plays an significant role in worldwide food security, mostly in developing Asian countries where approximately 90% of worldwide rice is grown (Cosslett and Cosslett, 2018). Most rice production necessitates extensive fertilizer inputs, which have undesirable environmental costs such as increased water contamination and GHG emissions. Nitrogen, among other macronutrients, is vital for higher crop production, and its use in rice systems accounts for 16% of total N fertilizer consumption globally (Tilman *et al.* 2002; Wood *et al.* 2004). Only 28% of this massive quantity of N input is absorbed by the rice crop, with the remainder ending up in leaching, runoff, gaseous emissions, and remain in soil as legacy N (Yao *et al.*, 2017; Guo *et al.*, 2010), highlighting the need for greater N use efficiency.

On the other hand, 50% of global aquatic food production comes from aquaculture while 13% of overall protein output (Bush *et al.*, 2013) from animals. Freshwater aquaculture is fast gaining popularity among farmers worldwide, particularly in Asian countries, due to rising fishing expenses and the risk for water contamination in natural fishing grounds (Troell *et al.*, 2014). In this situation, rice fields contain a lot of water, offering a favorable environment for aquaculture (Fernando, 1993; Brooks and Sieu, 2016). Rice-fish/rice-animal co-culture is the common term used to describe the culture freshwater aquatic creatures in the paddy fields. By utilizing land and water resources in tandem, the co-culture system allows us to manage numerous issues related to the one/ single aquaculture system and provide us with an adequate supply of aquatic creatures for sustenance (Hu *et al.*, 2016). This system was first implemented in Asia, and subsequently scaled-up in most part of the world (Erwin and Mendoza, 2008).

Depending on custom and farmers' interests, the aquatic animals opted in the integrated/ co-culture method may differ from site to site (Halwart and Gupta, 2004; Zhang *et al.*, 2017). According to earlier research, co-cultivating rice with fish helped manage weed growth and pest infestations in rice ecosystems (Hu *et al.*, 2016; Frei *et al.*, 2007). Furthermore, because raised animals may help suppress weeds and eat a variety of pests and insects as food, co-culture practice may improve the production of safe / nontoxic food with little to no application of pesticide (Rahman, 2016). Furthermore, in a rice-fish system, the N and P output from the meal of aquatic animals are accessible for plant uptake. As a result, less N fertilizer is required for the rice crop, which will lessen environmental pollution (Feng *et al.*, 2016; Zhang *et al.*, 2016).

The integrated co-culture practice of rice, fish, and animals creates a distinct agro-landscape globally, particularly in tropical and sub-tropical Asia (Haroon and Pittman, 1997). It is now an agro-cultural pattern rather than just a method of agro-production. The FAO and UNESCO have listed rice-fish farming as one of the Globally Important Ingenious Agricultural Heritage System (GIAHS) due to its exceptional contribution to food and livelihood security, significance in terms of biological diversity and genetic resources, landscape diversity, aesthetic beauty, cultural values, and indigenous knowledge of land and water management. In Asian countries, rice-fish farming is an age old practice of over 2000 years (Islam *et al.*, 2015; Anita *et al.*, 2014; Lu and Li, 2006) and well recognized as a promising alternative to rice mono-cropping (Suloma and Ogata, 2006). Earlier research works have already been demonstrated that such integrated system could enhance productivity (Mohanty *et al.*, 2009; Xie *et al.*, 2011; Saikia *et al.*, 2015), soil fertility by increasing the availability of N and P (Giap *et al.,* 2005) and other nutrients, nutrients use efficiency and reduce nutrients loss to environment (Feng *et al.*, 2016; Li *et al.*, 2008a; Oehme *et al.*, 2007) due to complementary use between fish and rice. Rice-fish integration is thus a main choice for developing ecological agriculture that maximizes system benefits, avoids negative consequences, and strives for maximum output utilizing available energy and materials.

Rice, a prominent staple crop in tropical Asia with a long history of domestication, has a wide range of farmed ecotypes based on three *Oryza sativa* varieties: indica, japonica, and javanica. There are four fundamental rice agro-habitats, each with its own set of edaphic conditions: irrigated ecosystems, upland and lowland rainfed ecosystems, and flood-prone (deep water) ecosystems, as discussed in Chapter 3 of the book. In this environment, fish farming is carried out concurrently or in rotation with rice at four intensities: traditional (capture), low intensity (without feed and fertilizer), medium intensity (just fertilization), and high density (with feed and fertilizer). The incorporation of fish into rice production provides important protein, particularly for subsistence farmers managing rain-fed agricultural systems. Rice fields provide shade and organic matter for fish, which then oxygenate the soil and water, eat rice pests, and promote nutrient recycling (Mohanty *et al.*, 2009; Brahmanand *et al.*, 2006). Rice-fish farming systems, as part of an integrated ecosystem tailored to local cultural, environmental, and economic conditions, are made up of complementary sub-agricultural ecosystems that offer essential ecological services such as bio-control, nitrogen fixation, and landscape combination.

In most countries, rice-fish/animal co-culture farming is characterized by intensive rice cultivation and extensive fish cultivation, with fish yields typically ranging between 0.3 and 2.5 t ha^{-1} (Nhan *et al.,* 1998; Mohanty *et al.*, 2008), though higher production levels of up to 7.03 t ha^{-1} have also been recorded (Chen *et al.,* 1995). Rice yields in these systems typically range between 2.60-3.25 t ha^{-1} (Mohanty *et al.*, 2009), but can reach up to 4.84 t ha^{-1} (Feng *et al.*, 2016), above the global average yield of 4.66 t ha^{-1} (FAOSTAT, 2023). Usually, farmers adopt different types of field design for integrated rice-fish culture in low and medium-land ecosystems, with or without integration of other components like duckery, poultry, on-dyke horticulture etc. In low-land situations, an in-field refuge with a peripheral trench is generally practised, where fish move all around the field and when water recedes, they aggregate in the refuge. In a medium land ecosystem, an out-field refuge downstream of the field is preferred, where excess water from the rice field is collected in the refuge for fish culture separately. However, yield performance seems to be higher in the medium land ecosystem due to less water submergence in the rice field, more water available to fish in the refuge and better operational management (Mohanty *et al.*, 2004; Brahmanand *et al.*, 2006; Mohanty *et al.*, 2009).

6.2. Rice-fish integration: Indian scenario

Approximately two-thirds of the population in India depend on agriculture, making it the backbone of the country's economy, accounting for nearly one-fourth of the country's GDP. By 2050, food grain output is anticipated to rise by 185% in order to feed the world's 1.5 billion people. This can be accomplished through increasing productivity, crop variety, and cropping intensity. However, because of the current rate of urbanization and industrialization, there is little scope for the horizontal spread of aquaculture or agricultural. The amount of land available per person in India has decreased from 0.5 ha to 0.15 ha. Further India's population has crossed 1.4 billion including undernourished population of 270 million. Under this scenario, meeting the food requirements of the burgeoning population has turned out to be a challenging task. Vertical extension by integrating well-matched farming components requiring lesser space and time, and ensuring periodic income to farmer therefore, seems to be a viable alternative which assume greater importance for sound management of farm resources. This would definitely enhance farm productivity, lessen environmental degradation, enhance quality of resource poor farmer's life and maintain the sustainability. However, the most significant barrier to aquaculture technology is the high expense of inputs, particularly fertilizer and fish feed. Moreover, one crucial management strategy to improve fish output is to stock healthy fingerlings that are more than 100 mm in length. Unfortunately, the planned stocking program is hampered by the lack of suitable land-based nursery ponds already in existence and the financial limitations associated with building new infrastructure facilities. Because of these limitations and the resources at hand, the ecosystem of rice fields offers a feasible chance for advanced fingerling rearing on a big scale as part of a stocking program.

Additionally, of India's 46.38 million hectares of rice-growing land, 20 million ha are ideal for the implementation of a rice-fish integration system, primarily in low-lying, rain-fed medium-lands, and canal commands. But as of right now, just 0.23 million hectares are used for rice-fish farming (Mohanty *et al.*, 2004; Brahmanand *et al.*, 2006; Mohanty *et al.*, 2009). The main causes of this low level of adoption, exploitation, and yield are the introduction of high-yielding rice varieties that involve the use of pesticides and other chemicals, which has significantly hindered fish culture in paddy fields; other factors include shorter fish rearing times; a lack of water management strategies; and an unpredictable monsoon. It is therefore urgently necessary to increase the productivity from these underutilized, highly prospective areas, especially in the eastern region. By implementing an integrated rice-fish system and employing appropriate scientific interventions together with water conservation and monitoring methods, these lands could potentially offset the financial losses incurred in rice production due to natural disasters. For smallholder rice farmers in low-lying areas with structures for conserving water and soil, as well as a steady supply of water for irrigation, this method may be a good option for diversification (Ofori *et al.*, 2005). Because two or more subsystems are involved rather than a single commodity farming system, this IFS system not only accommodates crop diversification, enhances productivity, creates employment opportunities, increases income, and provides nutritional security to resource-poor farming communities, but it also distributes risk (both biological and economic) (Mohanty *et al.*, 2009; Brahmanand *et al.*, 2006). Therefore, when developing ecological agriculture that maximizes system benefits, minimizes negative consequences, and aims to maximize output utilizing available energy and materials, rice-fish integration is a main alternative. Additionally, including fish in the ecology of rice fields improves social, economic, and ecological outcomes while also boosting yield.

The idea of using or recycling farm resources for production is the same for rice-fish farming worldwide, despite regional variations in practice. Variations in field design, species composition, stocking density, sizes, crops, crop rotation, etc. are the key differences between different agricultural methods. There are usually three different types of field designs: the deep water rice field system used in India; the pond refuge next to the rice field on one side; and the shallow trench system within the rice field, as observed in the Philippines, Indonesia, and China. Several kinds of land changes are made in these systems in order to keep and raise the fish. In Eastern Indian states, rice-fish farming is primarily done in ponds and trenches, whereas the Andamans use a large bed-farrow technique. The other traditional modified methods suiting to the source of water and land condition are discussed as follows:

6.2.1. Filtration system

This system of rice-fish farming is practiced in the coastal rice field. Seeds of brackish water fishes enter during the saline water ingress to the rice field at high tide through an inlet. The seeds are filtered into the rice plots by using bamboo mats or net materials. The fish are allowed to grow in this system for a certain period ranging from 4-5 months. The system is further protected from the entry of predatory fishes in the next high tide by using the nylon screen or bamboo mat. Sometimes supplementary feed is given for higher and quicker fattening of the stocked fished. This type of farming is common in the *Pokkalli* fields of Kerala, *Gajani* rice fields of Karnataka, *Khajan* area of Goa and also in West Bengal as the *Bhasabada* farming.

6.2.2. Flow-through system

In northeastern India, many rice fields are located in valleys where water accumulates from the adjoining hills and flows down the valley by gravity. Dwarf varieties of rice are generally cultivated in such plots integrated mainly with *Cyprinus carpio*. Fish production from such plots ranges from 200 to 1,000 kg ha^{-1} during the monsoon. Recently, such integrated farming is gaining popularity in these areas. Flow-through rice-fish farming system is also practiced in the rice fields of hilly terrain by diverting the stream water to the rice fields. Since the rice fields are arranged in steps in these areas, the system of practice is almost similar to that of running water aquaculture.

6.2.3. Lowland system

In India, waterlogged lowland constitutes the most important potential area for rice-fish farming. The rice-fish farming is generally practiced in the medium to lowland rice fields where at least 30 cm water accumulates and stagnates for more than 4-6 months. Generally two types of fish farming system is practiced in the rice field. It may be the capture-based system or culture system. In capture system, the bunded lowlands or modified rice fields harbours a wide variety of fish fauna such as murrels, catfishes, feather backs, a variety of weed fishes, post larvae of prawns and even the fry and fingerlings of major and minor carps, if the areas is connected to the catchment area of river, canal or streams. The ecosystem provides an excellent breeding ground for catfish and weed fish, which enter into the system during overflow. Since most of these fishes are prolific breeders, a good population is auto-stocked leading to a considerable catch after 3-4 months. In this system, farmers do little or no management during retaining fish. After the rice harvest, the fishes are harvested immediately or left for further growth depending on the water availability. The production in this system is low depending on the auto-stocking and duration of culture. The culture system is used in lowland rice producing areas, with the development of pond refuges and trenches. The fish farming operations in this system are essentially identical to conventional ones, with the exception of the restricted use of herbicides and pesticides required for rice crops.

6.2.4. Research and developments

The summary of the rice-fish farming under natural and research environment in the country is summarized in Table 6.1. In the natural system, rice-fish farming has been carried out in brackish water, rainfed and waterlogged situations. In this system, the yield of fish/prawn ranges between 180 to 2135 kg ha^{-1}. Similarly, the rice yield ranges between 0.5 to 3 t ha^{-1}. In brackish water and low saline water, prawn culture is preferred. The traditional system of shrimp filtration in rice plots of Kerala and Goa has been modified into the current practices. Major changes in the management include the desalinization in *Pokhali* plots through various devices, namely field trenching in a crisscross manner for quick removal of runoff water carrying surface salt deposits that have accumulated during brackish water prawn culture. The aquaculture production from such rice-fish plots has been raised to 800-2150 kg ha^{-1} yr^{-1}. Juvenile mortality has been reduced considerably by the recent introduction of the built-in nursery system and supplementary feeding. In such a modified method of culture, entry of predator species is avoided by filtration of incoming water using bamboo mats and nylon nets. In northeastern India, many rice fields are located in valleys where water accumulates from the adjoining hills and flows down the valley by gravity. Dwarf varieties of rice are generally cultivated in such plots integrated mainly with *Cyprinus carpio.* Fish production from such plots ranges from 200 to 1000 kg ha^{-1} during the monsoon (Brahmanand *et al.*, 2006).

Table 6.1. Rice-fish farming (India experience)

Location	**Type of rice plot**	**Fish stocked**	**Fish** (kg ha^{-1}) **production**	**Rice** (t ha^{-1}) **production**
Natural System				
East Godavari (wild crop	Deep water (river basin)	Natural stocks	3 / year	1.0-1.5
Kerala (modified *Pokhali* system)	Shallow (brackish water)	*P.monodon, P. indicus* and natural stocks	785-2135 / year	1.5-2.5
NE Hill complex (mountain valley)	Waterlogged	*C.carpio*	Highly variable	1.0-1.5
Meghalaya (terrace system)	Shallow (hilly)	*C. carpio*	186 /2 months	0.5-0.8
West Bengal (modified *Bhasa-badha* system)	Shallow (2 aquatic crops	*P. monodon L. parsia* and Natural stocks	600-1000 / year	2.5-3.0

West Bengal (ordinary plot)	Rainfed (inland)	Wild stock (Tilapia, murrel, catfish, prawns etc.)	300 / year	2.0-3.0
West Bengal (sewage enriched)	Shallow (low salinity)	*P. monodon, L. parsia*, Tilapia and natural fry	1000 / year	1.5-3.0
Trends in research and development				
State fisheries plot, W.B.	Shallow (nursery)	Carp (1,457 nos/ ha)	112 / 3-4 months	2.5-3.7
Karnataka plot	Shallow (nursery)	Crap fry to fingerlings	153 /71 days	1.5-2.0
CSSRI, W.B (SR 26B)	Shallow (brackish water)	Carp and *M. rosenbergii* (in monsoon). *P.mononodon* and *L. parsia* (in summer)	870 / year	3.0
24 Parganas, WB (Developing)	Shallow (fresh water)	Carp with *M.rosenbergii*	630-930 / year	5.2
24 Parganas, WB (Developing)	Shallow (coastal)	Brackish water shrimp and fish with Tilapia	862 / year	2.7
WTCER (ICAR) Bhubaneswar (1999-2001)	Rainfed medium land (without pesticide application)	Fry of Indian Major Carp with *M. rosenbergii*	1156 / 120 days	3.6
WTCER (ICAR) Bhubaneswar (2001-2004)	Rainfed medium land	Fry of common carp and Indian Major Carp	1265 / 180 days	5.3
WTCER (ICAR) Bhubaneswar (2003-2004)	Raised and sunken bed system in canal command	Fingerling of Indian Major Carp	2740 /120 days	3.7
CRRI, Cuttack (2001)	Deepwater lowland	Indian Major Carp & Common Carp	600 /season	4.0
IIWM(ICAR), 2015)	SRI-fish	IMC	2600/ 180 days	6.16
Kuttanad, Kerala (2002)	Polders	*M. rosenbergii*	95-1297 / year	--
WTCER (ICAR) Bhubaneswar (2005-2006)	Deep water high-density rice-fish culture	Indian Major Carp with *M. rosenbergii*	3600 / 180 days	3.3

Source: Brahmanand *et al.* (2006)

About 55,000 ha of wetlands are available for paddy cultivation year-round in the derelict polders (embanked coastal flood plains) of Kuttanad, Kerala;

of these, about 5,000 ha are used for *Macrobrachium rosenbergi* culture as a follow-up crop. About 4,700–4,750 hectares are used for polyculture with Indian and exotic carps from November to June, while the remaining 250–300 ha are used for M. rosenbergii monoculture from March to October in fallow polder. In monocultures of *M.rosenbergii*, stocking density typically ranges from 15,000 to 60,000 ha; in polycultures with carps, it typically ranges from 5,000 to 20,000 per ha for prawns and 5,000 to 10,000 per ha for fish. Production from polyculture systems ranges from 50 to 500 kg/ha of prawns and 200–1200 kg ha^{-1} of fish, whereas production from monoculture systems ranges from 100 to 1300 kg ha^{-1}. The use of ploders for prawn and/or fish culture not only helps farmers increase their earnings but also creates jobs.

During the last four decades, experiments have been conducted in several states of the country to improve the production potential of fish/prawns in rice fields. Presently most research activities in rice-fish culture are confined to medium lands and deep-water rice fields in freshwater and coastal rice fields in brackish water habitats. The State Fisheries Department, Government of West Bengal, undertook nursery rearing of carp in a 279-ha rice field (Hora, 1951). Carp fry (19-64 mm) were stocked at 1,457 per ha and raised to 127-135 mm in size in three to four months. The total yield was estimated to be 112 kg/ha with an overall survival of 34%. An experiment was carried out in plots at the Hasserghatta and Visweswarya Farms in Karnataka to control insect pests in the rice fields through stocking of *Channa striata*. The average fish yield in four months was 112 kg/ha, showing a growth of 7-13%. In West Bengal, *H. fossilis* and *C. batrachus* were stocked at a 1:2 ratio at 10,000 fingerlings/ha in Randhunipagal rice fields. With no supplementary feeding, catfish yield was 199.4 kg ha^{-1} with 1.84 t/ha of rice yields. However, with supplementary feeding (fish meal and rice bran at a 1:2 ratio and given 5% of the fish biomass), yields were increased to 375 kg/ha for catfish and 1.88 t/ha for rice. The yield of rice was low (1.79 t ha^{-1}) in the control plot without fish.

A 1.1 ha rice plot in Bandipur, West Bengal, which was producing only one crop of rice (2.7 t ha^{-1} yr^{-1}) in a traditional system, was renovated, keeping 65% of the area in rice, 33% perimeter canal and the rest for dikes. Two crops of rice, *Ratna* in summer and *Jaya* in the monsoon, and one annual crop of carp were raised. Yields were 4.2 (*Ratna*) and 3.2 (*Jaya*) t ha^{-1} of rice and 237 kg ha^{-1} yr^{-1} of carp in the initial phase. In subsequent years, fish yields increased to 630 kg ha^{-1} yr^{-1}. Similarly, in freshwater rice plots with the excavation of deeper pools at Minakhan, fish yields ranged from 650 to 930 kg ha^{-1} $yr^{-1}$and rice yields, 5.1 to 6.4 t ha^{-1} yr^{-1} (Ghosh, 1992). In coastal saline areas of West Bengal, plots belonging to 35 farmers were developed and the average fish yield in these plots was 826 kg ha^{-1} yr^{-1} and rice yield of 2.7 t ha^{-1} during the monsoon season.

In low salinity plots near Basirhat in West Bengal, management practices were further developed and yields improved: 50 kg ha^{-1} of *P. monodon*; 250 kg ha^{-1} of mullet; 3,000 kg ha^{-1}of tilapia and 2.4 t ha^{-1}of rice (Ghosh, 1992).

Earlier, the Water Technology Centre for Eastern Region (presently ICAR, IIWM), Bhubaneswar had conducted integrated rice-fish farming in medium lands, lowlands and in modified alternate raised and sunken bed systems in four locations of Odisha covering 12 farmers' fields. Results indicated that construction of monolateral-type refuge (30-35% of total land area) acts as a drainage system and helps in lowering the water table, which can be adopted for rice-fish culture in reclaiming waterlogged area (Fig. 6.1). Ways to intensify fish production from the deepwater rice-fish unit can involve management strategies like high-density rearing followed by selective/ cull harvesting when the growth curve of fish/prawn starts to slow down. As density-dependent growth performance takes place at higher population density, culling helped in reducing size heterogeneity, weight distribution and stunting growth of fish. Under high-density deepwater rice-fish culture, record yield of rice and fish was achieved @ 3.3 and 3.6 t ha^{-1} per crop respectively, where culling enhanced fish & prawn yield by 16.3% and feed conversion ratio by 26% over control. The existence of fish & prawns enhanced rice yield by 22.7%. In this system, fish acts as a bio-controller of insects & aquatic weeds, thus reducing the input cost of insecticide and herbicide by 52-55%. In this system, net economic indices of water productivity (Rs m^{-3}) increased by 11.3-fold over deepwater rice mono-crop while farmers can get a net profit of Rs.1,08,500 per ha per crop with a B-C ratio of 4.1.

In the rainfed medium land ecosystem, two-stage rainwater conservation (Mohanty *et al.*, 2004; Mohanty, 2003), short duration rice-fish culture using the excess rainwater harvesting refuge (Mohanty, 2004), integration of on-dyke horticulture and utilization of conserved refuge water for growing low duty *rabi* crops (Mishra and Mohanty, 2004) seem to be a viable solution (Fig. 6.2 and 6.3) for increasing the net profit (Rs. 39,700 per ha per crop with a B-C ratio of 2.93) of small and marginal farmers. The addition of poultry/ duckery to the system and their droppings in water reduce the input cost of fish feed by 40% and fertilizer cost by 60-65% while the B-C ratio goes up to 3.68-3.87. This system is eco-friendly as it promotes synergism between different components leading to the recycling of wastes and generating additional farm employment of up to 220 man-days per year. In another experiment of rice-fish co-production using raised and sunken bed systems in canal command, fish yield (2740 kg per 120 days) increased by 29.8% against that of the mono-culture system.

In a medium land ecosystem, a new method of rice and fish cultivation was investigated to enhance the overall productivity at the ICAR-IIWM, Bhubaneswar (Thakur *et al.*, 2015). The system of rice intensification approach and triggering compensatory growth (CG) response in fish through cyclic feed restriction and refeeding (Turano *et al.*, 2008; Rubio *et al.*, 2010; Stumpf *et al.*, 2010; Mohanty, 2015) have been proven beneficial in terms of significantly higher yield output. In this system rice yield of 6.16 t ha^{-1}, fish yield of 2.6 t ha^{-1} and net water productivity of INR 18.9 m^3 t^{-1} were achieved (Thakur *et al.*, 2015). The strategy of adopting rice intensification i.e., low planting density, wider spacing, a small population and less fertilizer application helps in better root zone development, increasing the number of tillers and improving ecological conditions (ventilation and illumination) that prevent lodging and helps produce heavier grains, high and stable yields. Similarly, CG offers the possibility of improving the growth rates of fish by a careful choice of feeding protocol in which periods of feed deprivation are followed by periods of satiation feeding (Nikki *et al.*, 2004; Wang *et al.*, 2005; Montserrat *et al.*, 2007; Mattila *et al.*, 2009). Many fish can actively regulate their growth as a strategy against negative environmental changes and have the potential to elicit above-normal growth rates in recovery periods following stress (Mohanty, 2010a). If CG can completely make up for growth lost during starvation, there could be an opportunity to save on fish feed by starving the fish and making up for lost growth when feeding resumes.

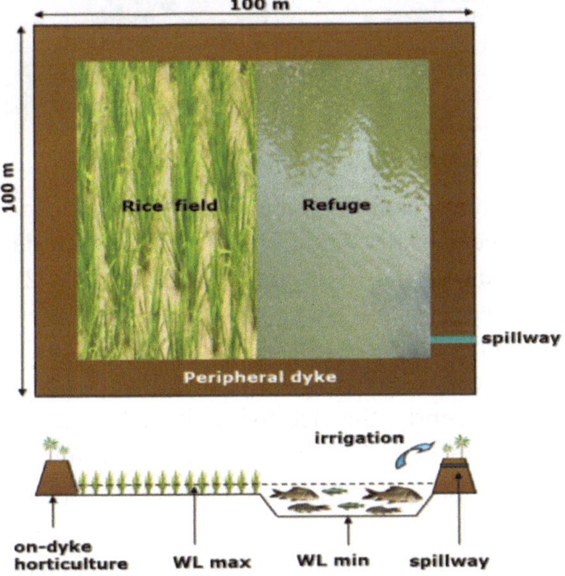

Fig. 6.1. Layout of the deep-water rice-fish system (1 ha unit)

Fig. 6.2. Rice-fish system in the medium land ecosystem without field refuge.

These developments serve as an impetus for promoting rice-fish farming in India. Interest in rice-fish farming over the years has waxed and waned among policymakers, scientists, extension workers and farmers in India.

Fig. 6.3. Two-stage rainwater conservation (*in-situ* in the rice field and *ex-situ* in refuge) and rice-fish system in a medium land ecosystem with on-dyke horticulture.

6.3. Rice-fish integration: global scenario

Rice-fish co-culture originated in China over 2000 years ago during the Han Dynasty, and has since spread to India, Bangladesh, Indonesia, the Philippines, Vietnam, Egypt, Thailand and Malaysia (Saikia and Das, 2015; Wang *et al.*, 2015; Little *et al.*, 1996). Rice - fish co-culture has been practiced in India for last 1500 years (Chakravartty *et al.*, 2017; Mohanty *et al.*, 2004). Co-cultivation systems were introduced in Japan in the past century, and in Java in the mid-nineteenth century. Recently, the rice-fish culture has garnered considerable attention due to its likely economic benefits by the farming community.

In this co-culture systems, a variety of aquatic species have appeared in addition to fish, such as crabs, prawns, shrimp, crayfish, ducks, and turtles. For instance, because of its high protein content, numerous medical applications, and significant economic worth, softshell turtles (*Pelodiscus sinensis*) are being co-cultured with rice more and more in China (Hu *et al.*, 2016; Zhang *et al.*, 2016; Haitao *et al.*, 2008). Another prominent approach in China is the co-culture of rice with crayfish, which in 2016 covered 4.2 - 105 ha^{-1}, or 70% of the total area (6 - 105 ha^{-1}) dedicated to crayfish breeding (Cao *et al.*, 2017). The use of rice-animal (co-culture) systems has grown quickly due to its benefits in producing grains in an environmentally sustainable manner, minimizing ecological disruption, and mitigating the effects of climate change (Datta *et al.*, 2009). According to earlier research, China achieves a comparatively high average fish output of almost 2.5 t ha^{-1}. Vietnam, India, Thailand, Bangladesh, and Indonesia follow with 2.2 t ha^{-1}, 2.0 t ha^{-1}, 1.1 t ha^{-1}, and 0.89 t ha^{-1}, respectively (Ofori *et al.*, 2005). While the highest recorded fish yields from rice-fish systems in India were 3.6 t ha^{-1} (Mohanty, 2009), Chen *et al.* (1995) reported the highest fish yield using a polyculture of *H.molitrix, C. carpio*, and crucian carp with 7 038 kg ha^{-1} for rice-azolla-fish system as opposed to only 4 119 kg ha^{-1} for rice-fish combination. Similar to most aquaculture operations, the amount of fish that can be harvested in rice-fish farms varies greatly. The harvest of aquatic animals from any rice field is a function of several factors such as water depth and water supply. Presence of predators, species, stocking density, whether monoculture or polyculture is practiced, size of fish at stocking, and the rearing period. Seasonal variations in natural productivity, and whether fertilization and/or supplemental feeding have been applied also affect fish production. Twenty to thirty percent of Bangladesh's ten million hectares of rice are appropriate for co-cultivating aquatic animals with rice (BRKB, 2010; Wahab *et al.*, 2008; Nhan *et al.*, 2007). While almost 20 million hectares of land in India can be used for rice-animal co-culture, only 0.23 million hectares of this practice have been implemented because of a lack of technical assistance and expertise (Chakravartty *et al.*, 2017; Mohanty and Mishra, 2003; Ali, 1990). Rice-fish farming is permitted on 1.0 million hectares of land in Indonesia (Dwiyana and Mendoza, 2008) and 0.09 million hectares in Thailand (Karapanagiotidis *et al.*, 2010).

Significant work on rice-fish farming has also been done in other countries such as the Philippines, Indonesia, Bangladesh, Vietnam, Thailand, Malaysia, Korea etc. Generally, three types of field designs for rice-fish farming predominate i.e., shallow trench within the rice field (in the Philippines, Indonesia and China), pond refuge adjacent to the rice field (in Indonesia, Thailand and China) and deep-water rice fields (Bangladesh). The concurrent culture of fish and rice as well as rotational systems where fish are grown after rice are common throughout Asia (Brahmanand *et al.*, 2006). An analysis shows that rice-fish systems presently occupy only a very small percentage of the potential area. The wide array of systems that exist can be broadly characterized by field design, growing period and fish species. Carp, tilapia, silver barb, snakeskin gourami and prawns are usually grown.

Combinations of rotational and concurrent rice-fish systems arranged in a one-year sequential cropping pattern are the best developed in Indonesia. When practised in combination, these systems can produce an annual fish yield of over 1000 kg ha^{-1}. Rice fields are being used both as fish nurseries and for table fish in Indonesia, China and Thailand. Interestingly, raising fingerlings in rice fields was more profitable than growing out. Published data on rice yields from China, India, Indonesia, Philippines and Thailand reveals that average percentage increases in rice yields ranged from 4.6 to 28.6% while in India, it is between 9 and 23%. Among the countries listed, Indonesia is second to China in rice area being used for rice-fish farming. Fish culture is concentrated in irrigated and coastal rice lands. Although it is much smaller in area, Thailand is next to Indonesia in the extent of rice-fish culture. Fish are grown in many different ways in both irrigated and rainfed environments. Bangladesh has more than 2 million ha of deep-water rice areas. The potential of this resource for stocking and feeding fish has not been tapped yet. The area devoted to rice-fish farming in the Philippines is still limited. Rice-fish culture techniques for irrigated rice using tilapia and carp have not yet been diffused to many farmers.

The most common fish species found in rice field systems are carps (Indonesia, India, China), tilapias (Philippines, China, Thailand), *Puntius gonionotus* (Thailand), and *Trichogaster pectoralis* (Thailand). Both mono and polyculture of these species are practised. In deep water and coastal areas of Bangladesh, India and Vietnam, shrimps are freshwater prawns (*Macrobrachium rosenbergii*) and finfishes are stocked during wet or monsoon months, while marine shrimps and fishes are culture during summer or saline water months. Thailand, Bangladesh and India have a seasonally important rice field capture fisheries system. Furthermore, Indonesian farmers integrate ducks and home gardens into their rice-fish culture systems (Brahmanand *et al.*, 2006).

6.3.1. Global experiences

China has the oldest archaeological and documentary evidence for rice-fish farming. In 1954 it was proposed that the expansion of rice-fish farming should be spread across the country (Cai *et al.*, 1995) and by 1959, the rice-fish culture area expanded to 6,66,000 ha. From the early 1960s to the mid-1970s there was a temporary decline in rice-fish farming. This was attributed to two developments: first, the intensification of rice production and second the ten-year cultural revolution (1965-75). By 1996, China had 1.2 million ha of rice-fish farms producing 3,77,000 t of fish (Halwart, 1999).

Rice-fish farming appears to be of small relevance in Japan, and there is little literature on the topic. Carp production on paddy fields peaked at 3400 t in 1943 as a result of war-time food production subsidies, but fell to 1000 t in the 1950s and is no longer practiced on a large scale, if ever. Rice-fish culture in Korea began in the 1950s and has never spread significantly since the inland fish supply was adequate to meet the limited freshwater fish demand. As of 1989, just 95 ha of rice fields were used for fish farming, and primarily to raise the most popular species of loach (*Misgurnus anguillicaudatus*).

Rice-fish farming in Indonesia started even before 1860 although its popularization started only in the 1870s. By the 1950s some 50 000 ha of rice land were already producing fish. The expansion of irrigation systems also contributed to the expansion of the area used for rice-fish farming. The average area of rice-fish farming increased steadily after Indonesia became independent in 1947 and rice-fish farms covered 72 650 ha in 1974, but declined to less than 49 000 in 1977. The decline was attributed, ironically, to the government's rice intensification program. Recent reports indicate that rice-fish farming is on the upswing. The 1995 figures from the Directorate General of Fisheries indicate a total area of over 138000 ha. At present rice-fish farming is practiced in 17 out of 27 provinces in Indonesia.

Rice-fish culture is believed to have been practiced for more than 200 years in Thailand, particularly in the Northeast where it was dependent upon capturing wild fish for stocking the rice fields. It was later promoted by the Department of Fisheries (DOF) and expanded into the Central Plains. The provision of seed fish and technology helped in popularizing the concept. However, during the 1970s, Thailand, like the rest of Asia, introduced the HYVs of rice and with it the increased use of chemical pesticides. This resulted in the near collapse of rice-fish farming in the Central Plains as farmers either separated their rice and fish operations or stopped growing fish altogether. However, rice-fish farming has recovered, particularly in the Central Plains, North and Northeast Regions. Presently, more than 25 500 ha is under rice-fish farming.

Rice-fish farming began in Malaysia in 1928, and rice fields have long been a significant source of freshwater fish. Fish output in rice fields began to fall with the adoption of the double-cropping technique and the widespread use of pesticides and herbicides (Ali, 1990). Vietnam has a rich history of mixing aquaculture and agriculture. Le (1999) describes five main rice-fish cultivation systems used in Vietnam. There are five systems: fish-cum-rice for nursery and growout, fish-cum-rice for growout alone, shrimp-cum-rice, fish/rice rotation, and shrimp/rice rotation.

In the Philippines, fish/ prawn are usually allowed to enter the paddy fields with the irrigation water and are later harvested with the rice. Despite the lower rice yields, in 1979 the government proceeded to promote rice-fish farming nationwide. The decision was based on the results of the economic analysis that even with reduced rice production, the farmer would still be economically ahead due to the additional income from the fish. After a peak of 1397 ha involving 2284 farms in 1982, the program was discontinued in 1986. At that time the average production of rice from rice-fish farms was above the national average. However, it was re-lunched in 1999 with a more modest rice-fish program in the Philippines.

Fish harvesting from rice fields has long been a practice among Bangladeshi farmers. Bangladesh is among the few nations that actively supports rice-fish farming and is engaged in a robust R&D effort. In Bangladesh, non-governmental organizations are also becoming more interested in rice-fish farming. More recently, CARE has become the most active NGO involved in rice-fish farming. The traditional *bheri* system is used wherein the rice fields are enclosed by small embankments complete with inlet channels and sluice gates. Fields vary in size from 3 to 50 ha. Both rotational and concurrent systems are practiced. Thousands of farmers have experimented with rice-fish culture in Bangladesh and have developed the package of practices to suit their farming systems. Initially, the adoption rate was lower among females, but the activity is reported to be gaining popularity among both male and female groups.

In Africa, the earliest report on rice-fish culture comes from Madagascar as early as 1928. Both concurrent and rotational systems relying on the entry of natural fish stock were practiced. Only in 1979, sufficient progress was made for the government to promote rice-fish culture. However, an average yield of 80 kg ha^{-1} indicates that culture techniques at the farm level still need to be improved (Randriamiarana *et al.*, 1995). A country with almost 900 000 ha of rice fields does have a great potential for rice-fish farming as about 150 000 ha cold be suitable for rice-fish farming. A potential annual production of 300,000 t of edible fish has been projected from the said areas. Farmers in Malawi are just beginning to grow rice and fish together as well as fish and vegetables. Although not specifically mentioned, the fish involved are tilapia, the principal species in the country.

Egypt, the Middle East's largest rice producer, began with a capture-type of rice-fish system that relied solely on infrequent fish stock brought in with irrigation. The rice-fish farming area extended significantly using reclaimed salt-affected soils, peaking at 225000 ha in 1989 before declining to 172,800 ha by 1995. In 1995, fish production from rice fields accounted for 32% of overall aquaculture out put in the country. Since then, 58000 acres of farmland have been added, yielding 7,000 t of *C. carpio* in 1997 (Wassef, 2000).

Iran began rice-fish culture trials in 1997. Chinese major carps are used concurrently with rice, sometimes with supplementary feeding. Productions over 1.5 t of fish per ha together with 7 t of rice have been achieved with a high survival rate (96%), despite an average water temperature of only 23 ^{0}C during the culture period. In addition, 70 farms have adopted a rotational rice-fish farming system where the rice field is stocked with trout during the winter months when the average water temperature is 12 ^{0}C, yielding 640 kg ha^{-1}. A concurrent culture of *M. rosenbergii* with rice is also being tried.

The rice-fish co-culture was introduced to Italy at the end of the 19th century and progressively became important in the subsequent 40 years. The main species were *C. carpio, C. auratus* and *Tinca tinca*. The rice fields were used to produce fingerlings that had a ready market. The practice gradually declined and by 1967 it was no longer considered an important activity. Although rice is produced in nine countries in South America and eight countries in the Caribbean, fish culture using rice fields is not widespread. Experiments on integrating fish culture with rice production are or were being conducted in Brazil, Haiti, Panama and Peru, but only Brazil appears to have had some degree of commercial success. In the northeast, farmers became interested in semi-intensive rice-fish culture using native fish species. Experiments on intensive rice-fish culture were also conducted in the Paraiba basin using the *C. carpio* and Congo tilapia (*T. rendalli*). Because of the region's excellent climate and irrigated lands, the outlook for rice-fish co-culture appears to be positive. Rice-fish farming used to be important in the United States. However, as an alternative to finfish, crawfish are now alternated alongside rice. Two species of crawfish are popular due to their resilience and adaptability: the red swamp crawfish (*Procambarus clarkii*) and, to a lesser extent, the white river crawfish (*P. zonangulus*). The life cycle of crawfish, as well as their environmental requirements, makes them ideal for rotation with rice. The majority of crawfish produced in the United States presently comes from the southern states' rice farms.

6.4. Eco-agriculture: agronomic and aquaculture management

Agriculture, the backbone of the Indian economy, accounts for almost one-fourth of national GDP and supports the livelihoods of around two-thirds of the people. It is anticipated that food grain output must increase by 185% to cover the food needs of 1.5 billion people by 2050, which is achievable through increased productivity, crop diversification, and cropping intensity. However, presently, there is little scope for horizontal expansion of agriculture/ aquaculture due to rapid industrialization and urbanization. Our per capita availability of land has also dropped from 0.5 ha to 0.15 ha. Further India's population has crossed

1050 million including an undernourished population of 270 million. Under this scenario, meeting the food requirements of the burgeoning population has turned out to be a challenging task. Vertical expansion by integrating compatible farming components requiring lesser space and time, and ensuring periodic income to farmers, therefore, seems to be a viable alternative which assumes greater importance for sound management of farm resources. This would enhance farm productivity, reduce environmental degradation, improve the quality of life of resource-poor farmers and maintain sustainability. On the other hand, the most significant restraint in aquaculture technology is the high cost of inputs, particularly fish feed and fertilizer. Furthermore, stocking healthy fingerlings of more than 100 mm is an important management tool for increasing fish productivity. However, the current lack of land-based nursery ponds, as well as financial limits in creating new infrastructure facilities impede the planned stocking programme. With these limits and sufficient supplies, the rice field ecology offers a realistic option for large-scale fry to advanced fingerling rearing as part of a stocking operation.

Furthermore, of India's 46.38 million hectares of rice cultivated land, 20 million ha are appropriate for the implementation of rice-fish integration systems, primarily in rainfed medium lands, waterlogged/low lands, and canal command areas. However, only 0.23 million hectares are currently under rice-fish cultivation. This low level of adoption, exploitation, and yield is primarily due to the introduction of high-yielding rice varieties involving pesticide use, which has greatly hampered fish culture in paddy fields, shorter fish rearing duration, insufficient water availability, a lack of water management measures, and erratic monsoons. Achieving improved output from these underutilized high-potential areas is thus an urgent requirement, especially in the eastern region. If these lands were put under an integrated rice-fish system with appropriate scientific intervention and water conservation/monitoring methods, it would help to compensate for economic losses in rice production caused by natural disasters. Integrated rice-fish farming not only allows for crop diversification, increases productivity, generates employment opportunities, raises income, and provides nutritional security to resource-poor farming communities, but it also distributes risk (both biological and economic) because two or more subsystems are involved rather than a single commodity farming system.

6.4.1. Eco-agriculture

Eco-agriculture is an ecologically balanced, highly efficient and rational agricultural system. It is assessed not only in terms of grain output but also in terms of total biological output and profit. There is unity in economics and ecological objectives. Rice-fish integration is therefore a primary option when trying to develop ecological agriculture that exploits all-out benefit from the system, avoids harmful effects and strives for maximum output using existing energy and materials. Further, introducing fish to the paddy field ecology helps increase production and achieves social, economic and ecological benefits.

6.4.2. Rice field ecosystem

Non-biological factors (water, soil, heat, air and light) and biological factors (plant, animal and microorganisms) are always interrelated and interdependent. They form an ecosystem of unilateral functions. Change in one factor triggers a chain of changes within the ecosystem. The rice land ecosystem is an anthropogenic system where rice is predominant. In the biological community of the rice field ecosystem weeds, plankton, humus and bacteria are the primary producers. These primary producers of rice undertake energy conversion and storage similarly and competitively. Both rice and primary producers absorb solar energy, carbon dioxide and nutrients from soil and water to manufacture organic matter by photosynthesis and they convert, transport and store energy. This competition among the biological communities degrades the growing environment and increases factors that are unfavourable to rice growth.

6.4.3. Mutualism of rice and fish

Some of the ecological requirements of rice and fish are similar and this provides the basis for their synchronized growth under one ecosystem (Fig. 6.4). Water is a prerequisite for raising fish and for the growth performance of rice. Water, as a component of plant cytoplasm, is indispensable for the synthesis of organic matter in plants and for the absorption and transfer of nutrients. Further, when fish, especially herbivorous and omnivorous fish are introduced into the rice fields, they add a new link to the existing food chain. They feed on primary producers and therefore reduce energy losses, improve the use of photosynthetic products and promote transformations in the rice field ecosystem that increase the carrying capacity of the rice field.

Rice-fish fields maintain water temperature and quality, creating an environment conducive to the growth of natural fish food organisms. These primary producers convert solar energy into food energy that is required by fish for their survival. The sequential relationship in the distribution of rice and fish is apparent. Fish in rice fields feed on available plankton (that compete with rice for fertilizer), weeds (that compete with rice for nutrients), insects and bacteria (that harm rice plants) and mosquito larvae (harmful to humans).

Fish assimilate only 3-4% of these feeds and discharge the rest into rice field that acts as manure. Because fish devour phytoplankton, zooplankton, and rice-competing weeds, they play a vital role in energy recovery and flow adjustment. When they swim in water, they release carbon dioxide and that increases the volume of carbon available to the plants. They also break the soil surface and oxidize layers of soil, which increases oxygen supply to promote root growth and tillering capability of rice plants. It is reported that under this system, paddy yield has been increased by 8-47% (8-14% in India). The increase in the yield of paddy probably results from improved aeration and a better tillering result triggered by the presence of fish. The excreta of fish, leftover supplemental feed and additional fertilizer used also help in increasing the soil fertility and productivity.

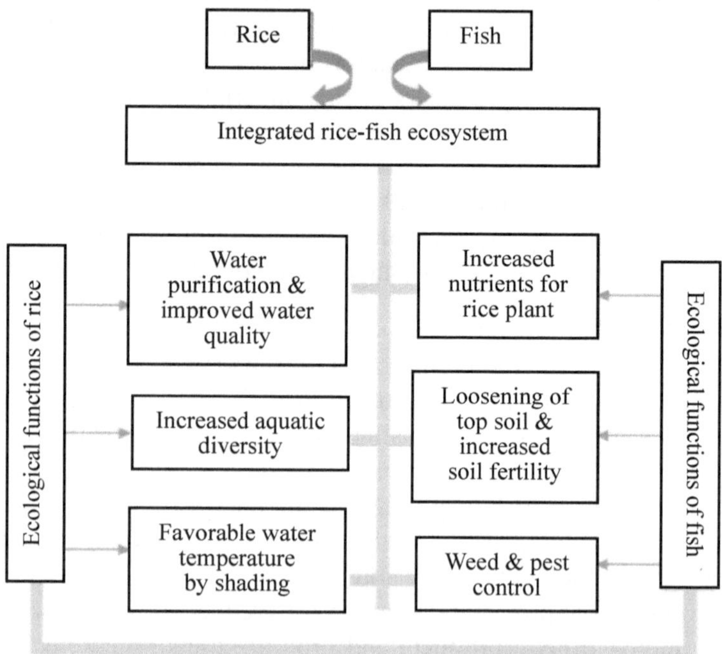

Fig. 6.4. Inter-component interaction in an integrated rice-fish ecosystem.

Rice-fish mutualism is therefore, the best way to maximize the output of the ecological system that turns material and energy into fish production, accelerates the growth of rice and increases the solar energy fixation, thus raising the productivity of paddy fields. Major components, pathways of material flow and benefits in the integrated rice-animal co-culture system are depicted in the next section as Fig. 6.5. Integrated rice-fish farming not only increases productivity but also generates employment opportunities, increases income and provides nutritional security to resource-poor farming communities and distribute the risk (both biological and economical), since two subsystems are involved instead of a single commodity farming system.

6.4.4. Effect of fish on the growth and yield of rice

When fish are reared in the paddy field, the nitrogen, phosphorous, and potassium (NPK) content of the water and soil increases dramatically. Fish movements in low-lying waters breach the surface membrane generated by microorganisms that cover the soil. This raises the dissolved oxygen level in the soil, increasing its oxidation and reduction potential during the rice development cycle. These changes increase the oxygen content of the soil and the pace at which nutrients are utilized. Fish in the rice-fish system promotes more efficient NPK utilization and distribution, minimizing fertilizer loss while enhancing soil fertility. The rice-fish system is typically used to measure the following: the total dry weight of the rice plant increased; the NPK content of the leaves and culm of the rice plants; the surface area of the leaves; the chlorophyll content of the rice plants at every developmental stage; and the strong root system activity that increases the uptake of nutrients from the soil. The efficiency of photosynthesis is increased by the bigger surface area and higher chlorophyll content of the leaves. This results in the accumulation of more carbohydrates, which in turn increases the number of effective ears, the number of grains per ear, and the weight of the grains. A prerequisite for higher yield of rice is an increase in the dry weight of the entire rice plant.

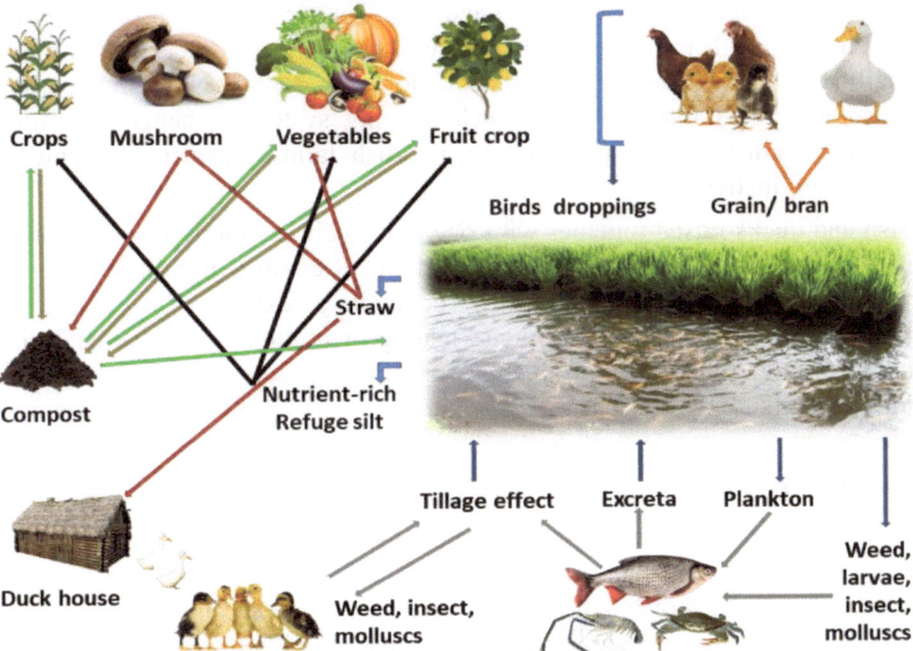

Fig. 6.5. Schematic representation of major components, pathways of material flow and benefits in integrated rice-animal co-culture system

6.4.5. Production intensification approach

The strategy for rice cultivation under the rice-fish system, is low planting density (20-30% lower than the density used in regular fields), a small population and less fertilizer application. This helps in better root zone development, increasing the number of tillers and improving ecological conditions (ventilation and illumination) that prevent lodging and help produce heavier grains and high and stable yields. The principles for selection of fish species for stocking into the rice-fish farming system should be based on the capability of fish species to filter and feed on plankton (bacteria, phytoplankton, and zooplankton) and to tolerate low levels of dissolved oxygen. An ideal stocking density of fish/prawn species is critical in attaining high cumulative fish yields and in reaching the upper carrying capacity of the system.

Ways to increase fish production from an integrated rice-fish farming system involve management strategies like high-density rearing (stocking with higher initial fish biomass) followed by phased/selective harvesting when the growth curve of stocked fish/prawn starts to slow down. The productivity of the current practices of rice-fish culture has a great prospective for improvement with further integration with duck and nitrogen-fixing aquatic fern *Azolla*. *Azolla* can be utilized not only as organic fertilizer for crops but also as feed for livestock and fish. Fish, *Azolla* and ducks integrated with a rice farming system can result in nutrient enhancement, pest (weed, insect, golden apple snail) control, feed supplementation and biological control. Nutrient recycling in an integrated rice-fish-azolla-duck farming system is better and more efficient compared to a rice-duck or rice-fish farming system resulting in greater productivity.

Fish and ducks raised together with rice have their economic value resulting in increased overall productivity of the farm. Rice-duck farming integrated with fish and *Azolla* increases the production potential of this traditional farming practice. Natural resources management in tropical wetland rice fields through integrated production systems is a promising approach to introducing sustainability in rice farming and addressing ecological issues of the conservation of rice field ecology and aquatic biodiversity.

6.5. Rice-fish integration practice in India

As mentioned earlier rice and fish sometimes have conflicting requirements. Growing fish in the paddy field does require some modifications to the management to ensure that the fish meet their requirements and to facilitate fish survival and growth during certain critical periods. Whether the modification is in the form of trenches, or lateral ponds of higher and wider dikes, nothing suggests that one form of modification can be considered superior to others.

The type of modification used is based on a combination of different factors: the terrain, soil quality, water supply, traditions, exposure to other methods, past experiences, the relative importance given to either rice and fish, whether fingerling or food fish is desired and the financial resources available. Although generally, rice is the main crop in any rice-fish farming activity, there are exceptions where rice is planted or rationed to provide forage for the culture organism.

For rice-fish integration in the waterlogged ecosystem, rice fields with infield refuge of 1.5 - 2.0 m depths, peripheral trench of 0.5 m depth and 1.0 m width and moderate slope towards refuge are most preferred, in which rice and fish are reared under one ecosystem. The refuge area can be 20-30% of the total available land area. Alternate raised and sunken bed technology through topographic modification in canal command also provides various options for crop diversification and short-duration fish culture. In sunken beds of 45-60 cm deep (70-75% land) followed by 1.0-1.4 m deep at one end (25-30% land) or rice field with dyke height of 50 cm, infield refuge (1.2-1.5 m depths) of 15-25% area of rice field, peripheral trench of 0.5 m depth and 1.0 m width with a moderate slope of 0.5% towards the refuge is most preferred, for adoption of rice-fish farming system. Water and soil quality variables generally determine the production potentiality of this system (Table 6.2), as several *biotic* and *abiotic* factors play a key role in the enhancement of productivity. Therefore, intensive hydro-biological studies before site selection are essential. The texture and properties of soil, its water retentive capacity and suitability as a constructional material to build up the surrounding dykes should be determined. Generally, soil with a higher percentage of clay and pH ranging from 6.5 to 7.5 is considered suitable. The lands having higher water tables and with suitable drainage systems may be selected.

Table 6.2. Classification of a productive rice-fish system based on water and soil characteristics.

Parameter	Unproductive	Low productive	Highly productive
1. Water			
pH	6.0 – 7.0	7.0 – 7.5	7.5 – 8.5
Total alkalinity (ppm)	10 – 30	30 – 90	90 – 170
Inorganic nitrogen (ppm)	< 0.1	0.1 – 0.3	0.3 – 0.7
Orthophosphate (ppm)	< 0.05	0.05 – 0.2	0.2 – 0.6
Dissolved oxygen (ppm)	2.0 – 4.0	4.0 – 6.0	6.0 – 9.0
Plankton (number / l)	5 – 35	35 – 70	>70
2. Soil			
pH	6.0 – 7.0	7.0 – 7.5	7.5 – 8.5

Available Nitrogen (mg / 100g)	5 – 15	15 – 35	35 – 90
Avl. Phosphorus (mg / 100g)	3 – 6	6 – 20	20 – 50
Exchangeable calcium (mg/ 100g)	20 – 50	50 – 150	150 – 300
Bottom fauna (g / m^2)	1 - 15	15 - 20	>20

Source: Brahmanand *et al.* (2006)

6.5.1. Species suitability

Fish species should be adaptable, compatible, resistant to environmental changes, high-yielding and able to tolerate heavy doses of fertilizer. Since the water column in the refuge, perimeter canal and the paddy field in the renovated system are suitable for rearing of carps, three Indian major carps, i.e., *Catla catla, Labeo rohita and Cirrhinus mrigala,* prawn (*Macrobrachium rosenbergii* and *M. malcolmsonii*), exotic carps like silver carp and common carp and tilapia may be stocked for culture in the integrated system. Improved high-yielding, tall, long-duration, submergence and pest resistance variety of rice with in-built characteristics of photoperiod-sensitivity, and strong seedling vigor can be tried. Some of the promising medium-deep to deep water varieties of rice in different states of India are listed as follows:

Odisha:	*Durga, Tulasi, Gayatri, Sarala, Panidhan, Jayanti dhan, Jalamani, Utkalprava*
West Bengal:	*Mahananda, Sabita, Purnendu, Sasi, Mandira, Kalashree*
Assam:	*Prafulla, Joymati, Rupsundari, Prasant, Kalashree*
Andhra:	*Badava Mahsuri*
Bihar:	*Santosh, Janaki, Sudha, Vaidehi*
Tamil Nadu:	*TNR1, TNR2*
Uttar Pradesh:	*Jal-Lahari jalapriya, Madhukar, Chakia-59, Jalamagna*
Chhatisgarh:	*Jaldubhi*

6.5.2. Rice cultivation

Land preparation for rice crops is done at the beginning of the season with minimum tillage and levelling for uniform crop establishment. Sowing of rice is usually preferred over transplanting, as sowing helps in the establishment of a good crop stand of rice by the time the field is waterlogged during monsoon. Sowing also helps in avoiding puddling (for transplanting), which would otherwise lead to siltation of refuge, besides creating turbidity problems for the fish. Normally line sowing and skip-row methods of sowing are followed to provide fish passage and to encourage movement of fish inside the rice field when it is flooded with water. Interestingly, the skip-row method reduces the plant population; it does not affect the overall rice yield significantly. When sowing of rice is not feasible due to unprecedented rainfall or some other reason, transplanting is done with 25-30 days rice seedling, after proper land preparation, preferably when the water level is around 5-10 cm in the rice field.

Sowing can be carried out in dry conditions well before monsoon, with a spacing of 20 cm between rows. Transplanting if required can be done early by using aged and healthy seedlings. Fertilizer @ 40:20:20 kg NPK ha^{-1} at seeding should be applied (apply 50% N at seeding and the rest 50% after weeding). FYM @ 5 t ha^{-1} at the time of land preparation is also essential. Avoid insecticides and herbicides and use of sex pheromone trap to monitor and control yellow stem borer, neem extract-based composition such as Nethrin or Nimbecidine @ 1% for control of stem borer. To ensure better productivity, higher planting density must be avoided to provide greater ventilation and illumination, which helps in enhancing both plant and plankton growth which ultimately helps fish growth and yield. The yield of rice for a specific variety correlates positively with planting density.

6.5.3. Application of fertilizer and chemicals

The application of fertilizers, organic or inorganic, benefits both rice and fish. The presence of adequate nutrients increases the growth of phytoplankton, which may be consumed directly by the fish or indirectly through supporting zooplankton production. The kind, amount, and application technique of fertilizers have a major impact on the growth performance of fish and paddy. Planktonic and benthic creatures, which are fish's natural food source, also need the same nutrients—nitrogen, phosphorus, and potassium—that the paddy need. Fish are also poisoned by excessive amounts of inorganic fertilizer. The optimal method of fertilization is to apply as little inorganic fertilizer and as much nutrient-rich organic manures (Table 6.3) as feasible. Applying organic manure is recommended after fermentation. Seventy percent of the total manure should be treated as basal, and the remaining portion should be applied sparingly but periodically as supplemental manure.

Experiments in China indicate that the organic nitrogen, alkaline nitrogen and total nitrogen in the soil are consistently higher in fields with fish than in the control fields without fish (Wu, 1995). It was attributed that fish in the rice field consumed weeds and were able to assimilate 30% of the weed biomass. The rest was excreted which helped maintain soil fertility since nutrients, otherwise locked up in weeds, were released. Further experiments showed that rice-fish plots required less fertilizer than rice-only plots. On average the control plots used 23% more fertilizer than the rice-fish plots (Li *et al.*, 1995). In summary, the Chinese experiments indicate that less, not more, fertilizer was required in rice-fish farming.

Fertilizer applied in rice-fish farms by incorporating the nitrogen fertilizer thoroughly in the soil during land preparation results in higher rice yields than when broadcast on the surface. No difference has been found between applying the phosphorus fertilizer on the surface or incorporating it in the soil. However, surface application is believed to be better for promoting plankton growth in the water. Split applications of phosphorus may be better for sustained plankton production without hampering rice production as long as they are made before tillering. If applied at a later time, this should be on top of the normal requirements for rice. An application rate of 30-50 kg P_2O_5 per ha is often reported as optimum for algal growth.

Organic fertilizers benefit both rice and fish. In addition to nutrients, the particles can also act as substrates for the growth of epiphytic fish food organisms. Animal manure should be considered as an input to benefit the fish in addition to inorganic fertilizers applied primarily for the rice (Sevilleja *et al.*, 1992). Manure should be applied several weeks before transplanting and the fields kept flooded for complete decomposition and to avoid any toxic effects. Fertilization is a complex issue and varies greatly depending on the particular location. Providing general statements runs the risk of oversimplifying the issue, but there is evidence that nutrients are more efficiently utilized in rice-fish systems compared to rice-only systems, this effect being more enhanced, particularly on poorer and unfertilized soils where the effect of fish may be greatest (Halwart, 1998).

Table 6.3. Nutrient composition of different organic manures used for aquaculture.

Manure	**Nutrient composition (%)**			**Manure**	**Nutrient composition (%)**		
	N	P_2O_5	K_2O		N	P_2O_5	K_2O
Animal Origin				**Plant Origin**			
Cattle dung	0.5	0.4	0.2	Groundnut cake	6.5	1.0	1.0
Pig dung	0.6	0.6	0.4	Mustard cake	4.5	1.5	1.6
Poultry dropping	1.7	0.9	0.7	Mahua cake	2.5	0.8	1.9
Duck dropping	0.9	0.5	0.6	Neem cake	5.2	1.1	1.5
Farm Yard Manure	0.6	0.5	0.1				

Fish in rice fields can consume some pests and help with biological control, but they cannot replace pesticides, thus chemical control is required. However, chemical plant protection should be avoided to prevent fish/prawn mortality. In fact, in a waterlogged situation, insecticide hardly works as high-standing water dilutes the concentration. However, compounds with minimal toxicity, little residue, high efficacy, and a wide range of applications can be used in an emergency. Spraying should wait until after the light disappears and applying chemicals in powder form should be done in the early morning while there is still some dew present. The splashing approach is used nowadays, and it works well, especially when the rice gets tall. It is always economical and advisable to reduce the water level before the application of fertilizer and chemicals.

6.5.4. Fish culture

After proper refuge preparation, liming @ 500-750 kg ha^{-1}, manuring with raw cattle dung @ 5000-7000 kg ha^{-1} as basal dose should be carried out at the onset of monsoon during June. Before fingerlings are released in the rice field refuge, it is essential to clear it from aquatic vegetation and predatory fishes. The floating and emergent weeds may be removed manually instead of using chemical/weedicide. It is better to use MOC (mahua oil cake) @ 250 ppm at the onset of monsoon during June when rainwater starts accumulating in the perimeter canal. MOC not only helps in eradicating predatory/unwanted species such as catfish, *Channa punctatus, C. Orientalis, Glossogobius giuris, Puntius ticto, Esomus danricus, Ambasis* spp. and *Barilius* spp., but also acts as a manuring substance.

During the month of July-August, when the rainwater starts accumulating in the refuge and paddy field, early fingerlings of Catla, Rohu, Mrigal, Silver carp, Common carp and prawn juveniles may be stocked with a composition of 30:25:45 (surface feeder: column feeder: bottom feeder) for optimum utilization of different food niches (Fig. 6.6). *Labeo bata* can also form a stocking component in this system. As the culture duration is short, fingerlings (>15g size) should be stocked at a higher density of 15,000 – 20,000 per ha for continuous rearing for 4-5 months or advanced fry (<1.5g size) at a higher density of 75,000-1,00,000 per ha for advanced fingerling production, based on the principle of phased harvesting. An optional stocking density of fish/prawn species is critical in attaining high cumulative fish yields and in reaching the upper carrying capacity of the system. Ways to intensify fish production from an integrated rice-fish farming system, therefore, involve management strategies like high-density stocking (stocking with higher initial fish biomass) followed by phased harvesting, when the growth curve of stocked fish/prawn starts to slow down. This will help in reducing the population pressure and enhance the growth of the remaining stock. To augment growth, a supplementary feed comprising mustard oil cake/ground nut oil cake and rice bran in a 1:1 ratio may be given to fishes daily @ 3% in the initial two months and then a 2.5% rate of mean body weight of stocked fish/prawn. In this culture system, the fish/prawn will grow for a period of 3-4 months in the entire area and then 2-3 months in the confined area of the infield refuge/ perimeter canal. Under this system, production range between 1400-1800 kg ha^{-1} (if stocking size <1.5g) or 2800-3200 kg ha^{-1} (if stocking size >15g) per crop with a survival rate of about 65-90 per cent.

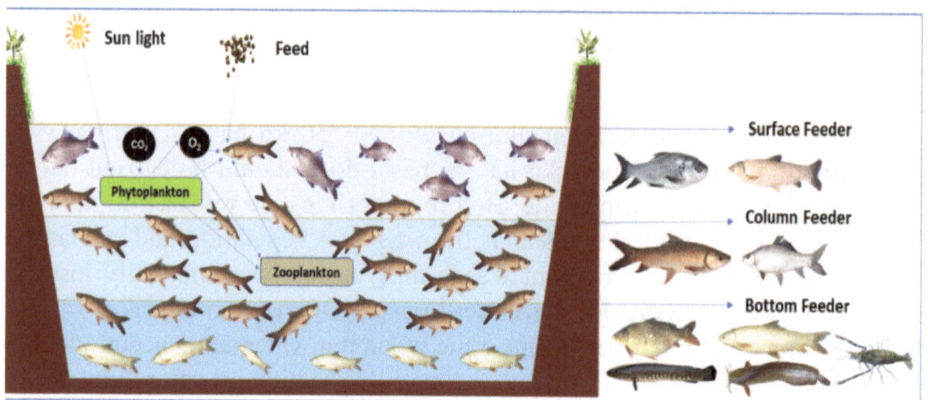

Fig. 6.6. Fish species composition in rice field refuge of rice-fish co-culture system

6.5.5. Prawn culture

Among various intrinsic and extrinsic factors associated with the culture of scampi, heterogeneous individual growth (HIG) happens to be a serious threat to the farmers resulting in heavy economic loss. *M. rosenbergii* is known to exhibit a complex social organizational hierarchy comprising morphologically distinct dominant, sub-dominant and subordinate groups (Mohanty, 2010). The predominance of a definite social hierarchy among the male morphotype increases the differential growth pattern within these prawns. The males show a high degree of HIG while the size distribution of the female population is rather homogeneous. Usually, three male morphotypes (SM-stunted male, OCM- orange-clawed male and BCM- blue-clawed male) represent three developmental stages of the male maturation process and are known to transform SM >> OC >> BC. SMs are subordinate, not territorial and in the initial stage of the developmental pathway. OCMs are sub-dominant and represent high somatic growth. BCMs represent the terminal stage in the morphotypic transformation pathway. Once a set of prawns reaches the terminal morphotype stage, it inhibits the transformation of other morphotypes to successive stages. This leads to a wide range of variations in growth patterns (Mohanty, 2010). In ponds/refuges, generally, 50% of the male population comprises the SM having a weight range of 5-20 g, while OCM (40%) and BCM (10%) are characterized by a wide weight range of 30-180 g and 180-250 g respectively. To overcome the relative proportion of SM, OCM and BCM morphotypes in the harvested population, the following suggestions may be considered.

- Adopt two-stage (nursery and grow-out) rearing management
- Stock post larvae (PL_{15-20}) in nursery for 50-60 days till they attain 4-5 g size
- Sex segregation should be done at the nursery site and only males should be stocked in the rice-fish system (all-male culture)
- Cull harvesting should be carried out from 2nd month onwards (BCMs should be targeted first followed by OCMs) in the grow-out phase.

6.5.6. Water management in rice-fish culture

Water quality is a dynamic property of a system affected by chemical, biological and physical factors, which ultimately influences the aquatic environment and production of rice-based farming systems. Water quality parameters of the most sensitive component of the farming system are required to be realized. Management of water quality is a daily monitoring process, which includes the behaviour of fish, the nature of water and inputs of diligence from farmers. Since scientific water quality management and maintenance of tolerance limit (Table 6.4) of hydrological parameters cannot be expected at farmers' end, the minimum adaptable techniques for water quality monitoring at farmers' end should be encouraged. This includes periodic liming @ 100-150 kg ha^{-1}, and phased manuring with raw cattle dung @ 2000-3000 kg ha^{-1} to maintain plankton bloom, which in turn regulates dissolved oxygen and pH of the ecosystem. As these two parameters are too critical for fish growth and survival, utmost care must be taken for bloom management. As the optimum ratio of phyto and zooplankton is 10:1, periodic estimation of plankton is essential. In case of plankton crash, re-inoculation should be carried out followed by fertilization with Urea + SSP (1:1) @ 3-4 ppm or systematic water replenishment in case of excessive bloom build-up. Further to maintain a clear aquatic environment, meal-to-meal feed management is essential in reducing the organic load in the system that affects water quality.

Table 6.4. Suggested water quality criteria for fish rearing

Chemical parameters	Tolerance ranges
Ammonia (NH_3)	<0.125 ppm (un-ionized form)
Calcium	10.0-160.00 ppm
Carbon dioxide	0.00 to 10 ppm
Chlorine	0.03 ppm
Hydrogen sulphide	0.002 ppm
Iron (total)	0.0 to 0.15 ppm
Nitrate (NO_3)	0.0 to 3.0 ppm

Nitrite (NO_2)	0.1 ppm in soft water, 0.2 ppm in hard water
Dissolved oxygen	5.0 -9.0 ppm
pH	7.5 to 8.5
Phosphorous	0.01 to 3.0 ppm
Total suspended solids	80.0 ppm or less
Total Alkalinity	90-170 ppm
Water colour	Green to brown
Temperature	26-32^0 C

6.5.7. Integrated pest management

Pest management includes many options falling into four major categories: mechanical, chemical, cultural and biological. The first is the most widely used and the one with the longest tradition, together with natural control is considered part of biological control. Weeding is perhaps the best example of this, but also includes cultural techniques such as water level control. Chemical pest management is relatively new and widespread, particularly popular for its perceived effectiveness and for the fact that it is not labour-intensive. Unfortunately, insecticide applications in rice have been proven to become a major problem because they destabilize the ecosystem and trigger pest resurgence thus creating an even more critical situation than without their use. Biological control of pests has a range of applications from favouring certain organisms that are predators of certain pests, to the use of disease-resistant rice varieties. Particularly when pesticide-related health impairments are included, natural control is the most profitable option for farmers. An integrated approach using various management options termed Integrated Pest Management or IPM is the preferred choice for plant protection in rice, and in fact, has been adopted as the national plant protection strategy by most rice-producing countries.

Integrated pest management encompasses all four management options outlined above and attempts to optimize their use. The use of chemicals is often cited as one of the major constraints in the popularization of rice-fish farming. Yet stocking fish in rice fields reduces pest infestation, and thus also reduces if not eliminates the need for the application of herbicides and insecticides, particularly molluscicides where snail predatory fish are cultivated. The practical and economic advantages of using fish instead of chemicals are often obvious. The effectiveness of fish as a bio-control agent depends on how well they are distributed within a rice field. If fish stay mostly in the pond refuge, then they cannot be effective in controlling rice pests. In such a situation, the use of pesticides as well as other control methods should be considered based on the potential costs and losses in terms of rice yield and fish harvest. The important characteristics to be considered in the selection of any pesticide to be applied in a rice-fish farm can be summed up as follows:

i) Relative safety to fish and effective against the target insect species
ii) Should not accumulate or persist in rice and should be metabolized into non-toxic compounds and excreted by fish
iii) Should either volatize, bio-degrade or chemically degrade shortly after its application

6.5.8. Weed management

There are several practical options for controlling weeds in rice fields: land preparation, water depth variation, mechanical weeding, herbicide use and stocking of herbivorous fish. Mechanical weeding is perhaps the most frequently used way of controlling weeds. It is, however, a very labour-intensive way of controlling weeds and as such often carries a high opportunity cost. Herbicides are used extensively but are not considered a serious problem in rice-fish farming. If a herbicide is applied, it is normally done immediately after transplanting. Fish are stocked 10 to 14 days after application. Further, it is also possible to select a herbicide which can be tolerated by fish even at relatively high levels.

Introducing fish to the rice field can reduce the amount of weeds in several ways. To the herbivorous species of fish, weeds are part of their diet. Weeds just happen to be in the way to bottom-feeding species. In the process of looking for food, the muddy bottom of a rice field is tilled giving little chance for the submerged weeds to anchor their roots in the soil thus affecting their growth and proliferation. In China, fish are more effective in weed control than either manual weeding or the use of herbicides. The introduction of fish reduced the amount of weeds in one rice field from 101 kg to only 20 kg after five weeks, while in an adjacent rice field with no stocked fish, the weed biomass increased from 44 to 273 kg during the same period (Wu, 1995).

6.5.9. On-dyke horticulture

The dykes to be constructed for preventing the escape of fish from the integrated system may be used for growing vegetables and other fruit trees like papaya and banana to make the system more economically viable. Vegetables such as gourd, radish, brinjal, and leafy vegetables during pre and *Kharif* season and vegetables such as tomato, French beans, radish, bitter gourd, cucumber, cauliflower, cabbage, brinjal, pumpkin and leafy vegetables (coriander, *Amaranthus* and Indian spinach) can be grown during winter. Vegetables such as snake gourd, bitter gourd, ridge gourd, bottle gourd or ash gourd can be grown throughout the year on raised platforms.

6.5.10. Advantages of rice-fish integrated system

i) Fish/prawn culture in paddy fields can turn material and energy into fish/ prawn production, accelerate the growth of rice and increase the solar energy fixation, thus raising the productivity of paddy fields.

ii) Fish acts as a bio-controller of insects, thus helps in reducing the input cost of insecticide.

iii) Weeds compete with rice, as they need CO_2, H_2O and other nutrients for photosynthesis. Fish helps in the biological control of aquatic weeds, thus minimizing the input cost of herbicide

iv) Fish in a paddy field transforms insoluble N in the soil into a soluble state, which rises soil fertility. Cultivation of fish in rice fields increases the amount of N in soil and the amount of N absorbed by rice plants, thus more N is transported to the rice grains, improving the quality of rice.

v) Fish disturb the soil, which improves the porosity and structure of the topsoil. This raises the soil's oxidation and reduction potential and dissolved oxygen content during the rice-growing season.

vi) Rice plants protect fish/prawns from predation by birds.

vii) Fish get sufficient oxygen released by rice and phytoplankton for survival and growth.

viii) Fish diseases are rare in rice-fish culture, due to a clear aquatic environment, high oxygen content and rich natural food that produces strong and disease-resistant fish/prawns.

ix) After harvesting of rice, the roots and remaining parts of straw (straw contains 9-13% cellulite, 1.5-3% potassium and 30-40% cellulose) provide organic matter and favours the growth of microorganisms, the ultimate natural food of fish/ prawn.

x) The voiding of poultry/duck/animals could be recycled as fish feed and this could also increase the biological productivity of water. The addition of animal/bird droppings in water reduces the fish feed cost by 40% and fertilizer cost by 60-65%.

6.6. Impact of rice-fish/ animal co-culture

The impact of rice-fish/ animal co-culture can be measured in many ways, but this section exclusively focuses on the direct economic impact followed by its impact on household nutrition, public health, the environment, biodiversity, water resources and sustainability.

6.6.1. Ecological impacts and social efficiency of rice-fish/ animal co-culture

Rice-fish culture makes various uses of rice fields to enhance land and water resource utilization, as well as increasing rice field production value. Profits can be significantly higher if the fry are raised on rice fields. The fish have a high value, which increases economic efficiency. In rice-fish farming, hazardous insects and pests are significantly decreased. As a result, pesticide application can be reduced or discontinued, minimizing toxicity accumulation. This benefits both human health and the ecological balance of the ecosystem. The introduction of fish into the rice field ecosystem causes changes in the quantity and composition of aquatic creatures, as well as their connections. Rice-fish culture has the potential to change the direction of energy flow in the ecosystem. In the rice field, stocked fish convert stationary energy (e.g., weeds) and possibly wasted energy (e.g., phytoplankton, zooplankton, and aquatic insects) into useful products (fish and rice). Figure 6.7 depicts the ecological components and material flow channels in both the traditional and integrated rice-fish co-culture systems.

Rice-fish culture coordinates the interactions between the biotic and abiotic environments, reducing infectious illnesses in animals and humans. Mosquito larvae, maggots, snails, and leeches, which serve as intermediate hosts for malaria, encephalitis, dysentery, blood fluke, and filaria, multiply quickly in rice fields. Fish, particularly common carp and other omnivorous fish, devour and remove these harmful parasites, reducing human infection rates and improving farmers' living conditions and health. Farmers must weed rice fields that do not include fish on a regular and labor-intensive basis. As a result, soil fertility and solar energy are significantly reduced, and production costs rise. Farmers in some areas do not plough the fields when rice-fish culture is practiced. This further minimizes the inputs required for rice planting, lowering production costs and increasing the economic efficiency of rice farming.

Within the framework of rural development, rice-fish culture has great potential in India and Southeast Asia. Rice-fish culture is not only considered a profitable enterprise but also an ideal method of water and land use patterns since both are utilized for the production of plants and animals. Viewed from a broader perspective, the system allows farmers to diversify their production and enables them to fully utilize their farm labour, especially during periods of low labour demand and off-season. Thus, the system results in the maximum utilization of farm resources. In addition, from an environmental point of view, rice-fish culture can play a large role in the bio-management of natural resources in the rice agro-ecosystem. Rice-fish culture requires small cash outlays and could provide significant increases in rural employment and family incomes. Rice-fish systems could also improve the nutritional status of rural people directly by additional consumption of fish, or indirectly by improving incomes. Therefore, rice-fish farming should be given sufficient attention in its role in fostering the development of a vibrant rural aquaculture production network.

When there is a large-scale adoption of rice-fish integration involving an entire community, the social impact can be quite profound. The use of fallow rice lands for fish culture by landless farmers in Indonesia is one such case. The situation prevailing in Indonesia in the past was that landless tenants were allowed to use the rice fields for fish culture during the fallow season (*Palawija* system). Nowadays, the use of rice fields for fish production during the fallow season is not limited to landless tenants but involves fish breeders requiring a larger area for raising fingerlings. Although the Indonesian example may be unique, in general adoption of rice-fish farming should result in job creation. Physical modifications of rice fields to accommodate and harvest fish require extra labour. In the Philippines, ancillary activities connected to tilapia fingerlings production in rice fields are: diking and excavation, making hapa-nets, harvesting seines and other fish culture accessories, renting out water pumps, harvesting nets, oxygen tanks, repair of pumps, harvesting, sorting and packing of fingerlings and transport of fingerlings. Each type of activity is done by a different person, thus creating job opportunities.

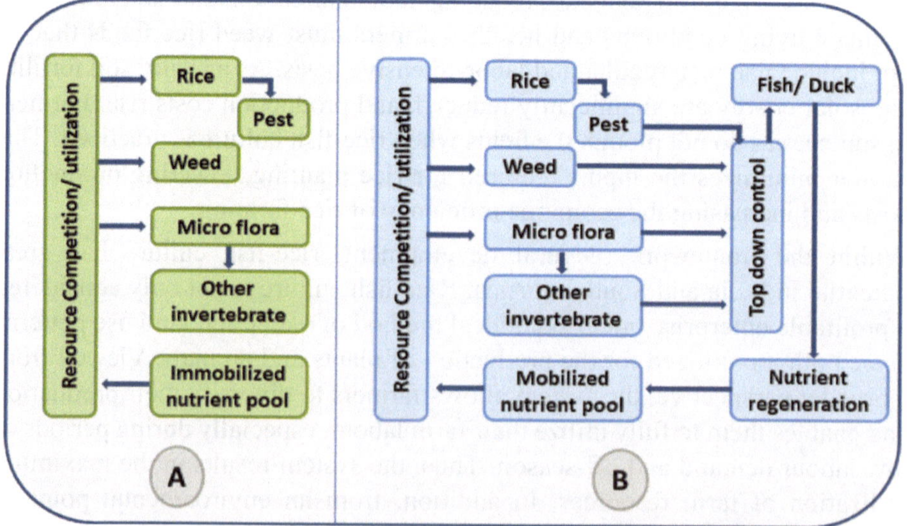

Fig. 6.7. Schematic representation of ecological components and pathways of material flow in conventional **(A)**, and integrated rice-fish co-culture system **(B)**

Rice-fish co-culture systems have the potential to provide variety of benefits to farmers, including agricultural (i.e., enhancing food security), ecological, social and economic. This system is a helpful approach for improving field water ecology by changing the temperature and oxygen concentrations. Similarly, animal excreta are a source of organic fertilizer that increases C content in field systems, resulting in a higher C:N ratio (Paudel *et al.*, 2015; Desta *et al.*, 2014; Ahmed, 2013; Halwart and Gupta, 2004; Edwards *et al.*, 1988). This co-culture systemprovides ecological benefits in two ways: i) biodiversity, and ii) resource and nutrient usage.

i) *Biodiversity*: According to reports, rice-fish co-culture systems effectively preserve biodiversity, as seen by the variety of aquatic animals and plants that may be found in paddy fields (Halwart, 2008). According to research by Dey *et al.* (2005), there was no discernible impact of rice field farming on fish populations or the biodiversity of aquatic creatures in co-culture systems. Furthermore, Bunting *et al.* (2015) asserted that the implementation of rice-fish systems, which also produced resilient and productive farming, might boost the biodiversity of aquatic organisms (Nhan *et al.*, 2007).

ii) *Resource and nutrient utilization:* A rice-fish co-culture is an integrated system in which many organism groups interact with one another in intricate ways, either directly or indirectly. Co-culture systems offer significant chances for plants and animals to share water resources, significantly improving the efficiency with which water resources are used. In co-culture methods, rice plants provide shade from the sun, especially during the sweltering summer months. They also aid in bringing down the field's standing water temperature, which is better for aquatic animal life. The movement of fish/ prawn loosens the soil, thus increases the accessibility of plant nutrients, which are partially fixed with clay particles. Soil loosening improves oxygen concentration and soil permeability, as well as shifts the breakdown patterns of fish feed and other organic compounds from anaerobic to aerobic. Aquatic creatures make excellent predators as well. This technique aids in the control of insect and pest attacks on rice crops while also reducing pesticide consumption (Berg and Tam, 2018; Ahmed *et al.*, 2014; Wahab *et al.*, 2008; Nhan *et al.*, 2007). Rice-animal co-culture offers a valuable possibility for the efficient and long-term usage of nutrients. In a co-culture system, excess nitrogen given to rice can be utilized by aquatic creatures (fish, prawn, crab, carp, shrimp, etc.) for their growth (Wongkiew *et al.*, 2018; Sha *et al.*, 2017); otherwise, this amount of

N may go as run-off during the rainy days (Yue *et al.*, 2014). In turn, the animals' excreta provide nutrients to rice plants. Co-culture creates a vibrant, fast-tracked nutrition cycle. Furthermore, fish use surplus nitrogen from rice fields to produce proteins, which are then sold in the nearby market to improve a rural community's socioeconomic condition (Costa-Pierce, 2002; Hu *et al.*, 2013; Das *et al.*, 2014). In Bangladesh, prawn/ shrimp farming in rice field is seen as a technological solution for the integrated management of existing land, water, and aquatic species resources and a critical component of the green economy (Ahmed and Garnett, 2011; Hu *et al.*, 2016).

6.6.2. Economic impact

Rice-fish co-culture is an essential technique to enhancing farmers' socioeconomic standing by making efficient use of existing resources and gaining fish/ prawn as surplus income producers alongside rice crops. Several studies were carried out in Bangladesh to determine the viability of an integrated rice-fish system and its economic significances. These experimental results suggest that the co-culture method provided more net benefits than rice monoculture, with reduced production costs and higher fish and rice yields (Dey *et al.*, 2005). In the rainy season, net farm revenue in the co-culture practice was 64.4% higher than in rice monoculture, while in the dry season, it was 98.2% higher. Furthermore, the price rate of rice grown in a co-culture system was higher than that of a traditional rice monoculture system, because cultured fish have a inclination to eat aquatic weeds and insects available in the rice fields, reducing herbicide and pesticide requirement that favors food safety (Feng *et al.*, 2016). The adoption of the rice-fish/animal co-culture system lowered pesticide and fertilizer expenditures by 50.0% and 79.5, respectively (Cao *et al.*, 2017), while increasing the NEEB (net ecosystem economic budget) by 26.9-75.6% when compared to monoculture rice (Sun *et al.*, 2019). The average revenue from a rice-fish farming system was found to be 50% greater than that from a rice-monoculture system, probably due to cheaper production expenses with dual income sources like fish and rice (Gupta *et al.*, 1998; Ahmed *et al.*, 2014). Co-culture systems not only reduce GHG emissions and nitrogen losses, but they also increase farm income and improve the livelihoods of rural poor / progressive farmers.

According to the published report, the net return from rice fish was over 50% greater than that from rice monoculture in Bangladesh. The higher net returns were probably due to the lower mean costs of rice cultivation and higher rice yields in addition to the fish yield from integrated farms. In China, the increase varied from 45 to 270%. Growing fish was almost three times more profitable than rice alone. Rice yields in rice-fish culture were 8-26% higher, labour input 19-22% lower, and material costs were 7% lower (savings in the cost of controlling diseases and pests). Additionally, fish production increased the net income. Indonesian figures show that having two crops of rice-fish and using the rice field for a short intermediate crop or *Penyelang* of fish has a 116% higher return than having two crops of rice and leaving the rice field fallow for two months or so. It is concluded that the rice-fish system is a profitable technology and that its adoption is likely to increase farm household income, labour absorption, and better liquidity. In the Philippines, rice-fish farms yielded a 27% higher net return with fish compared to a single crop of rice (Sevilleja *et al.*,1992). In addition, it has been demonstrated that it is possible to achieve a three-fold increase in the profitability of rice farming by culturing fish as well as rice. However, Thailand, in contrast to the previously mentioned countries, showed lower net returns in the rice-fish fields than in the rice-only fields. The Thai figures indicate that profitability in the rice-fish fields was only 80% of that of rice monoculture due to the high initial investment in rice-fish culture. In Vietnam, the net returns from the rice fields with unfed shrimps were 52% higher than that of rice monoculture and 176% higher in the rice fields where shrimps were fed with rice bran and decomposing animals.

Except for Indonesia, all the other countries consistently showed an increase in the overall labour requirement when fish are raised in the rice field, with the amount of increase varying from only 10% to as high as 234%. This was mainly due to the need to prepare the rice field for fish stocking as well as for fish harvesting. However, in some specific activities connected with the rice crop such as fertilizing, weeding and pesticide applications, the presence of fish lessened the labour required. Again, the amount varies from activity to activity and from one area to another. In terms of fertilizer expense Bangladesh, Indonesia and the Philippines showed from 4% to 14% lower fertilizer costs in rice-fish fields, while the costs of chemical pesticides in rice-fish farms were significantly lower (44-86%).

The immediate beneficiaries of the production of fish and rice yield in rice-fish farming are the farmers who adopt the technology. Although it seems obvious, the existence of such a relationship has not been demonstrated unequivocally. However, the fact that many farmers in different countries continue to practice it year after year, even without any government program, would seem to be proof enough of the benefits derived from this type of rice farming. Models developed using linear programming techniques on a 2.3 ha farm in the Philippines, infers that the rice-fish farming approach can generate an additional 23% more farm income by raising fish as well in 0.5 ha. This increases to 91% if the entire 2.3 ha area is stocked with fish, even if rice production remains constant and farm requirements for cash and labour increase by 22 and 17% respectively (Ahmed *et al.*, 1992).

One indication that fish farming in paddy fields must be satisfactory (economically or otherwise) from the farmers' perspective is that in many cases farmers their continue or even expand the extent of their rice-fish farms after having tried the technology. For example, Zambian farmers wanted to continue with rife-fish farming although researchers had found it to be uneconomical. In Northeast Thailand, the total rice field area stocked with fish increased each year from 1985 to 1987 despite a dismal showing in the first year. It has been pointed out that nutritional benefits and lowered risk of production may provide strong motivation for rice farmers to diversify and that rice-fish farming can be "profitable" in many ways including from social, environmental, or ecological points of view.

6.6.3. Environmental impact of rice-fish co-culture

The animals in rice-fish co-culture systems need favorable/satisfactory environmental settings to grow faster resulting in higher productivity. Overall, the integrated co-culture system requires a specific range of water depth, temperature, pH, DO, CO_2 and transparency. The first key aspect for good co-culture system management is water depth, which is essential for both aquatic animals and rice (Mohanty *et al.*, 2009). Other key environmental parameters in rice habitats include nutrient status (N-P-K), water pH, unionized NH_3, light penetration, DO, CO_2, water temperature etc. (Roger, 1996). DO concentrations are heavily influenced by the photosynthetic actions of rice, macrophytes and algae as well as the respiration of cultured organisms. High quantities of NH_3 and CO_2 are dangerous to aquatic animals. In rice field ecosystems, CO_2 is mostly produced by organism respiration, organic matter decomposition and agitation, whereas NH_3 is primarily produced by N fertilizer applications. Notably, co-culture practice can change environmental settings by adjusting substance concentrations (DO and CO_2 levels) and energy fluxes. The available increased concentration of C serves as a substrate for bacteria capable of controlling GHG and N loss. Along with the benefits, co-culture systems raise some difficulties, such as environmental concerns over nitrogen fertilizer in the co-culture system.

6.6.4. Impact of rice-fish farming on water quality

Rice cultivation practices are recognized as a significant source of pollutants in downstream water bodies, owing to the direct use of pesticides and nutrients to standing water in the fields. As a result, improving paddy field water quality has become a critical environmental goal for sustainable / viable rice cultivation. Crop absorption and environmental losses have the greatest influence on nutrient utilization efficiency (particularly nitrogen). In rice mono-culture, part of the applied nitrogen is absorbed by plants as a nutrient, while the remainder is released into the environment (Smith and Siciliano, 2015). Rice-animal co-culture, on the other hand, increases N use efficiency due to the presence of additional user/ consumer such as fish, shrimp, crab, carp, etc. Because of the additional consumers in the co-culture system, less nitrogen is lost to the environment. Furthermore, this co-culture can significantly improve the fertility status of soil and long-term crop productivity by increasing oxidizable organic carbon, total P and N, and total potassium (Yan *et al.*, 2014; Cao *et al.*, 2017). It has also been observed that rice-fish culture can cut inputs of chemical fertilizer by 24% compared to rice mono-culture while producing the same quantity of rice yield (Xie *et al.*, 2011a, 2011b). When compared to fish monoculture, rice-fish co-culture reduced nutrient concentrations in water samples by 70–79% for different N (total N, NH_3-N, NO_3-N, and NO_2-N) and total P and dissolved organic P (Feng *et al.*, 2016). Additionally, because animals naturally prey on weeds, co-culture systems can minimize pesticide application by as much as 68% (Xie *et al.*, 2011a, 2011b). According to studies, co-culture lowers pesticide applications to rice fields by 65%, while pesticide applications increase by 40% in paddies without co-culture (Haroon and Pittman, 1997; Berg, 2001; Su, 2006). To sum up, co-cultivating rice with animals can have a beneficial effect on delaying the eutrophication of surrounding watercourses and environmental contamination.

6.6.5. Impact on water resources

Freshwater is a limited resource and the integration of fish with rice is one way of using water more efficiently by producing both aquatic animals and rice. With fish in the rice field, a greater water depth has to be maintained and more water may be required, an issue raised half a century ago. Even without fish, rice farming consumes large volumes of water. For rice culture in general, the crop needs a minimum of 1 000 to 1 800 mm of water. If one ha of rice field produces 10 million t of rice, it still takes from 1-1.5 m^3 of water to produce 1 kg of paddy. Fish, are a non-consumptive user of water and while they can degrade the water, they do not use it up. If cleaned, the same water can be returned and reused by the fish. The increased water use is due to percolation (P) seepage (S) and leakage (L), which increase with rice-fish culture due to the deeper water maintained, a purely physical process that takes place with or without the fish. It is estimated that the average water requirement for rice culture is 1662 mm while rice-fish culture requires up to 2 100 mm, or 26% more than rice monoculture. The main water losses are attributable to P&S (67%), followed by L (21%). Thorough puddling during land preparation, good maintenance of the dikes and proper sealing of inlets and outlets may reduce water losses.

6.6.6. Impact of rice-fish co-culture on GHG emissions

Greenhouse gases (CHG) that absorb heat from sunlight, reduce the backflow of hazardous radiations from the earth's surface into space, and significantly contribute towards global warming. Rice systems are significant producers of GHG emissions in the agriculture sector (Pimentel *et al.*, 2006; Liu *et al.*, 2014). The primary greenhouse gas emissions from agricultural sectors include CH_4, N_2O, and CO_2 (Zhang *et al.*, 2018). Because of their movement, which mixes water at different levels and raises the DO level in the standing water, aquatic animals in rice-fish/animal co-culture systems have the potential to reduce greenhouse gas emissions (Wu *et al.*, 2018; Ghosh and Mazumdar, 2016; Liu *et al.*, 2016; Mohanty *et al.*, 2009; Cai *et al.*, 2002; Mustow, 2002).

Methane production: Anthropogenic happenings in paddy fields account for approximately 20% of worldwide CH_4 production (Weller *et al.* 2016; Sebastian *et al.*, 2015; IPCC, 2007; Lightfoot *et al.* 1993). Methane emissions are caused by the anaerobic decomposition of organic compounds such as organic debris, plant wastes, and organic fertilizers in submerged settings with low oxygen availability. Bottom feeder aquatic animals, particularly (crabs, prawn and carp), disturb soil layers by their movement or, in some cases, foraging, influencing CH_4 generation processes. Aquatic animals have the potential to improve DO in field water and soil, shifting anaerobic to aerobic digestion and so reducing CH_4 emissions (Ding *et al.*, 2013; Zhang *et al.*, 2017). A rice-crayfish co-culture system lowered CH_4 emissions by 18.1-19.6% and global warming potential by 16.8-22.0% when compared to monoculture (Sun *et al.*, 2019). The accumulation of organic substances in aquatic animals' feces may increase C levels, which serve as food for methanogenic organisms (Paudel *et al.*, 2015; Ahmed *et al.*, 2014). Co-cultivating rice with animals may also indirectly lower CH_4 emissions. The DO in the water tends to rise as co-culture reduces eutrophication and enhances water quality. According to Feng *et al.* (2016), this may improve aerobic conditions and reduce the generation of CH_4.

Nitrous oxide emission: Nitrous oxide (N_2O), having 298 times more global warming potential than CO_2, is one of the primary N loss channels to the environment (World Meteorological Organization, 2010; Butterbach-Bahl *et al.*, 2013). The rice field is a source of anaerobic digestion under submerged conditions, which influences microbial nitrification and denitrification, resulting in N_2O emissions (Wu *et al.*, 2018). According to this framework, water drainage, microbiological activity, and the degree of N fertilization all contribute to gaseous N emissions, which in paddy systems can reach 70%

(Braker and Conrad, 2011; Syakila and Kroeze, 2011). Rice-animal co-culture, as opposed to traditional rice monoculture, would be a significant factor in balancing greenhouse gas reduction with increased agricultural profitability and more efficient use of water and land. In co-culture systems, fish, crayfish, crabs and ducks significantly lower N_2O emissions, lessen the amount of nitrogen lost by NH_3 volatilization and nitrate leaching, increase the nitrogen use efficiency and reduce the quantity of nitrogen fertilizer requirement (Wang *et al.*, 2019; Cheng-Fang *et al.*, 2008).

6.6.7. Impact on nutrition

One benefit that is often assumed, but never supported by solid evidence is that farmers who culture fish in their rice fields have improved nutrition. In the case of rice-fish farming, there are no figures available as to how much the caloric and protein intake or the per capita fish consumption of farmer families has been increased by the availability of fish once these are grown in their own rice fields. Income augmentation was the most frequent reason provided for engaging in rice-fish, additional food only ranked next. In Bangladesh, it was pointed out that extra income was the most appreciated benefit from growing fish (70%) followed by increased food for the family (59%).

Improvements in a farming household's nutrition as a result of culturing fish in the rice fields may just be an incidental and perhaps even indirect effect, such as being able to buy meat or chicken as a result of the extra cash earned from fish. The main benefit of rice-fish farming is often seen as providing an opportunity to earn cash. Improvement in the local community's nutrition has been cited as one of the benefits of rice-fish farming. With greater availability of fish, the local population of a rice farming community will have easy access to fish at affordable prices. However, in a free market, the farmer may opt to sell the fish to a trader at a higher price than what the neighbours can afford. The trader in turn may opt to bring the fish to the nearest urban center where prices are higher. This is a common situation in most fishing communities where fish can be difficult to find in the local market having been siphoned off to the cities.

Nevertheless, particularly in more remote areas and where the mixed forms of capture and culture are prevalent, it is estimated that fish and other aquatic organisms from rice fields provide a very important component of the daily diet. The nutritional contribution extends from micronutrients and proteins to essential fatty acids that are needed for visual and brain development. Recognizing this, the 20th Session of the International Rice Commission recommended its member countries pay increased attention to the nutritional value of fish and other aquatic organisms from rice fields (Halwart, 2003).

A recent FAO/IUCN study in Lao PDR confirms the urgent need for further focus on this issue.

6.6.8. Impact on public health

Mosquitoes are known carriers of malaria and dengue fever. Certain species of freshwater snails serve as hosts to trematodes (*Schistosoma* spp.) that cause schistosomiasis when it enters the human bloodstream. Fish in rice fields very well controls these two public health vectors. A third aspect is that rice-fish culture may reduce the use of agricultural chemicals that pose a health hazard to humans. Field surveys in China indicate that mosquito larvae densities in rice fields with fish were only 12,000 per ha as against 36,000 per ha in rice fields without fish. In Indonesia, fish were found to be even more effective in controlling mosquitoes than DDT. After five years of fish culture in rice fields, malaria cases decreased from 16.5 to 0.2% in a highly endemic area for malaria. In countries such as China, black carp (*Mylopharyngodon piceus*) is used to control snails that are intermediate hosts in parasite transmission. In Katanga, the majority of snails in rice fields were controlled by *Haplochromis mellandi* and *Tilapia melanopleura* stocked at 200 fish per ha and 300 fish per ha[1] respectively. Hence it is concluded that well-maintained aquaculture operations contribute, often significantly, to the control of insects and snails of agricultural and medical importance, and that integrated management programs should be pursued to keep vectors and pests at levels where they do not cause significant problems. Fishes are potentially a good biological herbicide and insecticide and stocking can greatly reduce, if not eliminate, the need for using chemical pesticides. The presence of fish discourages farmers from applying pesticides. The reduction or elimination of the need to apply chemicals cannot but result in an environment that is safer and healthier for the people.

6.6.9. Impact on sustainability

Wet rice cultivation has been practiced for at least 4000 years and its long history indicates that traditional rice farming is sustainable. What is less certain is whether the dramatic increases of rice production made possibly by the 'green revolution' are sustainable. Global warming, sea level rise, increased ultraviolet radiation and even availability of water are all expected to hurt rice production. However, such scenarios are far beyond the level and scope of this report, and for the foreseeable future it can be assumed that rice farming will continue. Further, it seems likely that the culture of fish in rice fields can enhance the sustainability of rice farming since indications are that the presence of fish makes the rice field ecosystem more balanced and stable. With fish removing the weeds and reducing the insect pest population to tolerable levels, the poisoning of the water and soil may be curtailed.

6.7. Constraints, challenges and future thrusts

The ultimate goal of any integrated farming system is to improve overall productivity and maximize the net return. In this context, several studies have been carried out in India (Table 6.5) and abroad. The lowest return is always obtained from conventional rice farming which goes on increasing as more and more components are integrated. To make this system more remunerative efforts are being made to integrate other plant and animal components with topographic modification of the existing land. Natural resource management in tropical wetland rice fields through integrated production systems thus seems to be a promising approach in introducing sustainability in rice farming and addressing ecological issues about the conservation of rice field ecology.

In addition to an increase in net return, the rice-based integrated farming system also has the potential to generate employment to engage family labour and check the migration of rural youth to urban areas. It is a fact that rice-fish-poultry/ duckery/ vegetable systems in a ha area can generate employment for a person throughout the year. The addition of more profitable enterprises such as apiary, mushroom, dairy, agro-forestry etc. can enhance employment opportunities about three times.

Table 6.5. Recent trends in water productivity and net return of different rice-based integrated production system.

Production system	**GWP** (Rs m^{-3})	**NWP** (Rs m^{-3})	**NR** (Rs ha^{-1})	**Location and year**
* *Kharif* rice (GMP)	2.76	1.06	15,294	Odisha, WTCER, 2004
* Rice+fish (rainfed medium land)	2.94	2.7	17,460	Odisha, WTCER, 2004
* Rice + fish	--	--	17,847	Annamalai Univ., 2005
* Rice + fish + ODH+ *rabi* crop	5.87	3.76	29,617	Odisha, WTCER, 2004
* Rice + fish + rabbit	--	--	43,932	Annamalai Univ., 2005
* Rice + fish + Vegetable in raised-sunken bed system	4.58	3.23	74,010	Odisha, WTCER, 2005
* Rice+fish+prawn+ODH followed by post-harvest 2nd crop (black gram) in waterlogged ecosystem	11.52	6.20	99,706	Odisha, WTCER, 2005

* Rice +fish + duck	--	--	98,936	Philippines, 2000
* Rice + fish + prawn + Vegetable	14.02	7.52	1.46 lakh	Odisha, WTCER, 2003
* Rice +fish + poultry	--	--	2.28 lakh	Annamalai Univ., 2005

GWP: gross water productivity, NWP: net water productivity, NR: net return, ODH: on-dyke horticulture, GMP: good management practice

The constraints that hamper the development of rice-fish culture are technical, socio-economic and management-related. Many farmers also lack motivation or find it difficult to change work and social habits to grow fish with the more important rice crop. Marketing is also a problem as freshwater fish must be sold live or fresh to command a high price. Further, the success of rice-fish culture is also affected by the lack of reliable sources of quality fish seed; reasonably-priced supplemental feeds; control of predators and pests; lack of sufficient water during the post-monsoon seasons and lack of technology on how to best store fish in the rice field when fish prices are low. Moreover, there is an urgent need to formulate ecological guidelines for fish cultivation in rice fields. This lack of guidelines is a major constraint to the development of rice-fish culture systems.

Rice-fish co-culture outperforms rice or fish monoculture in terms of biodiversity, productivity, resource usage, food supply, and production efficiency. However, the farming community does not employ this system as broadly as it should due of a lack of technical expertise, increased flood and drought risks, and the need for more comprehensive management (Sha *et al.*, 2017). Community-based techniques could help to control the hazards associated with co-culture systems; by using these technologies, the potential paybacks of this system can be fully realized (Brooks and Sieu, 2016). A rice-animal co-culture system is not widely recognized among farming systems, and it still faces many challenges, such as a lack of technical understanding and the requirement for initial financial support to cover the costs of paddy seeds and aquatic animals. To use this system, greater management is essential to boost profitability and eliminate limitations. Small farm sizes and deficient land tenure regimes in many nations exacerbate these limits, which are strongly tied to technical understandings of water management, labor requirements, plot preparation, and the market for aqua-products (Dey *et al.*, 2013). Instead of focusing on creating policies that are supportive of the co-culture system, the majority of nations continue to promote rice monoculture in their regions. A number of significant obstacles prevent the co-culture system from adapting, including inadequate knowledge, a lack of technical know-how, undecided field management practices (Dey *et al.*, 2013), a lack of extension services, the absence of specialized sales outlets for acquired aquatic animals close to the production area, and price maintenance.

References

Ahmed, M., Bimbao, M.A.P. and Sevilleja, R.C. 1992. The economics of rice-fish in Asian mixed farming system- a case study of the Philippines, p. 207-216. *In* Dela Cruz, C.R., Lightfoot, C., Costa-Pierce, B.A., Carangal, V.R. and Bimbao, M.P. (eds.) *Rice-fish research and development in Asia.* ICLARM Conf. Proc. 24, 457 p.

Ahmed, N. 2013. Linking prawn and shrimp farming towards a green economy in Bangladesh: confronting climate change. *Ocean Coast Management,* 75:33-42.

Ahmed, N., Bunting, S.W., Rahman, S. and Garforth, C.J. 2014. Community-based climate change adaptation strategies for integrated prawn-fish-rice farming in Bangladesh to promote social-ecological resilience. *Reviews in Aquaculture*, 6:20-35.

Ahmed, N. and Garnett, S.T. 2011. Integrated rice-fish farming in Bangladesh: meeting the challenges of food security. *Food Security*, 3:81-92.

Akinbile, C.O., El-Latif, K.A., Abdullah, R. and Yusoff, M.S. 2011. Rice production and water use efficiency for self-sufficiency in Malaysia: a review. *Trends in Applied Sciences Research,* 6:1127-1140.

Ali, A.B. 1990. Rice-fish farming in Malaysia: a resource optimization. *Ambio,* 19(8):404-408.

Anita, B., Manjeet, K. and Yadav, A.S. 2014. Fish culture in intensively cultivated rice fields: growth performance of *Cyprinus carpio. International Journal of Agricultural Innovation and Research*, 3:80-85.

Berg, H. 2001. Pesticide use in rice and rice-fish farms in the Mekong Delta, Vietnam. *Crop Protection*, 20:897-905.

Berg, H., Ekman Soderholm, A., Soderstrom, A.S. and Tam, N.T. 2017. Recognizing wetland ecosystem services for sustainable rice farming in the Mekong Delta, Vietnam. *Sustainability Science,* 12:137-154.

Berg, H. and Tam, N.T. 2018. Decreased use of pesticides for increased yields of rice and fish-options for sustainable food production in the Mekong Delta. *Science of the Total Environment*, 619:319-327.

Brahmanand, P.S., Mohanty, R.K. and Kumar, A. 2006. Integrated Rice-Fish Farming. Water Technology Centre for Eastern Region (ICAR), Bhubaneswar, Orissa, 78p.

Braker, G. and Conrad, R. 2011. Diversity, structure, and size of N_2O-producing microbial communities in soils-what matters for their functioning? *Advances in Applied Microbiology*, 75:33-70.

BRKB. 2010. Rice Statistics in Bangladesh. Bangladesh Rice Knowledge Bank (BRKB). Bangladesh Rice Research Institute, Gazipur, Bangladesh.

Brooks, A. and Sieu, C. 2016. The Potential of Community Fish Refuges (CFRs) in Rice Field Agro-Ecosystems for Improving Food and Nutrition Security in the Tonle Sap Region. WorldFish, Penang, Malaysia. Program Report: 2016-10.

Bunting, S.W., Mishra, R., Smith, K.G. and Ray, D. 2015. Evaluating sustainable intensification and diversification options for agriculture-based livelihoods within an aquatic biodiversity conservation context in Buxa, West Bengal, India. *International Journal of Sustainability*, 13:275-293.

Bush, S.R., Little, D.C., Bush, S.R., Belton, B., Hall, D., Vandergeest, P., Murray, F.J., Ponte, S., Oosterveer, P., Islam, M.S., Mol, A.P.J., Hatanaka, M., Kruijssen, F., Ha, T.T.T.,

Little, D.C. and Kusumawati, R. 2013. Certify sustainable Aquaculture? *Science*, 341(80):1067-1068.

Butterbach-Bahl, K., Baggs, E.M., Dannenmann, M., Kiese, R. and Zechmeister-Boltenstern, S. 2013. Nitrous oxide emissions from soils: how well do we understand the processes and their controls? *Philosophical Transaction of Royal Society of Biological Sciences*, 368:20130122.

Cai, G.X., Chen, D.L., Ding, H., Pacholski, A., Fan, X.H. and Zhu, Z.L. 2002. Nitrogen losses from fertilizers applied to maize, wheat and rice in the North China Plain. In: Nutrient Cycling in Agroecosystems. Kluwer Academic Publishers, pp. 187-195.

Cai, R., Ni, D. and Wang, J. 1995. Rice-fish culture in China: the past, present and future, p.3-14. *In* MacKay, K.T. (ed.) *Rice-fish culture in China*. International Development Research Centre (IDRC) Ottawa, Canada, 276 p.

Cao, C.G., Jiang, Y., Wang, J.P., Yuan, P.L. and Chen, S.W. 2017. Dual character of rice-ray fish culture and strategy for its sustainable development. *Chin. J. Eco-Agric.*, 25:1245-1253.

Chakravartty, D., Mondal, A. and Raychowdhury, P. 2017. Role of aquaponics in the sustenance of coastal India e aquaponics is a solution for modern agriculture in ecologically sensitive Indian mangrove Sundarbans: a review. *Int. J. Fish. Aquat. Stud.*, 5:441-448.

Chen, D., Ying, H. and Shui, M. 1995. Rice-*azolla*-fish in rice fields, p. 169-176. *In* MacKay, K.T. (ed.) Rice-fish culture in China. International Development Research Centre (IDRC) Ottawa, Canada, 276 p.

Cheng-Fang, L., Cou-Gui, C., Jin-Ping, W., Ming, Z., Wei-Ling, Y. and Ahmad, S. 2008. Nitrogen losses from integrated rice-duck and rice-fish ecosystems in southern China. *Plant and Soil*, 307:207-217.

Cosslett, T.L. and Cosslett, P.D. 2018. Introduction. *In*: Sustainable Development of Rice and Water Resources in Mainland Southeast Asia and Mekong River Basin, vols. 1-4. Springer, Singapore.

Costa-Pierce, B.A. 2002. Farming Systems Research and Extension Methods for the Development of Sustainable Aquaculture Ecosystems. Ecological aquaculture: the evolution of the blue revolution, pp. 103-124.

Das, T., Sarkar, P. and Prasad, N. 2014. Exploring the potential for concurrent rice-fish culture in wetlands of Assam, North East India. *Int. Res. J. Biol. Sci.*, 3:60-69.

Datta, A., Nayak, D.R., Sinhababu, D.P. and Adhya, T.K. 2009. Methane and nitrous oxide emissions from an integrated rainfed rice-fish farming system of Eastern India. *Agriculture, Ecosystems & Environment*, 129:228-237.

Desta, L., Prabha, L., Sreenivasa, V. and Amede, T. 2014. Studies on the ecology of the paddy and fish co-culture system at Dembi Gobu micro watershed at Bako, Ethiopia, 1, 49-53.

Dey, M.M., Prein, M., Haque, A.B.M.M., Sultana, P., Dan, N.C. and Van Hao, N. 2005. Economic feasibility of community-based fish culture in seasonally flooded rice fields in Bangladesh and Vietnam. *Aquacult. Econ. Manag.*, 9:65-88.

Dey, M.M., Spielman, D.J., Haque, A.B.M.M., Rahman, M.S. and Valmonte-Santos, R. 2013. Change and diversity in smallholder rice-fish systems: recent evidence and policy lessons from Bangladesh. *Food Policy*, 43:108-117.

Ding, W.H., Li, N.N., Ren, W.Z., Hu, L., Chen, X. and Tang, J. 2013. Effects of improved traditional rice-fish system productivity on field water environment. *Chin. J. Eco-Agric.*, 21:308e314.

Dwiyana, E. and Mendoza, T.C. 2008. Comparative productivity, profitability and efficiency of rice monoculture and rice-fish culture systems. *Journal of Sustainable Agriculture*, 29:145-166.

Edwards, P., Pullin, R.S. and Gartner, J.A. 1988. Research and Education for the Development of Integrated Crop-Livestock-Fish Farming Systems in the Tropics, vol 16. WorldFish.

Erwin, D. and Mendoza, Teodoro C. 2008. Determinants of productivity and profitability of rice-fish farming systems. *Journal of Sustainable Agriculture*, 29:119-130.

FAO. 2013. The state of food insecurity in the world 2013 [05 May 2019.] Available from: http://www.fao.org/3/i3434e/i3434e00.htm.

FAOSTAT. 2014. http://faostat.fao.org.

FAOSTAT. 2023. FAOSTAT_data_en_8-29-2023 https://www.fao.org/faostat/en/#data/QCL

Feng, J., Li, F., Zhou, X., Xu, C. and Fang, F. 2016. Nutrient removal ability and economical benefit of a rice-fish co-culture system in aquaculture pond. *Ecological Engineering*, 94:315-319.

Fernando, C.H. 1993. Rice field ecology and fish culture - an overview. *Hydrobiologia*, 259:91-113.

Frei, M., Khan, M.A.M., Razzak, M.A., Hossain, M.M., Dewan, S. and Becker, K. 2007. Effects of a mixed culture of common carp, *Cyprinus carpio* L., and Nile tilapia, *Oreochromis niloticus* (L.), on terrestrial arthropod population, benthic fauna, and weed biomass in rice fields in Bangladesh. *Biological Control*, 41:207-213.

Ghosh, A. and Mazumdar, A. 2016. A review on the role of chironomine midges (chironomidae: Diptera) in the paddy fields. *J. Environ. Sociobiol.*, 13:67-74.

Ghosh, A.1992 Rice-fish Farming Development in India: Past, Present and Future. *In*: *Rice-fish research and development in Asia* (Dela Cruz, C.R. *et al*., eds.), ICLARM Conf. Proc. 24, pp 27-43.

Giap, D.H., Yi, Y. and Lin, C.K. 2005. Effects of different fertilization and feeding regimes on the production of integrated farming of rice and prawn *Macrobrachium rosenbergii* (De Man). *Aquatic Research*, 36:292-299.

Guo, J.H., Liu, X.J., Zhang, Y., Shen, J.L., Han, W.X., Zhang, W.F., Christie, P., Goulding, K.W.T., Vitousek, P.M. and Zhang, F.S. 2010. Significant acidification in major Chinese croplands. *Science*, 80 (327):1008-1010.

Gupta, M.V., Sollows, J.D., Mazid, M. Abdul, Rahman, A., Hussain, M.G. and Dey, M.M. 1998. Integrating aquaculture with rice farming in Bangladesh: feasibility and economic viability, its adoption and impact. *ICLARM Tech. Rep*. 55, 90 p.

Haitao, S., Parham, J.F., Zhiyong, F., Meiling, H. and Feng, Y. 2008. Evidence for the massive scale of turtle farming in China. *Oryx*, 42:147-150.

Halwart, M. 1999. Fish in rice-based farming systems-trends and prospects. pp130-141. *In:* Dat van Tran (ed.) International Rice Commission-assessment and orientation towards the 21st century. Proceedings of the 19th Session of the International Rice Commission, 7-9 September 1998, Cairo, Egypt, 260 p.

Halwart, M. 2003. Recent initiatives on the availability and use of aquatic organisms in rice-based farming, pp195-206. *In*: Proceedings of the 20th Session of the International Rice Commission. Bangkok, Thailand, 23-26 July 2002. FAO, Rome.

Halwart, M. 2008. Biodiversity, nutrition and livelihoods in aquatic rice-based ecosystems. *Biodiversity*, 9:36-40.

Halwart, M. and Gupta, M.V. 2004. Culture of Fish in Rice Fields. Food and Agriculture Organization and The World Fish Center, Rome.

Haroon, A.K.Y. and Pittman, K.A. 1997. Rice-fish culture: feeding, growth and yield of two size classes of *Puntius gonionotus* Bleeker and *Oreochromis* spp. in Bangladesh. *Aquaculture*, 154: 261-281.

Hora, S.L. 1951. Fish culture in rice fields. *Current Science*, 20(7):171-173.

Hu, L., Ren, W., Tang, J., Li, N., Zhang, J. and Chen, X. 2013. The productivity of traditional rice-fish co-culture can be increased without increasing nitrogen loss to the environment. *Agriculture, Ecosystems & Environment*, 177:28-34.

Hu, L., Zhang, J., Ren, W., Guo, L., Cheng, Y., Li, J., Li, K., Zhu, Z., Zhang, J., Luo, S. and Cheng, L. 2016. Can the co-cultivation of rice and fish help sustain rice production? *Scientific Reports*, 6:28728.

IPCC. 2007. Climate Change 2007: The Physical Science Basis. Contribution of Working Group I to the Fourth Assessment Report of the Intergovernmental Panel on Climate Change. Cambridge University Press, Cambridge, United Kingdom and New York, USA.

Islama, A.H.M.S., Barman, B.K. and Murshed-e-Jahan, K. 2015. Adoption and impact of integrated rice-fish farming system in Bangladesh. *Aquaculture*, 447:76-85.

Karapanagiotidis, I.T., Yakupitiyage, A., Little, D.C., Bell, M.V., Mente, E. 2010. The nutritional value of lipids in various tropical aquatic animals from rice-fish farming systems in northeast Thailand. *J. Food Compos. Anal.*, 23:1-8.

Le, T.L. 1999. Small-scale aquaculture in the context of rural livelihood development in Vietnam. FAO/NACA Expert Consultation on "Sustainable Aquaculture for Rural Development", 25-31 March 1999, Chiang Mai, Thailand.

Li, C., Cao, C., Wang, J., Zhan, M., Yuan, W. and Ahmad, S. 2008. Nitrogen losses from integrated rice–duck and rice–fish ecosystems in southern China. *Plant and Soil*, 307:207-217.

Li, X., Huaixun, W. and Yontai, Z. 1995. Economic and ecological benefits of rice-fish culture, pp 129-138. *In:* MacKay, K.T. (ed.) Rice-fish culture in China. International Development Research Centre (IDRC), Ottawa, Canada, 276 p.

Lightfoot, C., Bimbao, M.A.P., Dalsgaard, J.P.T. and Pullin, R.S.V. 1993. Aquaculture and sustainability through integrated resources management. *Outlook in Agriculture*, 22:143-150.

Little, D.C., Surintaraseree, P. and Innes-Taylor, N. 1996. Fish culture in rainfed rice fields of northeast Thailand. *Aquaculture*, 140:295-321.

Liu, J., Liu, H.B., Liu, R., Mostofa, A., Zhai, L.M., Lu, H.M., Wang, H.Y., Zhang, B., Zhang, Y., Zhao, Y. and Ding, X. 2018. Water quality in irrigated paddy systems. *In:* Ondrasek (Ed.), Irrigation in Agroecosystems. Intech Open, London, United Kingdom.

Liu, M. cheng, Xiong, Y., Yuan, Z., Min, Q. wen, Sun, Y. hong and Fuller, A.M. 2014. Standards of ecological compensation for traditional eco-agriculture: taking rice-fish system in Hani terrace as an example. *J. Mt. Sci.,* 11:1049-1059.

Liu, S., Hu, Z., Wu, S., Li, S., Li, Z. and Zou, J. 2016. Methane and nitrous oxide emissions reduced following conversion of rice paddies to inland crab-fish aquaculture in southeast China. *Environmental Science & Technology*, 50:633-642.

Lu, J. and Li, X. 2006. Review of rice-fish-farming systems in China - one of the globally important ingenious agricultural heritage systems (GIAHS). *Aquaculture*, 260:106-113.

MacDonald, G.M. 2010. Water, climate change, and sustainability in the southwest. *Proceedings of the National Academy of Sciences USA*, 107: 21256-21262.

Mattila, J., Koskela, J. and Pirhonen, J. 2009. The effect of the length of repeated feed deprivation between single meals on compensatory growth of pikeperch Sander lucioperca. *Aquaculture*, 296:65-70.

Mishra, A. and Mohanty, R. K. 2004. Productivity enhancement through rice-fish farming using a two-stage rainwater conservation technique. *Agricultural Water Management*, 67: 119-131.

Mohanty, R.K. 2003. Feed intake pattern and growth performance of Indian Major Carps and freshwater prawn in a rice-fish integration system. *Journal of Asian Fisheries Science*, 16:307-316.

Mohanty, R.K., Verma, H.N. and Brahmanand, P.S. 2004. Performance evaluation of rice-fish integration system in rainfed medium land ecosystem. *Aquaculture,* 230:125-135.

Mohanty, R.K., Jena, S.K., Kumar, A., Sahoo, N. and Roy Chowdhury, S. 2008. Rice-fish culture: an ingenious agricultural heritage system. ICAR Water Technology Centre for Eastern Region. Research Bulletin No. 42, 54p

Mohanty, R.K. 2004. Density-dependent growth performance of Indian major carps in rainwater reservoirs. *Journal of Applied Ichthyology*, 20(2):123-127.

Mohanty, R.K. 2010. Impact of phased harvesting on population structure, feed intake pattern and growth performance of *Macrobrachium rosenbergii* DeMan (giant freshwater prawn) in polyculture with carps in concurrent rice-fish culture. *Aquaculture International*, 18:523-537.

Mohanty, R.K. 2010a. Dynamics of the compensatory growth response in fish. *Fishing Chimes*, 30:78-80.

Mohanty, R.K. 2015. Effects of feed restriction on compensatory growth performance of Indian major carps in a carp–prawn polyculture system: a response to growth depression. *Aquaculture Nutrition*, 21:464-473

Mohanty, R.K., Jena, S.K., Thakur, A.K. and Patil, D.U. 2009. Impact of high-density stocking and selective harvesting on yield and water productivity of deepwater rice-fish systems. *Agricultural Water Management*, 96(12):1844-1850

Mohanty, R.K. and Mishra, A. 2003. Successful rice-fish integrated farming adopting rainwater conservation measures. *Fishing Chimes*, 22:97-99.

Montserrat, N., Gabillard, J.C., Capilla, E., Navarro, M.I. and Gutierrez, J. 2007. Role of insulin, insulin-like growth factors, and muscle regulatory factors in the compensatory growth of trout (*Oncorhynchus mykiss*). *Gen. Comp. Endocrinol.*, 150:462-472.

Mustow, S.E. 2002. The effects of shading on phytoplankton photosynthesis in rice-fish fields in Bangladesh. *Agriculture, Ecosystems & Environment,* 90:89-96.

Nhan, D.K., Duong, L.T. and Rothuis, A. 1998. Rice-fish farming system research in the Vietnamese Mekong Delta: identification of constraints. *NAGA 20*, 107-111.

Nhan, D.K., Phong, L.T., Verdegem, M.J.C., Duong, L.T., Bosma, R.H. and Little, D.C. 2007. Integrated freshwater aquaculture, crop and livestock production in the Mekong delta, Vietnam: determinants and the role of the pond. *Agricultural Systems*, 94:445-458.

Nikki, J., Pirhonen, J., Jobling, M. and Karjalainen, J. 2004. Compensatory growth in juvenile rainbow trout, *Oncorhynchus mykiss* (Walbaum), held individually. *Aquaculture*, 235:285-296.

Oehme, M., Frei, M., Razzak, M., Dewan, S. and Becker, K. 2007. Studies on nitrogen cycling under different nitrogen inputs in integrated rice-fish culture in Bangladesh. *Nutrient Cycling in Agroecosystems*, 79(2):181-191.

Ofori, J., Abban, E.K., Otoo, E. and Wakatsuki, T. 2005. Rice-fish culture: an option for smallholder Sawah rice farmers of the West African lowlands. *Ecological Engineering*, 24:233-239.

Paudel, S.R., Choi, O., Khanal, S.K., Chandran, K., Kim, S. and Lee, J.W. 2015. Effects of temperature on nitrous oxide (N_2O) emission from intensive aquaculture system. *Science of the Total Environment,* 518:16-23.

Pimentel, D., Acquay, H., Biltonen, M., Rice, P., Silva, M., Nelson, J., Lipner, V., Giordano, S., Horowitz, A. and D'Amore, M. 2006. Environmental and economic costs of pesticide use. *Bioscience*, 42:750-760.

Pretty, J. and Bharucha, Z.P. 2014. Sustainable intensification in agricultural systems. *Annals of Botany*, 114: 1571-1596.

Rahman, M.A. 2016. Integration of aquaculture with rice farming: a way to increase farm productivity, food security, livelihood improvement and better environment. In: 6th Int'l Conference on Agriculture, Environment and Biological Sciences (ICAEBS'16), pp. 149-151. Dec. 21-22, 2016 Kuala Lumpur (Malaysia).

Randriamiarana, H., A. Rabelahatra and J. Janssen. 1995. Rice/fish farming in Madagascar: the present situation and future prospects and constraints, pp 353-371. *In:* Symoens, J.J. and Micha, J.C. (eds.) Proceedings of the Seminar on "*The Management of Integrated Agro-Piscicultural Ecosystems in Tropical Areas*", 16-19 May 1994. Brussels. Tech. Centre for Agric and Rural Co-op. (CTA) Royal Academy of Overseas Sciences, Brussels, 587p.

Roger, P.A. 1996. Biology and Management of the Floodwater Ecosystem in Rice Fields. International Rice Research Institute, Los Banos, Laguna, Philippines, p. 250.

Rubio, V.C., Sanchez, E. and Cerda-Reverter, J.M. 2010. Compensatory feeding in the sea bass after fasting and physical stress. *Aquaculture*, 298:332-337.

Saikia, R., Das, T. and Das, D.N. 2015. Physio-chemical qualities of water in high altitude rice fish farming system of Ziro valley Arunachal Pradesh. *Indian Journal of Environmental Biology*, 36:1153-1159.

Saikia, S.K. and Das, D.N. 2015. Sustainable aquaculture: agro-ecological role of periphyton in ricefish farming. *Reviews in Aquaculture,* 7:172-186.

Sebastian, W., Baldur, J., Lena, J., David, K., U, R.H.S., Reiner, W., Klaus, B.-B. and Ralf, K. 2015. Greenhouse gas emissions and global warming potential of traditional and diversified tropical rice rotation systems. *Global Change Biology*, 22:432-448.

Sevilleja, R.C., A.G. Cagauan, E.A. Lopez, C.R. de la Cruz and A.A. van Dam. 1992. Component technology research in rice-fish systems in the Philippines, pp 373-384. *In:* C.R. De la Cruz, C. Lightfoot, B.A. Costa-Pierce, V.R. Carangal and M.P. Bimbao (eds.) Rice-fish research and development in Asia. ICLARM Conf. Proc. 24, 457 p.

Sha, Z., Chu, Q., Zhao, Z., Yue, Y., Lu, L., Yuan, J. and Cao, L. 2017. Variations in nutrient and trace element composition of rice in an organic rice-frog coculture system. *Scientific Reports*, 7:1-10.

Smith, L.E.D. and Siciliano, G. 2015. A comprehensive review of constraints to improved management of fertilizers in China and mitigation of diffuse water pollution from agriculture. *Agriculture, Ecosystems & Environment*, 209:15-25.

Stumpf, L., Calvo, N.S., Pietrokovsky, S. and Lopez Greco, L.S. 2010. Nutritional vulnerability and compensatory growth in early juveniles of the "red claw" crayfish *Cherax quadricarinatus*. *Aquaculture,* 304:34-41.

Su Sin, T. 2006. Evaluation of different species of fish for biological control of golden apple snail *Pomacea canaliculata* (Lamarck) in rice. *Crop Protection*, 25:1004-1012.

Suloma, A. and Ogata, H.Y. 2006. Future of rice-fish culture, desert aquaculture and feed development in Africa: the case of Egypt as the leading country in Africa. *Japan Agricultural Research Quarterly,* 40:351-360.

Sun, Z., Guo, Y., Li, C., Cao, C., Yuan, P., Zou, F., Wang, J., Jia, P. and Wang, J. 2019. Effects of straw returning and feeding on greenhouse gas emissions from integrated rice-crayfish farming in Jianghan Plain, China. *Environmental Science and Pollution Research*, 1-9.

Syakila, A. and Kroeze, C. 2011. The global nitrous oxide budget revisited. *Greenh. Gas Meas. Manag*., 1:17-26.

Thakur, A.K., Mohanty, R.K., Singh, R. and Patil, D.U. 2015. Enhancing water and cropping productivity through Integrated System of Rice Intensification (ISRI) with aquaculture and horticulture under rainfed conditions. *Agricultural Water Management*, 161: 65-76.

Tilman, D., Cassman, K.G., Matson, P.A., Naylor, R. and Polasky, S. 2002. Agricultural sustainability and intensive production practices. *Nature,* 418:671-677.

Troell, M., Naylor, R.L., Metian, M., Beveridge, M., Tyedmers, P.H., Folke, C., Arrow, K.J., Barrett, S., Cr_epin, A.-S., Ehrlich, P.R., Gren, Å., Kautsky, N., Levin, S.A., Nyborg, K., €Osterblom, H., Polasky, S., Scheffer, M., Walker, B.H., Xepapadeas, T. and de Zeeuw, A. 2014. Does aquaculture add resilience to the global food system? *Proceedings of the National Academy of Sciences USA,* 111:13257-13263.

Turano, M.J., Borski, R.J. and Daniels, H.V. 2008. Effects of cyclic feeding on compensatory growth of hybrid striped bass (*Morone chrysops* X *M. saxitilis*) food fish and water quality in production ponds. *Aquacultural Research,* 39:1514-1523.

Wahab, M.A., Kunda, M., Azim, M.E., Dewan, S. and Thilsted, S.H. 2008. Evaluation of freshwater prawn-small fish culture concurrently with rice in Bangladesh. *Aquacultural Research,* 39:1524e1532.

Wang, A., Ma, X., Xu, J. and Lu, W. 2019. Methane and nitrous oxide emissions in rice-crab culture systems of northeast China. *Aquac. Fish.,* 4:134-141.

Wang, Q., Cheng, L., Liu, J., Li, Z., Xie, S. and De Silva, S.S. 2015. Freshwater aquaculture in PR China: trends and prospects. *Reviews in Aquaculture,* 7:283-302.

Wang, Y., Cui, Y. and Yang, Y. 2005. Partial compensatory growth in hybrid tilapia (*Oreochromis mossambicus* X *O. niloticus*) following food deprivation. *Journal of Applied Ichthyology,* 21:389-393.

Wassef, E.A. 2000. Status of aquaculture in Egypt. *World Aquaculture,* 31(1):29-32, 60-61.

Weller, S., Janz, B., J€org, L., Kraus, D., Racela, H.S.U., Wassmann, R., Butterbach- Bahl, K. and Kiese, R. 2016. Greenhouse gas emissions and global warming potential of traditional and diversified tropical rice rotation systems. *Global Change Biology,* 22:432-448.

Wongkiew, S., Park, M.R., Chandran, K. and Khanal, S.K. 2018. Aquaponic systems for sustainable resource recovery: linking nitrogen transformations to microbial communities. *Environmental Science & Technology,* 52 (21):12728-12739.

Wood, S., Henao, J. and Rosegrant, M. 2004. The role of nitrogen in sustaining food production and estimating future nitrogen fertilizer needs to meet food demand. In: Mosier, A.R., Syers, J.K. and Freney, J.R. (Eds.), Agriculture and the Nitrogen Cycle: Assessing the Impacts of Fertilizer Use on Food Production and the Environment. Scope 65. Island Press, Washington, DC, pp. 245-260.

World Meteorological Organization, 2010.WMO Greenhouse Gas Bulletin: The State of Greenhouse Gases in the Atmosphere Based on Observations through 2009.

Wu, L. 1995. Methods of rice-fish culture and their ecological efficiency, pp 91-96. *In:* MacKay, K.T. (ed.) Rice-fish culture in China. International Development Research Centre (IDRC), Ottawa, Canada, 276 p.

Wu, S., Hu, Z., Hu, T., Chen, J., Yu, K., Zou, J. and Liu, S. 2018. Annual methane and nitrous oxide emissions from rice paddies and inland fish aquaculture wetlands in southeast China. *Atmospheric Environment,* 175:135-144.

Xie, J., Hu, L., Tang, J., Wu, X., Li, N., Yuan, Y., Yang, H., Zhang, J., Luo, S. and Chen, X. 2011a. Ecological mechanisms underlying the sustainability of the agricultural heritage rice-fish coculture system. *Proceedings of the National Academy of Sciences USA,* 108: E1381-E1387.

Xie, J., Wu, X., Tang, J. jun, Zhan G, J. en, Luo, S. ming, Chen, X. 2011b. Conservation of traditional rice varieties in a globally important agricultural heritage system (GIAHS): rice-fish Co-culture. *Agricultural Sciences in China,* 10:754-761.

Yan, Y., Liu, M. da, Yang, D., Zhang, W., An, H., Wang, Y. jing, Xie, H. tu, Zhang, Dong, X. 2014. Effect of different rice-crab coculture modes on soil carbohydrates. *Journal of Integrative Agriculture*, 13:641-647.

Yao, Y., Zhang, M., Tian, Y., Zhao, Miao, Zhang, B., Zhao, Meng, Zeng, K. and Yin, B. 2017. Duckweed (*Spirodela polyrhiza*) as green manure for increasing yield and reducing nitrogen loss in rice production. *Field Crops Research*, 214:273-282.

Yue, Y., Sha, Z., Zhao, Z., Lu, X., Zhang, J., Zhao, Q. and Cao, L. 2014. Effects of rice cultivation patterns on nitrogen and phosphorus leaching and runoff losses. *Chin. J. Eco-Agric.*, 22:1424-1432.

Zhang, J., Guoming, Q., Benliang, Z., Kaiming, L. and Zhong, Q. 2017. Rice-Duck Co-culture in China and its Ecological Relationships and Functions, Vol. 111. Agroecology in China: Science, Practice, and Sustainable Management.

Zhang, J., Hu, L., Ren, W., Guo, L., Tang, J., Shu, M. and Chen, X. 2016. Rice-soft shell turtle coculture effects on yield and its environment. *Agriculture, Ecosystems & Environment*, 224:116-122.

Zhang, T., Liu, H., Luo, J., Wang, H., Zhai, L., Geng, Y., Zhang, Y., Li, J., Lei, Q., Bashir, M.A., Wu, S. and Lindsey, S. 2018. Long-term manure application increased greenhouse gas emissions but had no effect on ammonia volatilization in a Northern China upland field. *Science of the Total Environment*, 633:230-239.

7

Pond-based Integrated Farming System

7.1. Introduction

The integrated farming system (IFS) is becoming increasingly important for effective farm resource management in order to increase farm productivity, reduce environmental degradation, improve the quality of life for resource-poor farmers, and ensure sustainability. Integration happens when one production sub-system's outputs (typically by-products) are used as inputs in another within the farm unit (Radhammani *et al.*, 2003; Lightfoot and Minnick, 1991). IFS are a mixed farming system made up of at least two distinct but logically linked components of crop and animal enterprises. The integrated farming system is effective in reducing risk, increasing productivity and revenues, and improving the use of organic waste and agricultural leftovers.

Farm families use resources to effectively utilize the farm enterprises that already exist in order to increase the farm's production and profitability. This type of farming system is known as mixed farming. Agro-forestry, agri-horticulture, aquaculture, crops, livestock, and sericulture are some of these farming activities. The major goal of such diversified farming is to reduce risk, even while other components and crops coexist. When waste and crop residues are recycled efficiently, a complementary effect occurs, providing farmers with an extra revenue (Paramesh *et al.*, 2021). IFS activity is focused on a few selected interdependent, interrelated and interlinking production systems based on crops, animals and related subsidiary components. In addition to being a dependable means of achieving a respectable level of output with ample opportunity for resource recycling, an integrated farming system approach is an ecologically sound method that promotes sustainable agriculture.

Integrated crop and livestock production systems are highly efficient; crop residues can potentially be used as livestock feed; waste products (e.g., faeces and urine) are fed into biogas digesters, and the effluent is used to fertilize ponds for plankton/algae production, with fish farming serving as the final activity. These systems are well worth investigating as a means of supplying nutrients/fuel for the family, lowering fossil fuel burning and methane production, and so reducing environmental pollution (Preston, 1990). One of the best examples of mixed farming is integrated fish aquaculture. This sort of farming is an ecologically sustainable technology, and is practiced in many forms typically in East and South-East Asian nations. The technology comprises a combination of composite fish culture integrated with crop/ livestock production. The integration of aquaculture with livestock/or crop farming aids in the provision of high-quality protein food, resource use, agricultural waste recycling, job creation, and economic development (Shoko *et al.*, 2011; Megerssa Endebu *et al.*, 2016). Integrated aquaculture systems are used in various parts of the world and have a long history of improving fish production, lowering operational costs, minimizing the environment pollution and waste production, and increasing income by utilizing waste materials as fish feed (Prein 2002).

Several IFS models such as rice-fish-on dyke horticulture-vegetable (Mohanty *et al.* 2004; 2009; 2017b), integrated system of rice intensification (ISRI) with fish and horticulture crops (Thakur *et al.*, 2015), pond-based IFS (Srivastava *et al.* 2004), self-reliant agri-based IFS (Rautaray *et al.*, 2020) are most prevalent, lucrative and replicable system in Indian context. Aquaculture-based IFS integrating aquaculture (carp polyculture), agriculture (*kharif* rice followed by green gram and vegetables), on-dyke horticulture (banana and papaya), and poultry 'Vanaraja' was discovered to be one of the most water-saving, sustainable, productive, and replicable models in terms of water management. These systems as a whole, make a net profit of □ 110000- 170000 ha^{-1} with an output value - the cost of cultivation ratio of 2.7-2.8 and net consumptive water productivity of □ 9.8-16.2 m^{-3}. Further, when fish and prawns are included as key components, this system also helps to improve water production, water use efficiency, and minimize water footprint (Mohanty *et al.,* 2009; 2010). Since multiple subsystems are involved rather than a single commodity farming system, integrated aquafarming not only increases productivity but also creates employment opportunities, raises income, and provides nutritional security to farming communities with limited resources. It also distributes risk (both biological and economic).

7.2. Basic principles of integrated farming system

IFS is an interdependent, interconnected, and often interlocking production system. It is based on a few crops/ animals, and related subsidiary enterprises that are designed to maximize the use of nutrients in each system while minimizing the detrimental impact of these operations on the environment. The interdependent character of IFS entails the use of primary and secondary produce from one system as input to the other system, resulting in the components being mutually integrated as one whole unit (Parajuli *et al.*,

2018). Sometimes this system is called an integrated biosystem with the reason being involvement with the production of livestock and crops, or fish and livestock etc. This reduces costs and improves productivity and/ or overall farm income. New integrated methods include improved farming technologies such as integrated nutrient management, site-specific nutrient management, conservation technology, use of biofertilizers, crop rotation, zero tillage etc. (Behera *et al.*, 2016; Goverdhan *et al.*, 2020).

The basic principle for an integrated farming system is to enhance ecological diversity by several ways: i) selecting the appropriate cropping pattern, rotation, mixing, and intercropping so that there is less competition for water, nutrition, and space and adopting environmentally friendly methods, ii) by using a multi-story layout so that the total available area can be used effectively and has a high level of interaction between biological and abiotic components, iii) adopting a better cropping system based on rainfall and soil moisture availability and selection of suitable varieties of cereal crops and or agro-forestry tree species, and iv) by producing extra fodder leaves, crop residues during rainy season, storing them as silage or hay for the lean season (summer).

7.3. Aim and goals of integrated farming system

It is a combination of different agricultural activities in an unit area of land aimed at the following:

i) Maximum return from unit area i.e., maximizing yield output of all component to offer steady and stable income,
ii) Maintaining soil fertility and soil quality i.e., rejuvenation/ amelioration of the system's productivity and achieving agro-ecological equilibrium,
iii) Ensuring supplementary and complementary enterprise relationships to use the by-products of one component of the farming system as inputs to the other,
iv) Lessening weed population, insect pests, and diseases out-break, keeping them at low levels of intensity, and
v) Reducing environmental pollution by cutting down the use of agrochemicals (fertilizers and pesticides).

7.4. Components in IFS

Various components of IFS include agriculture, fish farming, fodder production, goat rearing, nursery, sheep rearing, horticulture, duck rearing, forestry, pigeon rearing, apiary, mushroom cultivation, sericulture, *Azolla* farming, dairy, kitchen gardening, poultry, seed Production, piggery, vermiculture, rabbitry, and value addition (Fig. 7.1)

These components can be divided into some categories:

- **A**: *Crop component*- cereals, pulses, oilseeds, fruits, vegetables, spices, planting crops, flowers, fodder/forage crops, agricultural forests, sugarcane, fibre crops
- **B**: Livestock and poultry *component* - cows, buffaloes, pigs, goats, sheep, poultry, duck
- **C**: Fishery *component* - fingerling production, composite fish culture, paddy cum fish culture
- **D**: Secondary agriculture *component* - bee cultivation, mushroom cultivation, food processing, vermicomposting, biogas production, *Azolla* cultivation, silk cultivation

It is feasible to create an integrated farming system by combining the four categories of components listed above (A + B, A + C, B + C, A + D, B + D, C + D, A + B + C, A + B + D, B + C + D, A + C + D, A + B + C + D).

7.4.1. Elements of integrated farming system

- Farm ponds
- Biological pesticides
- Biogas
- Bio-fertilizers
- Solar energy
- Vermicompost making
- Green manuring
- Rainwater harvesting
- Watershed management

7.4.2. Factors governing the implementation of IFS

- Soil and climatic features of a selected area
- Availability of resources, land and labour
- The current level of resource use
- Economics of proposed integrated farming
- Farmer's management skills

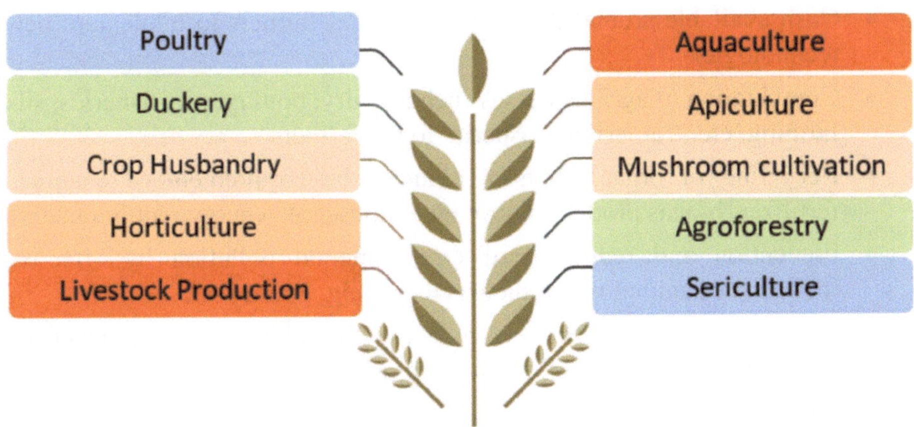

Fig. 7.1. Components of Integrated Farming System

7.5. Advantages of integrated farming system

In the integrated farming system, agriculture can be integrated with livestock, poultry, and fish kept in one place to create year-round employment and additional income. For example, keep poultry in the top layer and use their excreta. Pigs live in the lower layers, and the remaining water from the pond is used for agriculture and fodder crops. The emissions can be recycled as fertilizer for orchards as well as fodder crops, called integrated farming. Further, the system helps poor small and marginal farmers (Ayinla, 2003), and facilitates nutrient recycling (Velasquez, 1980), nature conservation, environmental safety, and land and water management (Fig. 7.2).

The major advantages of an IFS are:

- Maximum productivity and minimum utilization of external inputs with better economic yield per unit area per unit time by intensification of crop and allied enterprise,
- Increases profit by reducing input costs through recycling of wastes of one enterprise as energy inputs for other systems
- Greater sustainability in production due to the integration of diverse enterprises.
- Greater opportunity to solve the malnutrition problem due to the production of a variety of food products
- Recycling of farm waste reducing pollution potential.
- Year-round money flow to the farmer through sale of milk, eggs, vegetables, edible mushrooms, honey, silkworm cocoons, fish etc.

- Cash available around the year encourages farmers to implement new technologies
- Combination of agribusinesses such as dairy, poultry, pig, fisheries, silk farming, etc., can bring prosperity to the farmers.
- Recycling of organic wastes reduces the requirement of chemical fertilizer; biogas production can meet household energy requirements.
- Prevention and reduction of environmental pollution by effective recycling of animal waste such as dairy, pig, poultry, etc.
- Energy-saving by effectively reducing the additional dependence on fossil fuel-derived energy use in crop production
- Meeting the fodder crisis from perennial legume fodder trees; leguminous trees not only fix nitrogen for the field but also provide quality fodder.
- Solving the crisis of fuel and timber by introducing Silvi component in the system
- Employment generation around the year.
- This farming system improves the knowledge /literacy level of farmer.
- Increases input efficiency due to less dependency on external inputs such as fertilizers, food, agrochemicals
- Improves livelihood and living standard of the farmer
- Capability to adopt new technology as farmers will have investment capability once IFS is established.

Fig. 7.2. Integrated farming system and farm management approach

7.5.1. Possible impact of integrated farming system

IFS provide various ecosystem services in terms of food security, nutritional security, nutrient recycling, biodiversity conservation, increased employment, enhanced profitability and climate resilience (Fig. 7.3). It leads to the production of high-quality food and the strengthening of the food chain. It is necessary to eradicate nutritional disorders that have arisen as a result of mineral and vitamin deficiencies in the food being consumed. Vegetable and horticultural crops can produce 2-3 times more energy than cereal crops on the same plot of land, ensuring nutritional security through their inclusion in the existing system. Similarly, the inclusion of beekeeping, fisheries, sericulture, and mushroom in limited space provides additional high-energy food without interfering with food grain production. The integration of these enterprises in an ecosystem, helps in the production, consumption and decomposition realistically.

Fig. 7.3. Ecosystem services provided by integrated farming system

Similarly, it is necessary in farming systems to maintain the proper recycling of resources, particularly crop leftovers, because 80-90% of micronutrients remain in the biomass. Crop residue can be utilized as floor thatch in cattle shelters, compost, mushroom cultivation, and dry fodder. Multiple uses of water for crop, fruit, vegetable, and fisheries production may increase water productivity. Further, in communities/ villages, sewerage water can be cleansed using *Hydrilla* biomass before being released into fish ponds. Furthermore, the community land in the villages, which are more easily accessible, must be used productively. The implementation of social forestry, water harvesting and recycling fishery, and stall feeding of animals (piggery/ goatery) will increase the profit margin while also providing multiple secondary benefits such as employment and enhanced ecological conditions in the area. Such type of integration generates additional income ranging between Rs 50, 000 to 75,000 per ha under irrigated ecosystems and Rs 20,000 to 40,000 under rainfed ecosystems. It is observed that under IFS, income increased by 25-50% as a result of integrating processing and on-farm value addition, yield increased by 0.5 to 1.0 tonne per ha due to enhanced soil health, cost reduction by Rs. 2500 to 4,000 per ha, and employment generation increased by 50-75 man-days/ household. A multi-commodity agricultural system has more advantages than a single-crop approach. However, commodity integration must be tailored to the farmer's specific capabilities, resources, and needs, as well as the social, economic, and environmental aspects that surround him.

7.5.2. Limitations of integrated farming system

- Lack of awareness about the sustainable farming system.
- Unavailability of technical savoir-faire of various farming system models.
- Lack of easy and reasonable interest rate credit facilities.
- Unavailability of certain marketing facilities, especially for perishable products.
- Lack of storage facilities and lack of timely availability of input.
- Lack of education/knowledge/encouragement in farming communities especially rural youth.

7.5.3. Productivity of integrated fish farming

Productive capacity is defined as the biomass produced in kg ha^{-1}, which measures the system's "net community production" (Dalsgaard, 1995). This capacity is determined by the soil and pond water quality, as well as the type and quality of manure used. The biological productivity of any aquatic body is often assessed through the quantitative and qualitative measurement of plankton, a natural nutritious fish food (Ahmed and Singh, 1989). Animal feces also increases biological productivity of ponds via several pathways, resulting in higher fish production (Orhibhabor and Ansa, 2006). Singh (1996) asserts that the type of manure has an impact on the plankton community structure. The maturity of an ecosystem is determined by the biomass supported per unit of energy flow (Odum, 1989). This illustrates the system's capacity to transform available energy into something beneficial. By combining livestock such as cattle, pigs or poultry with fish, animal excreta and waste food serve to stimulate water productivity, natural feed production and fish growth, as well as delivering other symbiotic benefits (Fig. 7.4) for both animals (Barash *et al*, 1982). Furthermore, the importance of integrating aquaculture-based IFS into future rural development prog has been underlined by Edwards (2000) and FAO (2003). Aquaculture-based IFS is more profitable than the unitary system of fish farming (monoculture) that contributes to the economic empowerment of resource poor farmer, enabling them to be productive year-round and to fully maximize its production (Tokrishna, 1992, Nnaji *et al*., 2003; Tipraqsa *et al*., 2007; Gabriel *et al.,* 2007).

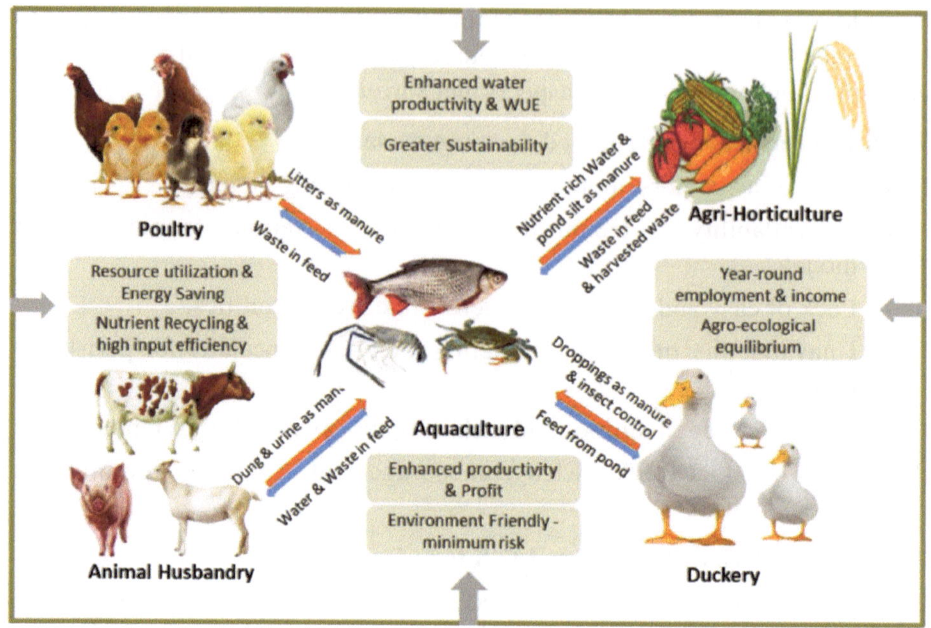

Fig. 7.4. Schematic representation of major components, pathways of material flow and benefits in an integrated farming system

7.6. Freshwater pond-based IFS models

Based on the ideas of making better use of the labor and resources that are available as well as the productive use of farm wastes, attempts have been made in India to construct affordable aquaculture-based integrated farming systems (Fig. 7.5) that are appropriate for Indian conditions. Consistent research has led to the creation of integrated farming systems that combine agriculture, animal rearing, and fish culture. The sets of procedures for fish-cum-duck, fish-cum-pig, and fish-cum-poultry farming have been carefully designed and their economic viability and practicality at the farmer's level has been tested. Livestock manure is rich in biomass and nutrients, and when it is added to a pond, it promotes the growth of zooplankton and phytoplankton, which provide fish with food. Manure contains indigestible feed elements that the fish directly eat. Via fish culture, leftover feed from feeding can also be recycled. Periodically, the pond's bottom layer of soil is removed and put to the field for biomass-rich crop or vegetable growing. There are numerous variations of integrated farming that can be implemented based on local environmental factors.

Fig. 7.5. An aquaculture-based integrated farming system model (Aquaculture, On-dyke horticulture, Agriculture, Duckery, Poultry, Agroforestry) developed by ICAR-IIWM, Odisha, India.

An example of integrated farming is the cultivation of fish and livestock as well as common farming methods that support each other. Fish can be fed from farm residues such as leaves, stalks, or other waste products. In addition, plankton, which is a major food source for fish, can be grown using livestock manure that grows around fish farms. Near fish farms, silk warms can also be increased (*i.e.,* sericulture). The silkworm can then be used to feed pupae and other waste fish. In turn, fish pond silt provides excellent manure and fertilizer for land crops and is highly demanded by farmers.

7.6.1. Fish-duck integrated farming

Ducks are quite compatible with farmed fish, therefore raising them in fish ponds works well with the fish culture method. Farmers can gain from this arrangement in a variety of ways. The ducks fertilize the pond with their waste when given free range at the pond's surface. Ducks are known as manuring machines because of their effective way of producing pond manure, which eliminates the need for pond fertilizer and supplemental fish feed, which make up about 60% of the whole expense of conventional fish culture. Ducks control aquatic plants, loosen the bottom of ponds with their dabbling, and aid in nutrient release from the soil, all of which boost pond productivity. Ducks are biological aerators because they aerate the water as they swim. On the pond's edge, duck houses have been constructed. Therefore, duck activities don't need any additional land. Ponds provide the majority of the food that ducks need in the form of aquatic grasses, insects, larvae, earthworms, and other such things. They require extremely little food, which is typically provided by farmers in the form of kitchen trash, molasses, and rice bran.

In this system, duck density usually varies between 100-3500 ha^{-1} (Sin, 1980). Woyanarovich (1980) reported that 300-500 ducks ha^{-1} gave fish yield of 500-800 kg ha^{-1} in 150 days in Hungary without supplemental feed where carp were cultured. In Vietnam, rearing 1,000- 2,000 ducks ha^{-1} on ponds enhanced the regular fish yield to as much as 5,000 kg ha^{-1} yr^{-1}, compared to 1,000 kg ha^{-1} yr^{-1} without ducks (Delmendo, 1980). The combination of duck and fish farming is currently thought to be a way to lower the cost of feed for the ducks on the one hand, and on the other hand, duck excrement acts as a cheap method of fertilizing ponds, which results in the development of fish food organisms. Such a mixed culture is very profitable since it significantly raises the amount of fish and duck protein produced per square meter. In addition, ducks consume a wide range of organisms, including weeds, snails, unwanted harmful insects and their larvae (some of which are vectors of fish pathogenic organisms), tadpoles, frogs, and water-borne disease-causing organisms that infect humans, the eradication of which is one of the key components of farm management).

In most Southeast Asian countries, egg-laying duck breeds are preferred rather than the white, Peking-type meat ducks, to scavenge on rice stubbles without formulated feeds. Duck raising can fully utilize the water surface of ponds. Ducks thrive in fish ponds' ideal environment, which shields them from parasite infestation. Ducks eat predators and aid in the growth of the fingerlings. The ducks are allowed to roam freely over the pond's surface from 9 am - 5 pm when they naturally manure the pond by dispersing their waste throughout it. Between 125 and 150 g of droppings are excreted daily by each duck. Every morning, the morning excretions are removed from the duck home and added to the pond. The rearing density of 200-300 ducks ha^{-1} in a one-hectare pond gives 10,000 to 15,000 kg of droppings which is recycled every year. On a dry matter basis, these droppings contain 81% moisture, 0.91% nitrogen and 0.38% phosphate. Duck raising in fish ponds reduces the protein demand by 2 – 3% in duck feeds. The daily waste of duck feed (around 20 – 30 g $duck^{-1}$) aids as fish feed or as manure in ponds, resulting in higher fish yield. Ducks' body weight and feed efficiency both rise and fish may be able to use the spilled meals. Additionally, because fish ponds are a clean environment, ducks raised in them have a 3.5% higher chance of surviving. The yield of fish can be increased to 35-50 kg ha^{-1} using duck droppings and the leftover food from each duck. This strategy provides great profit with little investment by producing a lot of fish, duck eggs, and duck meat per unit of time and water space.

For an integrated farming system, the pond should be water-retentive. Seasonal ponds that can hold water for 8 to 9 months can also be considered. A minimum of 1.0 m and 1.5 to 2.0 m water depth is optimal. The pH of the soil should fall between 6.5 and 7.5. If the soil pH is not at the desired level, it can be adjusted by adding lime in 3 to 4 split doses (about 400 kg ha^{-1} for pH 6.5 to 7.0, 200 kg ha^{-1} for pH 7.0 to 7.5). Lime helps in maintaining pH, and kills harmful parasites. The basal dose of lime and cow dung application in a one-hectare meter of water body is 400 kg and 5000 kg, respectively. There should be regular cleaning of ponds to ensure sunlight penetrates and oxygen circulates in the water and to ensure predators cannot hide in the pond.

Fish fingerlings can be stocked preferably during the June and July months. The stocking time varies depending upon the climate and the availability of optimum water level in the pond. In contrast, fingerlings may be stocked in April and harvested in November/December when water bodies remain functional for eight to nine months. Stocking density should range between 5000-6000 fingerlings ha^{-1}. The most suitable species of fish for integrated livestock-fish farming are those fishes that can filter and feed on phytoplankton, zooplankton and bacteria from water. To feed fish naturally, the goal of integrated livestock and fish farming is to produce as much protein-rich plankton as possible in the pond. Fish species with a macrophytic feeding habit, effective phyto- and zooplankton utilisation, and human consumption make excellent candidates for integrated livestock fish farming. Fishes are categorized into three groups based on how they feed: bottom feeders, column feeders, and surface feeders. Both indigenous and exotic fish species are advised for an integrated fish farming system. Indigenous species like *Catla catla* which is a zooplankton feeder and exotic species *Hypophthalmichthys molitrix* which is a phytoplankton feeder are the finest examples of surface feeders, whereas *Labio rohita* an indigenous omnivorous species that is a column feeder. The indigenous species *Cirrhinus mrigala*, *Labeo calbasu* are detritivorous and exotic common carp species (*Cyprinus carpio*) which is detritivorous/ omnivorous are bottom feeders. Exotic species like *Ctenopharyngodon idella* which is herbivorous cover the surface, column and marginal area of the feeding zone. Depending upon the availability of fingerlings, composite fish culture with 3, 4 or 6 species combinations can be adopted. In integrated livestock cum fish farming considering the surface, column and bottom feeder the ratio of fishes viz. Catla, Rohu and Mrigal - 4: 3: 3 (3 species), Catla, Rohu, Mrigal and Common carp - 3 : 3: 2: 2 (4 species) and Catla, Rohu, Mrigal, Silver carp, Grass carp and Common carp - 1.5 : 2.0: 1.5 : 1.5: 1.5: 2.0 (6 species), respectively.

7.6.2. Fish-poultry integrated farming

According to Ahmed and Lorica (1999), this approach uses chicken droppings from fully developed chicken litter for fish cultivation. To minimize expenses and increase returns, fish culture can be combined with the raising of broilers (chickens for meat) or layers (chickens for eggs). The chicken waste can be repurposed to manure the fishponds by raising chickens close to or over the ponds. Raising chickens over ponds makes the most efficient use of available space and reduces labour-intensive manure transportation. There are two methods for integrating fish and fowl. Installing the coops above the pond's surface is one approach to ensure that the droppings fall into the pond. The alternative is to construct deep litter chicken coops and recycle the waste daily at a rate of 50 kg ha^{-1} into fish ponds. Since fully built-up litter has nearly three times the manurial value of chicken droppings, the deep litter method is typically selected. Without the need for the addition of supplemental feed, poultry dung is a complete fertilizer that possesses the qualities of both organic and inorganic fertilizers (FAO, 2003). Poultry waste contains less moisture, fiber and compounds like tannins that discolors pond water when used as fertilizers. Fresh chicken dung contains 1.6% nitrogen,1.5% phosphorus and 0.9% potassium (Woynarovich, 1976). Every day, each bird produced about 40 g of dung. Like other organic waste, poultry manure can transform low-quality inedible materials into high-quality fish feed (Banerjee *et al.,* 1979; Schroeder, 1980). By releasing the nutrients and minerals that were once bound, they can promote microbial activity in the water column and at the pond bottom and can act as substrates for the photosynthetic production of essential fish food. For a 1.0-hectare pond, 500–600 birds are typically required to supply the necessary fertilizer. Two batches of broilers are made from chicks that are about two weeks old, and laying hens are made from chicks that are 4.5 to 5.5 months old. Laying hens are raised for 12 - 18 months to produce eggs. Without the need for any additional fish feed or ponds, this system produces 4500 - 5000 kg of fish, about 70,000 eggs, and 1,000 kg (live weight) of chicken meat from a 1.0-hectare water area in a single year. At low to moderate stocking densities, it has been claimed that ponds fertilized with poultry manure produce more fish (2.5–4.5 t ha^{-1}) than those treated with pig or cow manure (Fagberno and Sydenhan, 1988; Hatch and Engle, 1987).

Important points for poultry rearing under integrated poultry-cum-fish farming

- Dual-purpose poultry birds are suitable choice
- Brooding should be done in separate sheds
- Use a deep litter system of management instead of construction slatted floor over the pond
- Follow the vaccination and deworming schedule
- Don't allow the formation of cake in litter-turn at least once a day
- Use separate shoes to enter the shed
- Provide a good laying nest to get a clean egg
- Discourage visitors to the farm
- Maintain 14 hr light and 10 darkness schedules for layer
- Provide enough calcium to layer
- Maintain good record of egg production and other farm records

7.6.3. Fish-*cum*-cattle integrated farming

Fish-cum-cattle integration can cut down / save more fertilizers, and fish feeds and increase the income from milk, fish, and beef. In this system, cow dung is used as manure for fish culture. Farmers can apply 5,000 to 10,000 Kg of raw cow dung per ha to a fish pond over time. The growth of plankton, which serves as fish food, is encouraged by this cow dung. The wastewater from cleaning cow sheds can be diverted to a pond that has excellent manurial value, along with the cow dung, urine, and leftover feed (Athithan *et al.*, 2001; Moav *et al.*, 1977). One cow or buffalo will typically excrete 12000 kilograms of dung and 8000 liters of urine annually. The Filter-feeding and omnivorous fish benefit from the urine and faeces of cattle. The fish feed can be made from the leftover cattle feed. Building cattle sheds on the fish pond's broad embankment will allow waste and washing to run right into the fish pond. In addition to 9000 l of milk and 3000–4000 kg of fish annually, roughly 5–6 cows can supply enough manure for a 1-hectare pond. Building a cow house near a fish pond will make handling of cow manure easier. Black and white cow (for milk), Shorthorn (for beef), Simmental (for milk and meat), Hereford (for beef), Charolai (for beef), Jersey (for milk and beef), and Qincuan draft (for beef) are cow varieties that can be raised. This technique boosts revenue and lowers input costs because the farmer also receives milk, fish, and calves. The system gives a net profit of Rs.3-4 lakh per year from one-hectare land.

7.6.4. Fish-*cum*-rabbit farming

Fish-cum-rabbit farming benefits both. Rabbits are less susceptible to disease than chickens and ducks. The rabbit can be raised both on the ground and above the pond. The rearing density of rabbits varies per hectare of water. According to studies, 400-500 rabbits are sufficient for a one-hectare fish pond. This figure is determined by the pond's productivity. Every year, a rabbit excretes 60-70 kilograms (140-150 grams per day), which is turned into acceptable manure for fish farming. The faeces and excreta produced by rabbits are completely utilized in fish rearing as powerful organic fish food and manure.

Two species of rabbit are preferred for baby rabbit production (1) *European wild rabbit*: It is an exotic type rabbit and delivers 80 babies annually and (2) *American cotton tail rabbit*: It delivers 85 to 90 babies per year. One female rabbit produces around 8 offspring in 45 days; 400 female rabbits produce 3200 babies in 45 days; 3200 bunnies are ready after 3 to 5 months (0.5 to 5 kg), resulting in 9600-kilogram rabbits. The rabbit, like other animals, consumes cabbage, carrots, dry leaves, and hay. A mature rabbit (male or female) consumes only 140 to 200 grams of hay or vegetables per day. The fish pond requires 400-500 rabbits per hectare and does not require additional feed. To ensure that every rabbit is replaced every two years and that there are always rabbits available for farming, only 40 to 50 per cent of the entire rabbit population should be culled annually. Rabbits give birth to young between the ages of 3 and 6 years. To prevent the rabbit from contracting any diseases, a veterinarian should always take care of it, vaccinations should be given periodically, and the environment should be clean. The environment significantly affects rabbit reproduction. Heat stress accounts for 73% of rabbit mortality, but under a cooling regime, mortality is reduced to 6%, after accounting for losses owing to conception rate, litter weight, and pre-weaning death. To relieve heat stress in rabbits, cool drinking water should be provided. Temperature, humidity, and ventilation are the most critical factors influencing rabbit's physiological stress. In this system, fish yield of 3500 to 4000 kg by stocking 15000 fingerlings per hectare and rabbit production of 9000-9500 kg annually can be achieved with a gross income of Rs.6.0 lakh per year.

7.6.5. Fish-*cum*-pig farming

This system is very common in China, Vietnam, Taiwan, Malaysia, Thailand, Hungary and some European countries. The Hampshire, white Yorkshire, and Landrace are the popular breeds of pig for integration with fish. In integrated fish farming with pigs, the 'pig dung' is useful for conditioning the soil and providing the necessary nutrients required for fertilizing the pond water. Pigs are considered "costless fertilizer factories" in China. It has been reported that 100 pigs weighing 30 kg each were considered sufficient to fertilize and maximize total revenue from a 1 ha fish pond in Hong Kong and the Philippines, (Bardach, 1972). A scientifically solid and economically effective production system combining pig husbandry and fish culture has emerged for Indian circumstances. Pig stalls are built on or near the pond's back, and pig dung is either drained directly into the pond or transferred from the animal house to the pond. Pig excrement is an excellent fertilizer for the pond, increasing biological output of pond water and, as a result, fish production. Furthermore, fish consume directly from pig excreta, which is 70% digestible by fish. This technique does not require additional fish food or pond manure. Pig dung, which represent for 60% of the input, have greatly decreased the cost of fish farming.Usually, fish production of about 5.5-6.0 t ha^{-1} per year can be obtained through a fish-pig integrated farming system with a rearing density of 60-80 pigs and 20,000 fish fingerlings per ha pond (Singh, 1993; Francis and Pandurengan, 2004). In India, it is found that about 35-40 pigs are sufficient to fertilize 1 ha of fish pond, while fish production in pig-fish farming operations ranges between 2,000-5,000 kg ha^{-1} in six months. In some cases, total fish production of about 6.8 t ha^{-1} per year was obtained through a fish-pig integrated farming system with a stocking density of 80 pigs and 20,000 fish fingerlings per ha of water shad (Hopkin, 1982) whereas, in Thailand, the production of five tons of tilapia annually was obtained by integrating 150 pigs per ha of pond (Edward,1985). Farmers practising pig-cum-fish farming may get economic benefits up to 28-30% over normal pig farming (Ayinla, 2003).

The system has many advantages:

i) Pig dung is an excellent pond fertilizer, increasing the pond's biological productivity and thus fish production.

ii) Some fish feed directly on pig feces, which includes 70% digestible food for the fish.

iii) Supplemental feed is not required for fish culture, which typically accounts for 60% of total operational cost of aquaculture.

iv) The pond dikes make room for animal habitation modules.

v) Pond water is utilized to cleanse and bathe the pigs.

Due to religious restrictions, the system cannot be implemented throughout India; however, it is particularly important in the Northeastern region where it has the potential to raise the socioeconomic standing of marginalized rural communities, particularly the tribal people who enjoy eating fish and traditionally raise pigs in their backyards. They have no trouble starting a fish-pig farm.

7.6.6. Fish-*cum*-sheep/goat farming

An integrated goat and fish farm system offers several potential advantages, that is, increased productivity, income, cash flow, employment, a better diet for the farmer and his family, and the spread of both biological and economic risks since two subsystems are involved as opposed to one in a single commodity farming system. It is considered a poor man's cow and a goat's excreta is considered an excellent organic fertilizer. The goat excreta contains 60% organic carbon, 2.7% N, 1.78% P, and 2.88% K, while its urine is likewise rich in both N and P. At least 50-60 goats are essential to fertilize a 1 ha pond. The goats should be provided with a dry, safe, comfortable house protected from excessive heat. The suitable goat breeds are Jamunapari, Beetal, and Barbari for milk, and Black Bengal, Sirihi, and Deccani are used for meat purposes. Goats are selective feeders and consume Napier grass, berseem, cowpea soybean, mulberry, etc. This integration can provide 3500-4000 kg fish per ha per year without supplementary feeding and fertilizer.

7.6.7. Fish-*cum*-horticulture farming

A sizable portion of an aquaculture farm is often made up of dykes, some of which are utilized for regular farm operations and the remainder of which is left fallow all year round and overrun with terrestrial weeds with deep roots. This has a negative impact on aquafarming's economy, which may be greatly enhanced by using dykes wisely to produce vegetables and fish feed. An integrated approach to horticultural, agri-aquaculture production yields higher returns through better resource management. A family of four to five people typically needs 200 square meters of land to meet their daily fruit demands. A fish pond's embankments often offer more than 200 m^2 of space. Pond bunds represent between 25–33% of the total farmland. A farmer's family can cultivate enough fruits and vegetables to fulfil their daily needs while also earning extra money by cultivating produce on pond bunds. In ditches created between rows of banana or coconut trees serve as supply or drainage canals. These canals' abundant insect populations and constant water supply make them ideal for use as fish culture systems. Certain fish species are perfect for culture in this method, including tilapia, common carp, *Channa marulius* and *C. strialus*. The best fruit crops to grow on pond embankments include papaya, banana, lemon, coconut, and guava.

7.6.8. Fish-*cum*- vegetable farming

Kitchen gardening typically requires around 200 m^2 of ground, and the pond embankments provide more space than this. Annually two vegetable crops can be grown. Varieties of winter vegetables (tomato, brinjal, cabbage, cauliflower, turnip, radish, coriander, beans, spinach, potato, fenugreek, bottle gourd, and onion) and summer vegetables (amaranth, okra, bitter gourd, papaya, sponge gourd, ridge gourd, sweet gourd, ginger, chilly and turmeric) can be grown on pond dykes. Pond dykes can also be utilized to raise pulse and oil seed crops. Flowers such as marigold, rose, and jasmine can also be grown on pond embankments. Crops grown on pond dykes provide strength and stability to the pond embankment, prevent soil erosion, and keep the pond water from becoming excessively murky. Further, the pond helps the system by providing humus, which is pulled out of the pond after each harvest. Humus is an excellent source of fertilizer for crops.

Broad dykes (4 m and higher) allow for intensive vegetable growing and can be regularly ploughed and irrigated without causing damage to the dykes. The best practice for managing a dyke is to use the middle section, which makes up about two thirds of the total area, for intensive vegetable cultivation, and to use the remaining third of the dyke, which is along the length of the periphery, for papaya cultivation, leaving enough space on either side for netting operations. The production of 65 to 75 t ha^{-1} $year^{-1}$ can be achieved with the intensive cultivation of Indian spinach, water bindweed, radish, cauliflower, amaranth, okra, cabbage, sweet gourd, spinach, coriander, potato and papaya on pond dykes using the technique of multiple cropping with single or mixed crops throughout the year. Wider pond dykes (2 to 4 m), can support semi-intensive farming. Long-duration crops such as beans, papaya, mustard, okra, ridge gourd, tomato, brinjal, and chilli are considered appropriate for these types of dykes. Pond dykes (around 2 m wide) are suitable for extensive cultivation when mechanical plowing and irrigation are not feasible. After careful cleaning, weeding, and the creation of small pits along the dyke's length, these dykes can be utilized for the cultivation of papaya, citrus, sweet gourds, sponge gourds, bottle gourds etc. For shaded dykes, extensive turmeric and ginger cultivation is appropriate.

During harvest, a significant amount of leaves from turnips, radish, cauliflower, and cabbage are discarded. These can be used to complement grass carp feed profitably. Turnip, cabbage, and cauliflower leaves can be fed to grass carp in the winter, and amaranth and water-bind weed cut every two weeks can be fed as supplemental feed in the summer to help raise grass carp. An average of roughly 2 t ha^{-1} is produced annually by a monoculture of grass carp fed exclusively on vegetable leaves at a stocking level of 1000 fish per hectare. An average of 3 t ha^{-1} $year^{-1}$ can be achieved through polyculture of Indian major carps and grass carp, at a density of 5000 fish per hectare.

7.6.9. Integrated fish-water chestnut farming

Water chestnut (*Trapa bispinosa*), an annual aquatic angiosperm plant, found in tropical and subtropical regions, is suitable for integration with carp and air-breathing fish species. Further, water chestnut has a vast potential for its nutritional qualities for human beings. Moreover, it leaves all its biomasses in the same field where it grows. Henceforth, it contributes a lot of plant essential nutrients to the soil mass. Water chestnut is mainly grown in temperate and tropical Asia and central and south-east Europe. In India, it is extensively grown in Bihar, Madhya Pradesh, Uttar Pradesh, West Bengal, Odisha, Jharkhand, Karnataka, and Jammu and Kashmir. The plant is commercially cultivated for its nutritious nuts. The dried fruit contains about 70% carbohydrate, 13% protein, 2 to 7% sugar, 0.8% fat, 3% mineral elements and plenty of iodine including significant amounts of vitamins *viz.* thiamine, riboflavin, nicotinic acid, vitamin C, vitamin A and phosphorylases. The raw fruit contains 70% water, 23.3% starch (carbohydrate), 4.7% protein, 0.3% fat, 0.6% fibre and 1.1% minerals. The fruit can be consumed raw, boiled, roasted or dried to make flour. *Trapa bispinosa* is also an important plant of the Indian ayurvedic system of medicine which is used in the problems of the stomach, genito-urinary system, liver, kidney, and spleen. It is useful in diarrhoea, dysentery, ophthalmopathy, ulcers, and wounds. Water chestnut is also beneficial for women. Due to its nutritive, strengthening and cooling properties it increases fertility and chances of conception. It helps to stabilize pregnancy and prevent abortion.

In this dual production system, water chestnuts are transplanted @ 10,000-15,000 plants per ha. The 10% area in the middle of the pond/ waterbody is kept vacant as a refuge area for proper oxygenation and better growth of fishes. The fish species like catfishes (*Clarious batrachus*) @ 15000-20000 fingerlings per ha or murrels (*Channa* spp.) @ 10000-15000 fingerlings per ha or carps (rohu, catla, common carp and mrigal) @ 5000-6,000 advanced fingerlings per ha can be integrated in the ratio of 40:20:20:20, respectively between March and April, and harvesting of fishes gets completed between December and January. Water chestnut is harvested from October to November. Growth of fish is found appreciably higher in combination with water chestnut cultivation and 25 to 30% of the supplemental fish feed could be replaced with natural feed from the growing environment. The combined cultivation of the fish–water chestnut system can produce water chestnut @ 15-18 t ha^{-1}, carp @ 3.5 t ha^{-1} or air-breathing fish @ 1.7- 1.8 t ha^{-1} per crop. Net profit of Rs 1.2- 1.5 lakh ha^{-1} can easily be achieved from this co-production system. Coproduction of fish with water chestnut also has the potential to enhance the overall yield, water productivity, reciprocal advantage, and efficient utilization of shallow waterlogged areas for better productivity. Integrated cultivation of cat fish with water chestnut can reduce more than 65% of pest attacks by water chestnut beetle (*Galerucella birmanica*), as *magur* acts as a predator to *G. birmanica.*

7.6.10. Self-reliant agri-based IFS

An integrated farming system is the foreseeable management option for the small and marginal farmers of rainfed ecology to generate more income, food, and employment per unit area. Water and energy-efficient farming systems could be achieved through optimum cropping systems (rice-horse gram and rice-sunflower), less use of agro-chemicals (green manuring and vermicompost), mulching, ring method of irrigation (cucurbits), paired row bed planting (okra), multiple uses of water (pisciculture, horticulture, apiculture, poultry and mushroom cultivation), growing of widely spaced crop with large ground coverage (bottle gourd and watermelon) and red gram cultivation on field bunds. Residue recycling of poultry and duckery for pisciculture, paddy straw for mushroom cultivation, and bio-waste for vermicompost production saved energy on fish feed and fertilizer.

A self-reliant farming system model (3408.44 m^2) developed by ICAR-IIWM, Bhubaneswar, India was based on utilizing a water harvesting pond (30 m x 30 m x 2.8 m depth) for aquaculture, dyke around the pond (556 m^2) for on-dyke horticulture, field crop unit (1879 m^2) and field bund (73.44 m^2). Net return per ha and employment generation per ha were Rs 1,98,341 (without considering the cost of land modification) and 509.3, respectively under IFS as compared to wet season rice (Rs 27,982 and 157.7). Net water productivity under IFS was also very high (Rs 53.7 m^{-3}) due to multiple uses of water. Self-reliance on plant nutrients was achieved through green manuring, agri-waste recycling and the growing of legume crops. Growing paddy crops in the wet season under integrated nutrient management with in-situ green manuring resulted in higher energy output: input ratio (18.7) and net returns (Rs 27982 per ha) as compared to the use of chemical fertilizers alone (energy output: input ratio of 13.9 and net returns of Rs 23412 per ha). After paddy harvest, growing of legume (horse gram) under reduced tillage, and use of farm-generated vermicompost reduced energy requirement and resulted in high energy output: input ratio, water productivity, net returns and beneficial residual soil fertility (organic carbon, available N, P and K). This technology has the potential for implementation in around 12 lakhs ha in India as well as in other parts of Southeast Asia.

7.6.11. IFS using alternate raised and sunken bed system

The raised and sunken bed (RSB) is a land configuration specially designed by the cutting and filling method for low-lying areas. Such beds are prepared by removing the surface soil layer from an area and depositing the same on the adjacent area to a height as per the requirement. Adoptions of a permanent RSB system of cultivation with proper cropping sequence in rice paddies in lowlands or canal commands will not only increase the production and productivity of vegetables and cereal crops but also enhance cropping intensity, employment, and income substantially. This system helps in crop diversification and increases cropping intensity, year-round employment opportunities and profitability. This replicable system can very well be adopted in irrigated canal commands and medium and low-land rice ecosystems.

It is possible to diversify the cropping system with rice–pea or rice–lentil in sunken beds and vegetables such as tomato– okra–broccoli, tomato–brinjal–broccoli, etc., on raised beds with effective land utilization. Further, growing rice + fish in sunken beds of alternate raised and sunken bed systems (30 m length & 5 m width) instead of rice alone can give more yield (rice: 5.2 t ha^{-1}; fish: 3.7 t ha^{-1} 150 d^{-1}) and return. A combination of rice fish in sunken beds with cabbage (28.5 t ha^{-1}), pointed gourd (4.7 t ha^{-1}) and snake gourd (20.2 t ha^{-1}) crops in adjacent raised beds can give the highest rice equivalent yield (48.8 t ha^{-1}), water use efficiency (174.54 kg ha^{-1} cm^{-1}), net water productivity (Rs.13.3 m^{-3}) and benefit-cost ratio (4.78). In this system, fish fingerlings @ 5000 ha^{-1} is recommended. Fish culture in sunken beds helps in improving soil nutrient status while rice cultivation with fish improves water quality. This system has the potential to enhance production by 6-7-fold and water saving of up to 20%. The highest B: C ratio of 4.78 is recorded for rice-fish + snake gourd + pointed gourd + cabbage combination. The WUE as well as water productivity is also enhanced by about 3–4 times under RSB compared to rice monocropping. RSB technology has the potential to increase the productivity per unit of input including land and water while at the same time, it encourages self-employment and upliftment of the socio-economic condition of resource-poor farmers.

7.6.12. Integrated multi-trophic aquaculture (IMTA)

When inorganic (macroalgae/ aquatic weeds) and organic extractive (bivalves) species known as un-fed species are integrated with cultured or fed species in the same system, it is called integrated multi-trophic aquaculture (IMTA). IMTA is a way of farming multiple marine or freshwater organisms from different nutritional levels or trophic levels in the same system at the same time (Barrington *et al.*, 2009; Angel and Freeman, 2009). Complementary ecosystem functions in a way that allows one species' uneaten feed and wastes, nutrients, and by-products to be recaptured and converted into fertilizer, feed, and energy for the other crops taking advantage of synergistic interactions between species. The fed aquaculture species (e.g., finfish/shrimp) are combined with organic extractive aquaculture species (e.g., shellfish/herbivorous fish) and inorganic extractive aquaculture species (e.g., seaweed) to create balanced systems for environmental sustainability (biomitigation), economic stability (product diversification and risk reduction) and social acceptability (better management practices). It is a type of polyculture system with proper utilization of each trophic level and the generated wastes from one trophic level become the food for another. Better production with more desirable water quality is achieved compared to the age-old poly-culture practice (Biswas *et al.*, 2019). Typically, predatory fish or shrimp occupy the higher trophic levels. They excrete soluble inorganic ammonia, phosphorus (orthophosphate) and particulate organic carbons (fish faecal matter). Seaweeds and similar species can uptake these inorganic nutrients (inorganic nitrogen and phosphorous) directly from their environment. The organic nutrients viz. fish faecal matter and uneaten feeds, particulate organic matter (POM) is fed by shellfish and bottom feeders and their excretory inorganic products are again uptake by the seaweeds. Thus, the IMTA provides a new farming approach to increase food production by efficient utilisation of trophic levels in a sustainable way which ensures better biodiversity, biosecurity and ecosystem (Fig. 7.6).

Integrated Multi-Trophic Aquaculture = Fed Aquaculture (Finfish) + Extractive Aquaculture i.e., Organic (Shellfish) Inorganic (Seaweed)

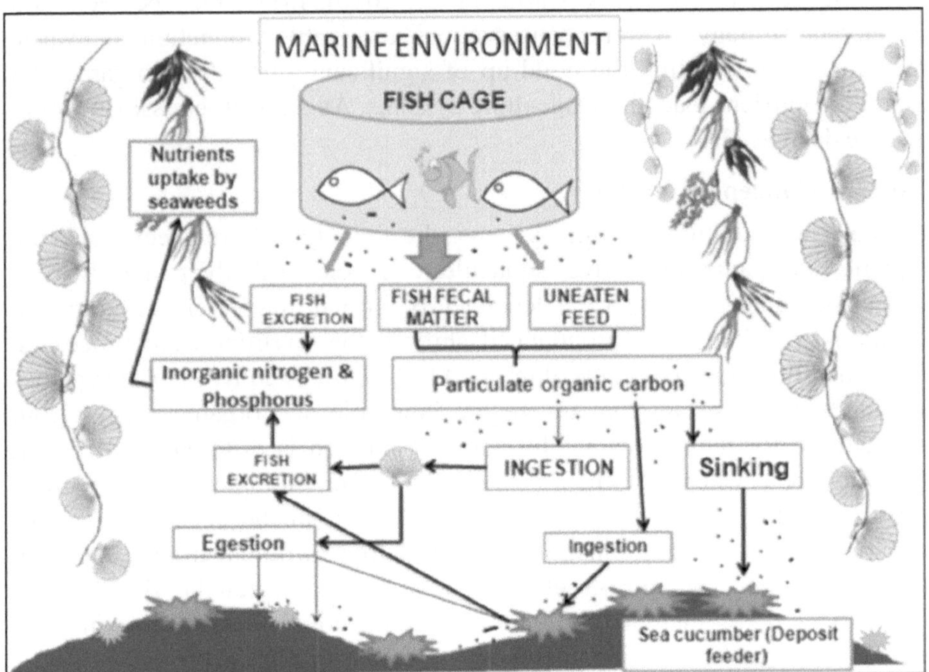

Fig. 7.6. Flow diagrammatic representation of Integrated Multi-Trophic Aquaculture system, IMTA

Theoretically, in an IMTA farm, low-trophic extractive species like bivalves, macroalgae, and detritivores might utilize metabolic waste and uneaten feed from fed species, reducing any potential environmental consequences. The natural habitat where IMTA farms are located, farming procedures, and the selected extractive species will all have a significant impact on how effectively they mitigate environmental change because of the way these species eat, which is essential to the movement of matter and energy within the farms. Although the configuration of an IMTA farm can vary greatly (Chopin *et al.*, 2012), three major types of extractive species have been considered in the literature: seaweeds, bivalves, and bottom feeders. According to Troell *et al.* (2009), seaweeds can lessen the amount of dissolved nutrients in the water column. Bivalve farms' ability to filter water could help to counteract the negative impacts of particulate organic matter (Troell *et al.*, 2009). Bottom feeders like sea cucumbers could be included in the IMTA system to reduce particle organic loading while being considerably less frequently discussed in the literature (Zhang and Kitazawa, 2016). IMTA has been promoted as a farming method to diversify products and reduce operational risks, even if this mitigation potential has been the main justification for creating and putting into practice IMTA systems (Carras *et al.*, 2020). IMTA may be seen as a developing strategy for water conservation, sustainable production, and waste recycling by preventing the build-up of significant amounts of organic and inorganic nutrient loads in aquaculture systems. As a result, emissions would also be decreased.

7.6.12.1. Concept of IMTA

In many monoculture farming systems, the fed-aquaculture species and the organic/ inorganic extractive aquaculture species (bivalves, herbivorous fishes and aquatic plants) are independently farmed in different geographical locations, resulting in a pronounced shift in the environmental processes. Integrated multi-trophic aquaculture (IMTA) involves cultivating fed species with extractive species that utilize the inorganic and organic wastes from aquaculture for their growth. This farming method is different from finfish "polyculture", where the fishes share the same biological and chemical processes which could potentially lead to a shift in the ecosystem. Multi-trophic refers to the combination of species from different trophic levels in the same system. The multi-trophic sub-systems are integrated in IMTA which refers to the more intensive cultivation of the different species in proximity to each other, linked by nutrient and energy transfer through water.

Selection of species and environmental sustainability is the major consideration in IMTA. Fed organisms, such as carnivorous fish and shrimp are nourished by feed, comprising of pellets or trash fish. Extractive organisms derive their nourishment from the environment. The two economically important cultured groups that fall into this category are bivalves and seaweed. A combination of co-cultured species will have to be carefully selected according to several conditions and criteria:

i) *Complementary roles with other species in the system:* Species should complement each other on different trophic levels. For example, species must be able to feed on the other species' waste for the newly integrated species to improve the quality of the water and grow efficiently. Not all species can be grown together efficiently.

ii) *Adaptability about the habitat:* Native species can be used that are well within their normal geographic range and for which technology is available. This will help to prevent the risk of invasive species causing harm to the local environment, and potentially harming other economic activities.

iii) *Culture technologies and site environmental conditions:* POM and dissolved inorganic nutrients should be considered, as well as the size range of particles, when selecting a farm site.

iv) *Ability to provide both efficient and continuous bio-mitigation:* Species be used that are capable of growing to a significant biomass. The organisms need to act as a bio-filter that captures many of the excess nutrients and that can be harvested from the water. The other alternative is to have a species with a very high value, in which case lesser volumes can be grown. However, with the latter, the bio-mitigating role is reduced.

v) *Market demand for the species and pricing as raw material or for their derived products*: Species be used those that have an established or perceived market value. Farmers must be able to sell the alternative species to increase their economic input. Therefore, they should establish buyers in markets before investing too heavily.

vi) *Commercialization potential:* Species be used for which regulators and policymakers will facilitate the exploration of new markets, not impose new regulatory impediments to commercialization.

7.6.12.2. Inorganic extractive sub-system in IMTA

Since aquatic plants' biofiltration involves assimilation, it increases the environment's potential to absorb nutrients. Plants photosynthesize new biomass using solar energy and extra nutrients, primarily C, N, and P. The process creates a miniature ecosystem in the culture system where, in a well-balanced system, fish and microbial heterotrophy are balanced out by plant autotrophy in terms of nutrients as well as oxygen, pH, and CO_2. As a result, plant bio-filters can preserve the cultural environment while drastically reducing the overall environmental effect of fish farming. Furthermore, relatively little input is required when farming animals that are low on the food chain and get their nutrition from the water. Because they can be inexpensively grown and have the highest production of any plant, seaweeds are the best choice for bio-filtration. As phycocolloids, feed additives, agrichemicals, nutraceuticals, and medicines, seaweeds are widely used for human nutrition. In marine cage culture zones, seaweed farming has long been advocated in China as a means of bio-extracting nutrients from the seawater. According to FAO aquaculture statistics, there are 37 different kinds of seaweed. Eucheuma seaweeds, which are grown in tropical and subtropical waters and account for 8.44 million tonnes, are dominant, followed by Japanese kelp (5.94 million tonnes).

A seaweed species is selected for inclusion in an integrated aquaculture system only after meeting a number of fundamental requirements, such as a high growth rate and tissue nitrogen concentration, ease of cultivation and life cycle control, resistance to epiphytes and disease-causing organisms, and a match between the ecophysiological characteristics and the growth environment. Because the introduction of non-native creatures may cause ecological harm, the seaweed should also be a native species. Beyond these fundamental requirements, the intended use will affect the choice of seaweed. The quality of the tissue and secondary compounds' added value will be taken into consideration in the decision-making process if the value of the biomass produced is the primary consideration. Nutrient intake, storage, and growth are the main variables if the bioremediation process is the major emphasis. The optimal system would include a seaweed species that incorporate both value and bioremediation. The 'thin sheet' morphology of seaweeds grows more quickly than that of squishy seaweeds. On nutrient sequestration, generalization is more challenging. A type of seaweed that is utilized as a biofilter needs to be able to flourish in conditions that are high in nutrients, particularly ammonium. This is a rare quality in seaweed applications. Seaweed that grows quickly should be able to absorb nitrogen quickly and build up a high biomass nitrogen content. Grown in eutrophic conditions, the typical bio-filter seaweeds accumulate a significant amount of total internal nitrogen. The potential of a small number of seaweeds for aquaculture and/or bioremediation has received much research. Probably the most comprehensive collection of studies pertaining to the genus Ulva. These flat sheet morphotypes are excellent candidates for cleanup because of their high nitrogen concentrations and associated rapid growth rates. In mid-to large-scale animal mariculture systems, Ulva has been successfully integrated because its life cycle and its regulations are generally well understood. The little market for harvested biomass may be the only negative aspect. While Laminaria and Porphyra cultivations are regarded to have the ability to produce sustainable seaweed mariculture and integrated aquaculture, Gracilaria has a history of mariculture research. Along the Indian coast, *Kappaphycus alvarezii* is employed in integrated open-sea aquaculture (IMTA) and has proven to be a promising species.

7.6.12.3. Organic extractive species sub-system in IMTA

Filter-feeding bivalves are cultured next to meshed fish cages in a conceptual open-water integrated culture system, reducing nutrient loadings by filtering and assimilating particulate wastes (fish feed and feces), as well as any phytoplankton production stimulated by introduced dissolved nutrient wastes. Instead of being lost to the surrounding environment as in typical monoculture, waste nutrients are collected during the harvest of cultured bivalves. Literature shows that 95% of particles released from aquaculture systems, fish farms, and closed recirculation systems are ~20 microns in diameter (5–200-micron range) and that they will settle. There is evidence that filter feeders are selective in extracting particles from the water column, rejecting the rest. Thus, it is important to know the particle size of wastes from an IMTA system and to choose from among the wide range of bivalves that will select the required particle size and type. The marketability of these secondary products is a factor, but it need not be an overriding consideration. Studies have shown that bivalves are capable of utilising fish farm wastes as an additional food supply. The bivalve mussel, *Perna viridis* and oyster *Crassostrea madrasensis* are commercially produced along the Indian coast, which can mitigate eutrophication in integrated aquaculture/ IMTA.

7.6.12.4. Fed-aquaculture species sub-system in IMTA

The only fed component in the majority of IMTA systems, finfish are the only source of nutritional energy supplied by humans to the system. Fish contribute dissolved and particulate nutrients, oxidation-reduction potential reducing chemicals, and revenue to the industry in an IMTA system. Species, size, and feed composition are only a few of the variables that affect the amount and form of these nutrients. The most obvious way to modify fish effluent for the extractive components of an IMTA system is probably through feed formulation. There are differences between IMTA systems that are semi-closed to the environment (recirculation aquaculture systems) and those that are open to the environment (cage-based). Most open systems provide environments that are both essential and adequate for raising extractive species, but semi-closed systems demand considerably tighter coupling of the various trophic levels that are being cultivated. For the industry to be lucrative, the choice of fish species for open and closed systems would probably vary to take use of each system's particular characteristics.

7.6.12.5. Case studies on IMTA

The countries viz. Canada, Chile, China, Ireland, South Africa, the UK Northern Ireland (mostly Scotland) and the USA have IMTA systems in temperate waters, which are near commercial scale. France, Portugal and Spain have ongoing research projects related to the development of IMTA. The countries of Scandinavia, especially Norway, have made some individual groundwork towards the development of IMTA, despite possessing a large finfish aquaculture network (Barrington *et al.* 2009). Studies have focussed on the integration of seaweeds with marine fish culture for the past fifteen years in Canada, Japan, Chile, New Zealand, Scotland and the USA. Recent IMTA research focuses on seaweeds, bivalves and crustaceans. Studies conducted in an IMTA system incorporating *Gracilaria lemaneiformis* and *Chlamys farreri* in North China have shown that a bivalve/seaweed biomass ratio from 1:0.33 to 1:0.80 was preferable for efficient nutrient uptake and for maintaining lower nutrient levels. Results indicate that *G. lemaneiformis* can efficiently absorb the ammonium and phosphorus from scallop excretion. In China, Seaweeds, *Gracilaria lemaneiformis*, grown over 5 km of culture ropes near fish net pens on rafts increased the density from 11.16 to 2025 g m^{-1} in a 3-month growing period. The scaling up of the culture area during the following 4 months to 80 km of rope, reported an increase in culture density on ropes to 4250 g m^{-1}. An increase in the biomass of *Gracilaria* (in the culture area) to 340 t wet-weight was estimated due to its culture near fish net pens.

Studies on IMTA have been conducted on Canada's east coast, where blue mussels (*Mytilus edulis*), kelp (*Saccharina latissima* and *Alaria esculenta*), and Atlantic salmon (*Salmo salar*) were all raised alongside one another at several IMTA sites in the Bay of Fundy. According to the study, kelp and mussels grown close to fish farms experienced growth rates that were 46 and 50% greater than those at control locations, respectively. The rapid growth of mussels and oysters raised close to fish cages has been the subject of numerous further researches. This is due to the finfish cages' increased availability of food and nutrients. Economic model results have also demonstrated that, when compared to monoculture, increased total net productivity at a specific IMTA site can increase farm profitability. Seaweeds can remove between 35% and 100% of the dissolved nitrogen generated by feeding species, according to studies from land-based systems. Based on the percentage of accessible nutrients that are bound by seaweeds at any given time, the ability of seaweeds in open-water cultures to extract nutrients from the water column can be assessed. For the complete elimination of the excess nitrogen associated with a commercial fish farm, experimental results and mass balance calculations suggested that a sizable area of seaweed cultivation, up to one ha for every ton of fish standing stock, would be needed.

Although the open-sea IMTA in India is relatively new, numerous studies on the advantageous polyculture of the various mariculture species have been conducted. To increase field productivity and efficiently utilize the biological niches present in the pond system, combined culture of suitable species of prawns and fish is of significant significance. High survival rates and growth rates were achieved by cultivating *Etroplus suratensis* in cages built inside the bivalve farms (racks). Co-cultivation of *Gracilaria* sp. at different stocking densities with *Feneropenaeus indicus* showed nutrient removal from shrimp culture waste by the seaweed. The ratio of 3:1 was found suitable for the co-cultivation. From the shrimp (200 g) waste, the seaweed (600 g) was able to remove 25% of ammonia, 22% of nitrate, and 14% of phosphate. Shrimp and mollusk polycultures aid in the efficient breakdown of organic matter, serve as a significant source of food for a variety of organisms, and either directly or indirectly create space or shelter for associated organisms, increasing the species diversity of the ecosystem. According to studies, a single mussel can filter 2 to 5 litres per hour, while a rope of mussels may filter more than 90,000 litres per day. Thus, mussel culture could be utilized to effectively remove phytoplankton and other debris and to lessen the eutrophication brought on by aquaculture. When combined with *Rachycentron canadum* finfish farming along the east coast of India, the use of IMTA in open sea cage farming resulted in a 50% increase in the production of the seaweed *Kappaphycus alvarezii*. When green mussels, *P. viridis*, and open-sea mariculture of fish were combined, the nutrients along Karnataka were somewhat but not significantly reduced. The beneficial effect of combining bivalves such as mussels, oyster and clams as bio-filters in utilizing such nutrient-rich aquaculture effluents has been documented in estuaries. The farming of bivalves (*Crassostrea madrasensis*) and finfish (*Etroplus suratensis*) in a tropical integrated aquaculture system had successful eutrophication management (Viji *et al.,* 2014; Viji, 2015). Oysters that were fed filters made the water in the agricultural area clearer, which decreased eutrophication. In this farming technique, the ideal ratio for co-cultivation of fish and oysters was observed to be 1:0.5.

Application of IMTA principle in brackishwater aquaculture involving (1) *Chanos chanos, Etroplus suratensis, Mugil cephalus and Penaeus indicus* as fed species and *Crassostrea madrasensis* as an extractive species (Balasubramanian *et al.,* 2018; Biswas *et al.,* 2019), (2) *Mugil cephalus, Liza tade and Peneaus monodon* as main fed-species, and oyster, *Crassostrea cuttackensis* and water spinach, *Ipomoea aquatica* as extractive species (Biswas *et al.*, 2020), (3) *Chanos chanos* and *Penaeus vannamei* as fed species and green seaweed, *E. intestinalis* as an extractive species (Naskar *et al.*, 2023) and (4) *C. chanos, P. vannamei* as fed species and estuarine oyster, *Crassostrea cuttackensis* as the extractive species (Naskar *et al.*, 2022) has confirmed that the BIMTA (Brackishwater Integrated Multi-Trophic Aquaculture) system is a better performer than brackish water polyculture in all aspects. For the development of a field application, this kind of small-scale BIMTA model can also be evaluated in a brackish water pond system with various species combinations and densities. To improve the BIMTA system, it is necessary to investigate edible aquatic weeds or seaweeds with commercial potential as inorganic extractive species, as well as their suitability for cultivation in brackish water ecosystems.

7.6.12.6. Benefits and advantages of IMTA

- *Effluent bio-mitigation*: Mitigation of effluents through the use of bio-filters which are suited to the ecological niche of the aquaculture site.
- *Increased profits through diversification*: Increased overall economic value of an operation from the commercial by-products that are cultivated and sold. By exploiting the extractive capacities of co-cultured lower trophic level taxa, the farm can obtain added products that can outweigh the added costs involved in constructing and operating an IMTA farm.
- *Improving local economy:* Economic growth through employment (both direct and indirect) and product processing and distribution.
- *Form of 'natural' crop insurance:* Product diversification may offer financial protection and decrease economic risks when price fluctuations occur, or if one of the crops is lost to disease or inclement weather.
- *Disease control:* Prevention or reduction of disease among farmed fish can be provided by certain seaweeds due to their antibacterial activity against fish pathogenic bacteria.
- *Premium prices and increased profits:* Potential for differentiation of the IMTA products through eco-labelling or organic certification programmes.
- Offers tremendous potential for food and nutritional security.
- Eco-friendly, promotes environmental sustainability by converting by-products, wastes and uneaten feeds, efficient utilisation of all the trophic levels, and reducing eutrophication.

7.7. Water budgeting in aquaculture-based IFS

Humans depend on water to survive, and social-ecological systems depend on it as a resource. Most living things are made of at least 50% water, which is essential to the ecosystem's ability to function. However, freshwater supplies are limited. Freshwater makes up just 2.5% of global water resources, and 70% of that is frozen in glaciers. 70 000 km^3 of the 110000 km^3 of freshwater that fall on Earth each year evaporate into the atmosphere. According to Mohanty *et al.* (2014), only 12500 km^3 of the 40000 km^3 are usable by humans. Amidst worldwide environmental and socioeconomic shifts, the scarcity of water remains a major concern in different parts of the world. Water disputes between users at many scales, ranging from the local to the transnational, may intensify due to changes in water availability. For a variety of reasons, the most fundamental resources for food production—land and water—have garnered a lot of attention in the discussion of climate change. The most pressing issue is how we will feed a population that is expanding and has an increasing need for higher-quality food and larger proportions of animal products (Kearney, 2010). Non-sustainable use also depletes freshwater and fertile land, two valuable resources. The availability of freshwater will also alter as a result of climate change. There is a great need to raise aquacultural water productivity since aquaculture production must be enhanced but there is a limited quantity of water. Therefore, the primary strategy for enhancing aquaculture's environmental performance is the application of improved management practices through intensification of currently operating systems with a focus on BMP. To increase aquacultural water productivity for a certain scenario or hydro-ecological state, a variety of technical methods are available. The management of inputs, particularly feed, and the availability of blue water for culture are the two main prerequisites for increasing aquaculture water productivity. Using the compensatory growth (CG) response and minimizing needless water exchange are also seen to be effective strategies for raising water productivity and profitability in aquaculture operations. Aquaculture's sustainability and rising production intensity are not mutually exclusive. On the other hand, increased production intensity is necessary for aquaculture sustainability.

The sustainability of aquaculture has been called into question on numerous occasions; consequently, regional and international organizations have put forth BMP (Best Management Practices) to improve sustainable production and make aquaculture environmentally responsible. Extreme weather, excessive freshwater use, organic pollution, disease, chemical contamination, and other dangers all pose a threat to the ecological viability of pond aquaculture. The availability of restricted land and water resources is now a constraint on the amount of area that may be used for aquaculture, even if production must rise to meet the increasing demand. Furthermore, fish production uses a lot of water—much more than conventional agriculture does—and the availability of freshwater will limit its potential to flourish in the future (Verdegem and Bosma, 2009). The unplanned, wasteful use of water in aquaculture is another factor impeding the industry's ability to grow. As is clear, aquaculture can have extremely high on-farm water usage, reaching up to 45 m^3 per kg of biomass produced in ponds (Verdegem *et al.*, 2006). To make the system more water-efficient, aquaculture production systems must be intensified in order to reduce the amount of water used per kilogram of biomass produced. Since inland aquaculture will no longer have unlimited access to water, increased efforts in input management—primarily feed—and water quality monitoring, along with the quantification and estimation of the water requirements of commercially significant, high-value fish and prawn species, will guarantee increased water productivity and profitability. Therefore, it is necessary to popularize new technologies and practices that will make fresh and brackish water aquaculture systems more resource-efficient and sustainable in the future. These include improving water use efficiency/productivity, nutritional water productivity, and lowering water footprint.

7.7.1. Importance of aquacultural water budgeting and water productivity

It is now essential to reduce operating costs by increasing water use efficiency due to the issue of low economic output in grow-out aquaculture (caused by increased price of feed, chemicals, aqua-drugs, fuel, power supply, etc.). In order to boost water productivity in the aquaculture industry, we have been obliged to reconsider the prudent use of water due to monsoon rain variability, shortage, and limited availability of freshwater resources. Water of days is more harder to get by and more expensive to obtain (pumping expense). The freshwater reserve and supply around the world, and in India specifically, are currently under jeopardy because of growing population pressure and rising water demands in home, industrial, and agricultural sectors. Because water is a scarce resource, water management should take a more comprehensive approach. Furthermore, the establishment of protocols for best water management practice (BWMP) in commercially significant grow-out aquaculture should start with water budgeting and its prudent application. The main water losses in static water ponds are evaporation, percolation, and seepage. This leads to low water productivity because of nutrient loss and water quality fluctuations. Resupply/exchange of water becomes crucial to replace/maintain this water loss, pond fertility, and the survival of stocked animals. Farmers frequently exchange water in an attempt to increase yield, but they frequently fail to account for the expense and necessity of doing so, which can occasionally have unfavourable effects. Therefore, determining the precise amount of water needed is crucial and depends on a number of variables, including species, rearing density, biomass, growth stage, plankton and nutritional status, water loss, agroclimatic conditions, etc. The amount of water needed depends on a number of factors, including the soil, climate, species to be cultured, culture technique, and management practices. In order to avoid the issue of water waste, it is very much required to determine whether replenishment or exchange is necessary and then quantify the amount of water needed for replenishment. The water budgeting for various species and production goals may serve as a beneficial tool for producing information that will help mitigate the water-related issues facing aquaculture production.

Aquaculture uses water in two ways: consumptive use and total use. Aquaculture's overall water usage varies substantially, mostly according on the cultivation technique employed. Cage, Biofloc and net pen culture use minimal water, while intensive pond and raceway cultures use the most. Fish production typically requires total water use of 4 - 8 m^3/kg fish in embankment ponds and 8 - 16 m^3/kg fish in watershed ponds, however, water use in ponds varies with the intensity of production, frequency and amount of water exchange employed (Boyd, 2005; Boyd *et al.*, 2007). As of right now, super-intensive recirculation systems use as little as 500–700 l of water per kilogram of biomass output, whereas extensive pond systems use as much as 45 m^3 of water per kilogram of produce (Verdegem *et al.*, 2006). The degree of water exchange is a major factor in figuring out how efficiently aquaculture uses water. But for the majority of pond aquaculture, water exchange is not required (Boyd and Tucker, 1998). Minimizing or doing away with water exchange conserves water and lowers pumping expenses. Additionally, the hydraulic retention time (HRT) in ponds rises with less water exchange. This lessens the amount of possible contaminants in effluent and enables natural mechanisms to digest wastes more thoroughly (Boyd, 2005). In static ponds, the HRT is typically weeks or even months; in ponds that exchange water, HRT is typically a week or slightly longer (Boyd *et al.*, 2007).

By 2050, increasing aquacultural water productivity at all levels would be necessary to offset decreased water diversion for aquaculture and rising food demands. In its broadest meaning, water productivity seeks to use less water to produce more food, money, improved livelihoods, and ecosystem services. The net return on a unit of water utilized, or the ratio of net benefits from livestock, forestry, fisheries, crops, and mixed agricultural systems to the volume of water required to generate such advantages, is known as water productivity.

Thus, "more crop per drop" refers to physical water productivity, which is the ratio of aquaculture output to water consumption, while "value derived per unit of water used" refers to economic water productivity. 'Increasing or improving water productivity' refers to the most efficient way to use the water that is now available to us to increase crop yields. More water is not needed when there is greater water productivity. A number of factors, including water withdrawal, water consumption, and virtual water usage, must be taken into account when evaluating the sustainability of water use. Water extracted for aquaculture purposes from streams, rivers, or aquifers is referred to as water withdrawal. Following withdrawal, some of the water is returned and can be recycled or reintroduced into the ecosystem. The portion that is not returned stands for water that has been consumed—that is, water that evaporates or becomes absorbed by products and living things. Indirect water use, such as water used to make fish feed, is referred to as virtual water use. The virtual water use should be included in the total amount of water used. Currently, 1.7 m^3 of water per kg of production of grains used in fish feed is indirectly used due to evaporation. As fish feeds contribute more to overall inland aquaculture production in the future, so will grain-associated water consumption; however, with current technology, this consumption will peak at 3 m^3 water per kilogram production. According to Verdegem and Bosma (2009), pond evaporation accounts an additional 5.2 m^3 of water per kg of production on average. In inland aquaculture, freshwater removal averages 16.9 m^3 per kilogram of production; however, if pollution is managed, infiltration losses (6.9 m^3) and water replacement (3.1 m^3) might be classified as green water (Verdegem and Bosma, 2009). The amount of infiltration, which is often estimated at 5 to 10 mm/d and depends on the soil type and topography of the ponds, is a function of both bottom percolation and lateral seepage. However infiltration losses from fish ponds will be limited to lateral seepage to nearby farms and waterways when the groundwater table is high, as it is during the monsoon season. Optimizing aquaculture can significantly lower the amount of evaporation loss per kilogram of production; therefore, studies should concentrate on raising pond water productivity while minimizing environmental effects.

7.7.2. Water productivity and economic efficiency

Water is the primary natural resource, thus conserving it and using it wisely is essential to improving productivity and preserving quality in the present-day context. By 2030, there will be a greater need for aquaculture water productivity due to rising water diversion for industrial, agricultural, and aqua-food sectors. The goal of providing more food, income, livelihood, and ecological advantages at a lower social and environmental cost per unit of water spent is reflected in aquacultural water productivity, which is defined as the ratio of the net benefits from aquacultural systems to the amount of water utilized (Molden *et al.*, 2010). Additionally, water productivity defines the relationship between the amount of water used in crop production and the quantity of biomass produced, and is an index of the economic value of water used (Boyd, 2005; Dasgupta *et al.*, 2008; Ali and Talukder, 2008). Increased water productivity thus saves operating costs while also lowering the requirement for more water.

7.7.3. Hydrological water balance study (water budgeting)

The study of hydrological water balance not only helps in the estimation of density-dependent water usage in aquaculture operations, but also contributes to increased productivity and efficiency in water use (Fig. 7.7). The hydrological water balance equation, which reads inflow = outflow ± change in volume (ΔV), is used to calculate the exact amount of water used in ponds. There are two types of water use in aquaculture: consumptive water use (CWU) and total water use (TWU). TWU, or likely inflows to ponds, is calculated as follows: runoff (R) + precipitation (P) + management additions or controlled inflows (I) + initial water filling (Wf). Potential outflows (CWU) are defined as follows: water content in harvested biomass (Wb) + overflow (Of) + evaporation (E) + seepage (So) + transpiration (T) + purposeful discharge or regulated discharge (D). Non-consumptive water usage (NWU) is the difference between total and consumptive water use. Commercial shrimp and fish ponds seldom receive direct input from creeks or streams. Moreover, water is hardly ever utilized for purposes other than farming, and aquatic weeds are forbidden from growing in ponds or along their margins. Consequently, transpiration and creek input are rarely important issues. Because embankment fish/shrimp ponds are small watersheds, groundwater inflow is rarely an issue and runoff is minimal. The harvested biomass's very small amount of water content (about 0.75 m^3/t) can be disregarded. Thus, the appropriate water balance equation is:

$$P+I = E + S_o + O_f + D \pm \Delta V \qquad (1)$$

A water level gauge can be installed in each pond to quantify the average water loss (evaporation + seepage) in order to estimate the CWU. When adding and withdrawing water, the outflow and inflow are also recorded using a water level gauge. Additionally, the following equation is used to estimate evaporation in order to distinguish it from regular water loss.

$$\textit{Evaporation (mm) from Pond} = \textit{Pond-pan coefficient} \times \textit{Class-A pan evaporation (mm)} \qquad (2)$$

In the preceding equation, a pond pan coefficient of 0.8 - which is best suited for ponds—may be utilized. The evaporation loss is subtracted from the total loss to determine the amount of seepage from a pond. The following is an estimate of the consumptive water usage index (CWUI), which indicates the volume of water consumed per unit of fish or shrimp production.

$$CWUI = CWU\ (m^3) / \text{total biomass (kg)} \quad \text{.........(3)}$$

7.7.4. Water use and economic efficiency

The gross total water productivity (GTWP), net total water productivity (NTWP), and net consumptive water productivity (NCWP) can be computed while accounting for the entire volume of water utilized, as shown below, to assess the effectiveness of water management.

$$GTWP = \text{Total economic value of the produce (Rs.)} / TWU\ (m^3) \quad \text{.........(4)}$$

$$NTWP = \text{Total economic value of the produce (Rs.) - Production cost (Rs.)} / TWU\ (m^3) \quad \text{.........(5)}$$

$$NCWP = \text{Total economic value of the produce (Rs.) - Production cost (Rs.)} / CWU\ (m^3) \quad \text{.........(6)}$$

To evaluate the efficacy of water management, total water use efficiency (WUE_t) and consumptive water use efficiency (WUE_c) are estimated as follows:

$$WUE_t\ (kg\ m^{-3}) = \text{Biomass production in kg } ha^{-1} / TWU \text{ in } m^3 \quad \text{.........(7)}$$

$$WUE_c\ (kg\ m^{-3}) = \text{Biomass production in kg } ha^{-1} / CWU \text{ in } m^3 \quad \text{.........(8)}$$

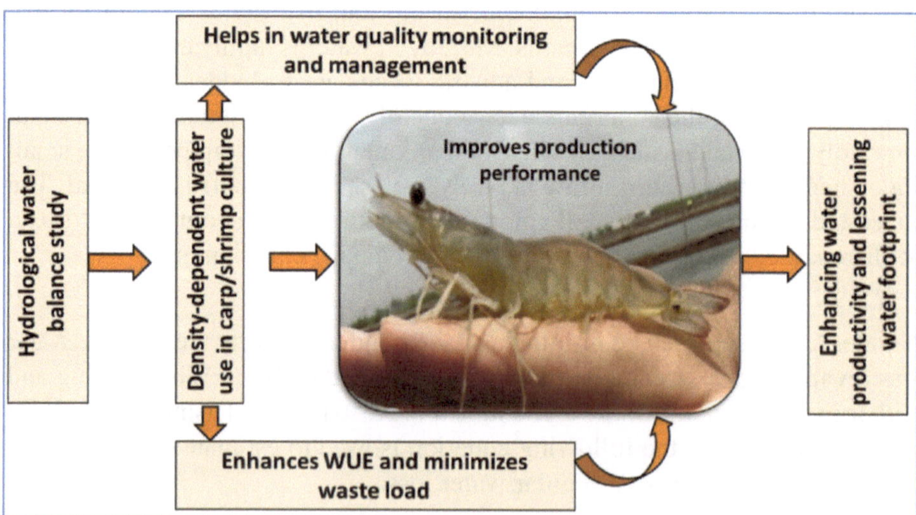

Fig. 7.7. Advantages of hydrological water balance study.

7.7.5. Water footprint

In general, blue and green water contribute to water footprint (cubic metre of consumptive water used per tonne of fish/shrimp produced, $m^3 t^{-1}$) in aquaculture. Water footprint is estimated considering all four components of water loss to the catchment such as (1) evaporation or evapotranspiration, (2) water content in harvested biomass, (3) contaminated water, and (4) non-returned water to the same area of withdrawn. However, seepage and percolation loss are not considered for the estimation of water footprint as these losses are not a loss to the catchment and later can be reused in the same area. Usually, blue (surface and groundwater) and green (precipitation) water contribute to water footprint in aquaculture. Thus, the suitable equation is:

$$\textit{Total water footprint } (WF_t, m^3 t^{-1}) = (D_i + O_f + E + W_b) / E_{cy} \quad \ldots\ldots (9)$$

Where, D_i = intentional or regulated discharge in m^3, O_f = overflow including other losses in m^3, E = evaporation in m^3, W_b = water content of harvested biomass in m^3 and E_{cy} = Economic crop yield in t ha^{-1}.

7.8. Conclusion

Integrated farming means combining crop production with livestock management that complement each other in a way that is a well-symbiotic relationship that is currently economically viable and profitable, environmentally friendly, and benefiting from the diversity of production. IFS are the need of the hour. The interplay of soil, water, animals, plants, and the environment makes the IFS a viable choice. IFS provides various ecosystem services in terms of food security, nutritional security, nutrient recycling, biodiversity conservation, increased employment, enhanced profitability and climate resilience. In the integrated farming system, agriculture can be integrated with livestock, poultry, and fish kept in one place to create year-round employment and additional income.

On the flip side, the aquaculture sector is expanding, but resource efficiency in production must be increased. The term "Blue Revolution" describes the swift expansion of aquaculture, which calls for "ecological aquaculture," which incorporates concepts of aquaculture water management and goes "green." Finding the ideal quantity of water required for aquaculture success is essential given the growing shortage of water resources and the risks of environmental contamination. Responsible practices are necessary for aquaculture's future growth in order to increase operational effectiveness, aid stop water waste, and maintain the pond water quality. To improve aquaculture performance, a hydrological water balance research to identify density-dependent water use is essential. Therefore, three key elements for a successful aquaculture under IFS or monoculture are water budgeting, density-dependent water usage, and water quality monitoring. Furthermore, aquaculture that uses demand-driven low to moderate water exchange helps minimize the water footprint, sediment load, and effluent outputs while simultaneously maintaining water quality that is conducive to growth.

References

Ahmed, M., and Lorica, M.H. 1999. Improving developing country food security through aquaculture development: socioeconomic and policy issues. Paper Presented at World Aquaculture 99, the Annual International Conference and Exposition of the World Aquaculture Society, Sydney, Australia, 26 April–2 May (No published proceedings).

Ahmed, S.H. and Sign A.K. 1989. Correlation between antibiotic factors of water and zooplanktonic communities of a tank in Patra, Bihar. In proceedings of National Seminar on forty years of Fresh water Aquaculture in India, Central Institute of Fresh water Aquaculture, Bhubneshwar. pp. 119-121.

Ali, M.H. and Talukder, M.S.U. 2008. Increasing water productivity in crop production- A synthesis. *Agricultural Water Management*, 95: 1201-1213.

Angel, D. and Freeman S. 2009. Integrated aquaculture (INTAQ) as a tool for an ecosystem approach in the Mediterranean Sea. *In*: Integrated mariculture: a global review (ed. Soto, D.). FAO Fisheries and Aquaculture Technical Paper, 529: 133-183.

Athithans, S., Francis, S. Ramanthan and Ramandas, U. 2001. Cattle urine as a benefitting source of nutrients for production. *Indian Journal of Fisheries,* 48:231-235.

Ayinla, O. A. 2003. Integrated fish farming: A veritable tool for poverty alleviation/Hunger eradication in the Niger Delta Region. *In*: Eyo, A.A and Atanda, J.O, (eds.) Conference Proceedings of Fisheries Society of Nigeria, Owerri, Nigeria. pp. 40-41.

Balasubramanian, C.P., Mhaskar, S.S., Sukumaran, K., Panigrahi, A., Vasagam, K., Kumararaja, P., Thigale, D., Sawant, R., Vijayan, K.K. and Vasudevan, N. 2018. Development of integrated multi-trophic aquaculture (IMTA) for tropical brackishwater species in Sindhudurg District, Maharashtra, west coast of India. *Indian J. Fish*, 65:59-64.

Banerjee,R.K., P.Ray, G.S.Sing and B.R.Dutta, 1979.poultry droppings and its manurial potentiality in aquaculture. *Journal of Inland Fisheries Societies,* 2:94-108

Barash, H., Plavnik, I. and Moav, R. 1982. Integration of duck and fish farming: experimental results. *Aquaculture,* 27:129-140.

Bardach, J.E. 1972. The farming and husbandry of fresh water and marine organisms. *Aquaculture,* John Wiley and Sons, New York, USA.

Barrington, K., Chopin, T. and Robinson, S. 2009. Integrated multi-trophic aquaculture (IMTA) in marine temperature waters. *In*: Integrated mariculture: a global review (ed. Soto, D.). FAO Fisheries and Aquaculture Technical Paper, 529: 7-46.

Behera, U. K. and France, J. 2016. Integrated farming systems and the livelihood security of small and marginal farmers in India and other developing countries. *Advances in Agronomy,* 138, 235-282.

Biswas, G., Kumar, P., Ghoshal, T.K., Kailasam, M., De, D., Bera, A., Mandal, B., Sukumaran, K. and Vijayan, K.K. 2020. Integrated multi-trophic aquaculture (IMTA) outperforms conventional polyculture concerning environmental remediation, productivity and economic return in brackishwater ponds. *Aquaculture*, 516:734626.

Biswas, G., Kumar, P., Kailasam, M., Ghoshal, T.K., Bera, A. and Vijayan, K.K. 2019. Application of integrated multi-trophic aquaculture (IMTA) concept in brackishwater ecosystem: the first exploratory trial in the Sundarban. *Indian Journal of Coastal Research,* 86:49-55.

Boyd, C. E. 2005. Water use in aquaculture. *World Aquaculturist,* 36(3):12-15 and 70

Boyd, C.E. and Tucker, C.S. 1998. Pond Aquaculture Water Quality Management. Boston, MA: Kluwer Academic Publishers.

Boyd, C.E., Tucker, C.S., Mcnevin, A., Bostick, K. and Clay, J. 2007. Indicators of Resource Use Efficiency and Environmental Performance in Fish and Crustacean Aquaculture. *Reviews in Fisheries Science*, 15:327-360.

Carras, M.A., Knowler, D., Pearce, C.M., Hamer, A., Chopin, T. and Weaire, T. 2020. A discounted cash-flow analysis of salmon monoculture and integrated multi-trophic aquaculture in eastern Canada. *Aquaculture, Economics and Management*, 24 (1):43-63.

Chopin, T., Cooper, J.A., Reid, G., Cross, S. and Moore, C. 2012. Open-water integrated multi-trophic aquaculture: environmental biomitigation and economic diversification of fed aquaculture by extractive aquaculture. *Reviews in Aquaculture*, 4(4):209-220.

Dalsgaard, J.P.T., Light, F.C. and Christensen V. 1995. Towards quantification of ecological sustainability in farming systems analysis. *Ecological Engineering*, 4:181-189.

Dasgupta, S., Pandey, B.K., Sarangi, N. and Mukhopadhyay, P.K. 2008. Evaluation of water productivity and fish yield in sewage-fed vis-a`-vis fertilized based carp culture. *Bioresource Technology*, 99: 3499-3506.

Delmendo, M.N. 1980. A review of integrated livestock-fowl-fish farming systems. *In*: Pullin, R.S.V. and Shehadeh, Z.H. (eds.), *Integrated Agriculture-Aquaculture Farming Systems*. ICLARM Conference Proceedings 4:59-71.

Edward, P. 1985. Pigs over fish ponds. *Pig International*, 15(9):10.

Edwards, P. 2000. Aquaculture and poverty: past, present and future prospects of impacts. A discussion paper prepared for the fifth fisheries development donor consultation, Rome, Italy, pp. 22-24 February, 1999.

Fagberno, O.A. and Sydenham, D.H. 1988. Evaluation of *Clarias isheriensis* (Sydenham) under semi intensive management in ponds. *Aquaculture,* 74:287-291.

FAO, 2003. The Role of Aquaculture in improving food Security and Nutrition. Committee on World Food Security, Food and Aquaculture Organization of the United Nations, Rome.

Francis, T. and Pandurengan, P. 2004. Review on integrated fish farming. *World Aquaculture,* 35(3):

Gabriel, U.U., Akinrotimi, O.A., Bekibele, D.O., Anyanwu, P.E. and Onunkwo, D.N. 2007. Economic benefit and ecological efficiency of integrated fish farming in Nigeria. *Scientific Research and Essay*, 2:302-308.

Goverdhan, M., Kumari, C.P., Reddy, G.K., Sridevi, S., Alibaba, M.D., Chiranjeevi, K. and Kumar, M.S. 2020. Evaluation of integrated farming system model for resource recycling and livelihood security of small and marginal farmers of Telangana State, India. *Current Journal of Applied Science and Technology*, 39(34):17-26.

Hatch, U. and Engle, C. 1987. Economic analysis of aquaculture as a component of integrated Agro aquaculture system. science evidence from Panama. *Journal of Aquaculture in Tropics,* 2:93-105.

Hopkins, K.D. 1982. Outstanding yields a profits from livestock tilapia integrated farming. ICLARM newsletter, 5(3):13.

Kearney, J. 2010. Food consumption trends and drivers. *Philosophical Transactions of the Royal Society, Biological Sciences*, 365:2793-2807.

Lightfoot, C. and Minnick, D.R. 1991. Farmer-first qualitative methods: Farmers diagrams for improving methods of experimental design in integrated farming systems. *Journal for Farming Systems Research and Extension*, 2:11-34.

Megerssa, E., Daba, T. and Tokuma, N. 2016. Fish growth performance in ponds integrated with poultry farm and fertilized with goat manure: a case in Ethiopian Rift Valley. *International Journal of Fishery Science and Aquaculture*, 3:040-045.

Moav, R., Wohlfarth, G. and Schroeder, G. 1977. Intensive poly culture of fish in fresh water ponds. I. substitution of extensive feeds by liquid cow manure. A*quaculture,* 10:25-43.

Modadugu, V. M. and Acosta, B.O. 2004. A review of global tilapia farming practices. World Fish Centre.

Mohanty, R.K., Jena, S.K., Thakur, A.K. and Patil, D.U. 2009. Impact of high-density stocking and selective harvesting on yield and water productivity of deepwater rice-fish systems. *Agricultural Water Management*, 96(12):1844-1850

Mohanty, R.K., Mishra, A., Ambast, S.K., Mandal, K.G., Panda, D.K. and Panigrahi, P. 2017b. Effects of various shrimp (*Penaeus monodon*) densities on their growth, water and sediment quality, and water budget *Aquaculture International,* 25:2161-2176.

Mohanty, R.K., Mishra, A. and Patil, D.U. 2014. Water budgeting in black tiger shrimp *Penaeus monodon* culture using different water and feed management systems. *Turkish Journal of Fisheries and Aquatic Sciences,* 14:487-496.

Mohanty, R.K., Thakur, A.K., Ghosh, S., Mohanty, S. and Patil, D.U. 2010. Performance evaluation of concurrent rice-fish-prawn culture with and without cull harvesting. *Aquaculture Research*, 41:1402-1412.

Mohanty, R.K., Verma, H.N. and Brahmanand, P.S. 2004. Performance evaluation of rice-fish integration system in rainfed medium land ecosystem. *Aquaculture*, 230:125-135.

Molden, D., Oweis, T., Steduto, P., Bindraban, P., Hanjra, M. A. and Kijne, J. 2010. Improving agricultural water productivity: Between optimism and caution. *Agricultural Water Management*, 97: 528-535.

Naskar, S., Biswas, G., Kumar, P., De, D., Banerjee, P.S., Das, S. and Roy, U. 2022. Effects of estuarine oyster, Crassostrea cuttackensis as the extractive species at varied densities on productivity and culture environment in brackishwater integrated multi-trophic aquaculture (BIMTA) system. Aquaculture, 554, https://doi.org/10.1016/j.aquaculture.2022.738128

Naskar, S., Biswas, G., Kumar, P., De, D., Das, S., Banerjee, P.S., Chadha, N.K. and Behera, P. 2023. The green seaweed, Enteromorpha intestinalis: An efficient inorganic extractive species for environmental remediation and improved performances of fed species in brackishwater integrated multi-trophic aquaculture (BIMTA) system. *Aquaculture*, 569, *https://doi.org/10.1016/j.aquaculture.2023.739359*

Nnaji, C.J., Okoye, F.C., and Ogunseye, J.O. 2003. Integrated Fish Farming practices with special reference to combination rates production figures and economic evaluation in: AA Eyo, JO Ayanda (eds). Conference proceedings of fisheries society of Nigerian (FISON) Owerri 8th-12th December, 2003. Pp.173-178.

Odum H.T. 1989. Self-organization transformation and information: *Ecol. Sci.* (242): 171-181.

Oribhabor, B.J. and Ansa E.J. 2006. Organic waste reclamation, recycling and re-use in integrated fish farming in the Niger Delta. *Journal of Applied Sciences and Environmental Management,* 10(3): 47-53.

Parajuli, R., Dalgaard, T. and Birkved, M. 2018. Can farmers mitigate environmental impacts through combined production of food, fuel and feed? A consequential life cycle assessment of integrated mixed crop-livestock system with a green biorefinery. *Science of the Total Environment, 619–620*, 127–143.

Paramesh, V., Chakurkar, E. B., Bhagat, T., Sreekanth, G. B., Kumar, H. B. C., Rajkumar, S., Gokuldas, P. P., Mahajan, G. R., Manohara, K. K. and Ravisankar, N. 2021. Impact of integrated farming system on residue recycling, nutrient budgeting and soil health. *Indian Journal of Agricultural Sciences*, *91*(1):44-48.

Prein, M. 2002. Integration of aquaculture into crop-animal systems in Asia. *Agricultural Systems*, 71:127-146.

Preston, T.R. 1990. Future strategies for livestock production in tropical third world countries, *Ambio*, 19:390-393.

Radhammani, S.; Balasubramanian, A.; Ramamoorthy, K. and Geethalakshmi, V. 2003. Sustainable integrated farming systems for dry lands: A review. *Agricultural Reviews*, 24:204-210.

Rautaray, S.K., Dubey, R., Raychaudhuri, S., Pradhan, S., Mohanty, S., Mohanty, R.K. and Ambast, S.K. 2020. Sustainable agriwaste management at farm level through self-reliant farming system. *Waste Management & Research*, DOI: 10.1177/0734242X20920350

Schroeder, G.L. 1980. fish farming in manure loaded ponds. pages 73-86.

Shoko, A.P., Getabu, A., Mwayuli, G. and Mgaya, Y.D. 2011. Growth performance, yields and economic benefits of Nile tilapia, Oreochromis niloticus and Kales Brassica oleracea cultured under vegetable-fish culture integration. *Tanzania Journal of Science, Vol.* 37.

Sin, A.W. 1980. Integrated fish husbandry systems in Hong Kong, with case studies on duck-fish and goose-fish systems. In: R.S.V. Pullin and Z.H. Shehadeh (eds.), *Integrated Agriculture-Aquaculture Farming Systems*. ICLARM Conference Proceedings 4, 113–123.

Singh, A.K. 1993. Fish cum pig farming. boom for tribal areas. *Fishing Chimes,* 12(10):47.

Singh A.K. 1996. Investigation on the effect of pig manure on growth and survivability of Indian major crop and exotic carps spawn in nursery a pond pp.6-8. In the fourth Indian fisheries forum. pp. 24-28.

Srivastava, R.C., Singandhupe, R. B. and Mohanty, R.K. 2004. Integrated farming approach for runoff recycling system in humid plateau areas of eastern India. *Agricultural Water Management*, 64:197-212.

Thakur, A.K., Mohanty, R.K., Singh, R. and Patil, D.U. 2015. Enhancing water and cropping productivity through Integrated System of Rice Intensification (ISRI) with aquaculture and horticulture under rainfed conditions. *Agricultural Water Management*, 161: 65-76.

Tipraqsa, P., Craswell, E.T., Noble, A.T. and Schmid V. D. 2007. Resource Integration for multiple benefits multifunctionality of Integrated farming system in Northeast Thailand. *Agricultural Systems,* 94: 694-703.

Tokrishna, R. 1992. Integrated livestock-fish farming systems I Thailand. In: Mukherjee, T.K., Moi, P.S., Panadam, J.M., and Yang, Y.S. (Eds.). Proceedings of the FAO/IPT Workshop on Integrated Livestock-Fish Production Systems, 16-20 December 1991. Institute of Advanced Studies, University of Malaya, Kuala Lumpur, Malaysia.

Troell, M., Joyce, A., Chopin, T., Neori, A., Buschmann, A.H. and Fang, J.G. 2009. Ecological engineering in aquaculture - potential for integrated multi-trophic aquaculture (IMTA) in marine offshore systems. *Aquaculture*. https://doi.org/10.1016/j.aquaculture.2009.09.010.

Velasquez, C.C. 1980. Health constraints to integrated animal-fish farming in the Philippines, pp-103-11 In: Pullins, RSV; Ziad, H; Shehadeh (eds). Integrated Agriculture- Aquaculture farming system. ICLARM- SEARCH, Manila, Philippines.

Verdegem, M. C. J. and Bosma, R.H. 2009. Water withdrawal for brackish and inland aquaculture, and options to produce more fish in ponds with present water use. *Water Policy*, 11, Supplement 1:52-68.

Verdegem, M.C.J., Bosma, R.H. and Verreth, J.A.J. 2006. Reducing water use for animal production through aquaculture. *Water Resources Development*, 22(1):101-113.

Viji C.S. 2015 Studies on integrated multi-trophic aquaculture in a tropical estuarine system in Kerala, India. *Ph.D. Thesis*, Central Institute of Fisheries Education, Mumbai, 128 p.

Viji, C.S., Chadha, N.K., Kripa, V., Prema, D., Prakash, C., Sharma, R., Jenni, B. and Mohamed, K.S. 2014. Can oysters control eutrophication in an integrated fish-oyster aquaculture system? *Journal of the Marine Biological Association of India*, 56 (2):67-73.

Woyanarovich, E. 1980. Raising ducks on fish ponds. In: R.S.V. Pullin and Z.H. Shehadeh (eds.), *Integrated Agriculture-Aquaculture Farming Systems*. ICLARM Conference Proceedings 4, 129–134.

Woyanarovich, E. 1976. The feasibility of combining animal husbandry with fish farming.

Zhang, J. and Kitazawa, D. 2016. Assessing the bio-mitigation effect of integrated multi-trophic aquaculture on marine environment by a numerical approach. *Marine Pollution Bulletin*, 110(1):484-492.

8

High-Value Aquaculture Under IFS

8.1. Biofloc aquaculture

Among the innovations in the field of aquaculture, biofloc technology (BFT) has attracted great attention due to its ability to foster the goals of sustainable aquaculture (Khanjani *et al.*, 2022a). If implemented correctly, BFT capitalizes on the existing organisms in water to significantly reduce water use and environmental impact and increase biosafety at farms (Avnimelech, 2012).

Biofloc organisms (BFOs) play three main roles in BFT: i) maintaining water quality by consuming nitrogenous compounds; ii) providing a source of food for the cultivated species (Wang *et al.*, 2016), and iii) pathogen competition and creating probiotic properties often by heterotrophic organisms (e.g., *Bacillus* spp bacteria). BFOs help in maintaining water quality by reducing harmful nitrogenous compounds such as ammonia and nitrite by metabolizing faecal matter and unconsumed food (Suita *et al.*, 2015; Wang *et al.*, 2016). The growth of BFOs is associated with the control of TAN in the environment. Bioflocs (BFs) contain organisms such as bacteria, microalgae, yeast, rotifers, ciliates, protozoans, nematodes, copepods and crustaceans (Monroy-Dosta *et al.*, 2013; Collazos-Lasso and Arias-Castellanos, 2015; Khanjani *et al.*, 2022a, 2022b). Greater diversity of BFOs can improve flocs' mineralization of waste (Martínez-Cordova *et al.*, 2015), protein utilization (Avnimelech, 2007), and pathogen control (Emerenciano *et al.*, 2013). The diversity and density of the floc community in a BFT system depend largely on factors e.g., the type of external sources of carbon (Deng *et al.*, 2018), different C/N ratios (Panigrahi *et al.*, 2019b), environmental conditions such as salinity, light, oxygen levels, and solids load and the type of cultivated species (Meenakshisundaram *et al.*, 2021; Khanjani and Sharifinia, 2021; Krummenauer *et al.*, 2021; Khanjani *et al.*, 2022c). Further, the dynamics in the BFT system are determined by the types of ecological relationships (e.g., commensalism, competition, and predation) between BFOs in a trophic network (Collazos-Lasso and Arias-Castellanos, 2015). The presence of BFOs in aquaculture, especially ciliates and nematodes, can make proteins, lipids, and vitamins available to the farmed species, thus promoting growth and reducing the need for these compounds in commercial feed (Khanjani and Sharifinia, 2020). Moreover, the identification of BFO composition can help fine-tune the applications of BFT and achieve increased production and reduced environmental footprint (Kumar *et al.*, 2021; Abakari *et al.*, 2022).

By recycling nutrients, especially nitrogenous waste, into microbial biomass that the cultured animals may use *in-situ*, BFT permits high-density culture and provides the opportunity to maintain adequate water quality with no or minimal water exchange (Fig. 8.1). Since nutrients may be continually recycled and reused in the culture medium while benefiting from little to no water exchange, BFT is regarded as a new "blue revolution." BFT is an environmentally friendly aquaculture method that relies on the production of microorganisms *in-situ*. The suspended growth known as "biofloc" in ponds and tanks is made up of aggregates of bacteria, phytoplankton, and other grazing bacteria as well as dead and living particulate organic matter. According to Wang *et al.* (2016) and Suita *et al.* (2015), it is the use of microbial processes that occur within the pond or tank to serve as a water treatment solution and a source of food for cultured organisms. As a result, another name for this system is a green soup pond or an active suspension / heterotrophic pond.

The technique's basic idea is to add sources of carbohydrates in order to maintain a greater C-N ratio (Panigrahi *et al.*, 2018; 2019a). Additionally, the process produces high-quality single-cell microbial protein, which improves the water quality. Under these circumstances, heterotrophic microbial growth takes place, assimilating the nitrogenous waste so that it can be used as feed for the cultured species and functioning as a bioreactor to regulate the water's quality. Because heterotrophs grow more quickly and produce ten times more microbes per unit of substrate than autotrophic nitrifying bacteria, hazardous nitrogen species are immobilized in biofloc more quickly. The basis of this technique is the idea of flocculation inside the system.

Fig. 8.1. A small-scale biofloc aquaculture unit.

According to Khanjani *et al.* (2022a, 2022b), biofloc is a heterogeneous aggregate of suspended particles and a range of microorganisms (including bacteria, algae, fungus, invertebrates, detritus etc.) associated with extracellular polymeric compounds. It is a protein-rich live feed that is created when leftover feed and excrement in a culture system are exposed to sunlight and intense aeration, turning them into a natural food. The loose mucus matrix that binds each floc together is released by bacteria and is held together by electrostatic attraction / filamentous microorganisms. The naked eye can see large flocs, but the majority are tiny. The size of floc is between 50 and 200 microns. Biofloc has a good nutritional value. Protein has a dry weight range of 25–50%, while fat has a range of 0.5–15%. It is an excellent source of minerals and vitamins, especially phosphorus. It functions similarly to probiotics. One can use the dried biofloc as a component in the feed in place of soybean meal or fishmeal.

BFT will enable aquaculture methods to adopt an environmentally beneficial approach, such as eco-based technologies. With the emergence of new disease concerns and rising energy prices, BFT can be a creative technique for disease prevention and control, in contrast to traditional tactics such as antibiotic, antifungal, probiotic, and prebiotic administration. BFT combined with biosecure modular systems may provide a solution for more efficient, sustainable, and economical aquaculture production. The clear benefits of this technique include reduced water usage, on-site recycling of nutrients and organic matter, and enhanced farm biosecurity due to disease exclusion, increased natural food production, improved feed conversion ratio, and stress-free surroundings. There would be little environmental damage if chemicals and other medications were not used. Because microbial biofloc is a naturally occurring food source, it makes up for aquatic animals' increased protein requirements.

Before or after stocking, microbial biofloc production is a crucial phenomena for shrimp and fish culture in the BFT system. Depending on the species to be cultivated, the intensity of farming, the buffering time for biofloc generation, and the system's design and compatibility, there are various ways to set up the system for biofloc-based fish and shrimp farming. Additionally, the system can run in either semi-biofloc or full biofloc mode, incorporating substrates for periphyton growth or not. Pond preparation is done in accordance with Standard Operating Procedures (SOP), which include drying, tilling, replenishing the ponds with old soil, sanitizing water that has been carefully sieved through a 60 to 100 μ screen, and following all biosecurity measures. Fertilizers high in phosphorus and nitrogen can be used alone or in combination with other biofertilizers to create an autotrophic bloom. Liming with hydrated lime or dolomite maintains the pH. In the beginning, nitrogen level is built up by using feed or any organic substance. After that, a CHO source, such as molasses, is added based on the amount of feed (the feed's CHO content should be considered when determining the CHO requirements) and the TAN level in order to maintain a high C: N ratio (15:1) while promoting the flocs (Khanjani and Sharifinia, 2020; Panigrahi *et al.*, 2019b). Usually, autotrophic algae take over ponds and, after a few days, floc forms, turning the water dark and frothy and turning the system heterotrophic with little to no algae. Ammonia levels rise to a peak and thereafter fall, following the rise of nitrite, which eventually stabilizes and can be managed by providing additional C. Molasses can be added at intervals of 1-3 days over a period of 2-3 weeks to encourage the transition from algae to bacteria. Based on the TAN level and feed quantity, the molasses or other carbohydrate supplements can be adjusted to maintain a high C: N ratio once the heterotrophic population has been established. If producing biofloc in a shorter time is the goal, the procedure may need to be altered. With the addition of a significant number of carbohydrates, the feed quantity is high enough to quick-start the bioflocs. Alternatively, biofloc inoculum in its preserved form or from other ponds might be added to speed up the process. Any commercial inoculum used for this purpose should, however, be used with caution. To maintain the floc in suspension, intense aeration is needed. By preventing dead spots and the buildup of inorganic debris from the pond's banks and walls brought on by excessive water circulation, pond liners facilitate simpler maintenance of floc in suspension.

8.1.1. Environmental requirements for BFT

BFT is a biosecure system as it works under zero water exchange conditions. In biofloc system, ammonia is consumed by bacteria and nitrite increases, with tolerance up to a higher level. There is a need to correct the declining alkalinity which must be maintained >75-150 ppm. Another important requirement in the biofloc-based system is to meet the high oxygen demand. There will also be the need to suspend the aerator of aspirator type at the sludge pile. Typical aeration requirements are 1 HP/350- 400 kg fish/shrimp biomass. The biofloc system requires high aeration/ mixing with aerators and blowers not only to provide dissolved oxygen (DO) and remove CO_2 but also to keep biofloc in suspension and prevent the production of nitrogenous metabolites, sulphides and organic acids.

8.1.2. Water quality management

During the culture period, the chemical composition of water in the system should be regularly monitored and maintained. Mainly few physico-chemical parameters should be monitored daily; others should be monitored periodically like once every 3 days. The DO, pH, temperature, and turbidity can be monitored twice a day, and NH_3, NO_3, NO_2 and TAN can be monitored daily. Algal community, alkalinity, hardness salinity can be seen once a week. Floc volume should be maintained at 12-20 ml L^{-1}. Total heterotrophic count can be monitored on every 3-day interval. The optimal water quality parameter levels must be maintained for retaining the fish/ shrimp's health. All the physicochemical parameters should be maintained at optimal levels for better production of fish/ shrimp.

8.1.3. Major cultivable species suitable for biofloc culture

The most suitable species for the Biofloc system are those that can gain some nutritional advantages from directly consuming floc. The biofloc system works best with species that can withstand large concentrations of solids in water and are generally tolerant of low water quality. Air-breathing fish, non-air-breathing fish (Common carp; rohu, tilapia, milkfish), and shellfish (Pacific white shrimp; tiger shrimp) are some of the species that are suitable for BFT.

8.1.4. How to prepare the inoculum?

About 150 L of inoculum is prerequisite for 15000 L of water for the floc development. In order to make this inoculum, fill a clean tub or can with 150 L of water, continue aerating vigorously, and add 3 kg of pond soil, 1.5 g of urea and 30 g of a carbon source (tapioca/ wheat flour / jaggery). The mixture gives enough aeration when thoroughly mixed with water in a tub. After 24 to 48 hours, the inoculum will be ready to be moved to the main tank. For floc to form, carbon sources must be added every day. To maintain a C:N ratio of 10:1, 600 grams of carbon source must be fed to the system for every kg of feed (containing 25% crude protein). Additional carbon is added after the floc volume reaches 15-20 ml.

8.1.5. Advantage of BFT

i. An eco-friendly culture system that lessens environmental effect thus higher biosecurity.
ii. Judicious/ Sensible use of land and water resources.
iii. It is a restricted or zero-water exchange system with higher productivity (In fish culture systems, it improves feed conversion, growth performance, and survival rate.).
iv. BFT reduces water contamination and alleviates the spread of pathogens
v. It reduces demand of protein-rich feed and cost of feed.
vi. It reduces the pressure on capture fisheries i.e., the use of cheaper food fish and trash fish for fish feed formulation.

Aquaculture could undergo a transformation thanks to BFT. This is still in its early phases, though, and further study will be needed before it can be modified, standardized, and put into practice.

8.2. Carp polyculture

Catla (*Catla catla*), rohu (*Labeo rohita*) and mrigal (*Cirrhinus mrigala*), known as Indian major carps (IMC) are the important carp species cultured traditionally in ponds due to their higher growth and consumer preference. Owing to the compatibility of food and habitat preference, these three species are cultured together in ponds under a polyculture system. While *catla* is a surface dweller, *rohu* and *mrigal* are column and bottom feeders, respectively. The exotic silver carp, grass carp and common carp having higher growth potential are also added along with these three IMCs for increasing the production under composite carp culture or polyculture (Fig. 8.2). The standardized carp production technologies have been based on the intensity of resource use with varied production levels. While the usual production level in carp polyculture in ponds range between 3-5 t ha^{-1}, higher production levels of 6-12 t ha^{-1} per yr are being achieved by many farmers and entrepreneurs, in the country. The success of the grow-out output is determined by the three essential inputs-the right kind of seed, feed, and fertilizers-and how well they are managed. Furthermore, factors pertaining to fish health and the maintenance of water and soil quality also have a significant role. For carp grow-out production, ponds that are between 0.4 and 1.0 hectares in size and with a water depth of 1.5 to 2.0 meters are thought to be ideal. Pre-stocking, stocking, and post-stocking operations are three broad categories for the environmental and biological manipulations used in carp polyculture management procedures. These activities are covered in more detail below.

8.2.1. Pre-stocking pond preparations

Aquatic weed control and removal of weed and predatory fishes are among the important pre-stocking operations for achieving greater survival of the stocked seed. Drying the ponds until soil cracks followed by ploughing not only helps in the eradication of undesired animals but also, removes the obnoxious gases from the bottom soil. However, drying of ponds often is a difficult proposition owing to the time requirement, non-availability of water for refilling and economic considerations. Such perennial ponds, thus, require necessary control measures against aquatic weeds, and predatory and weed fishes before stocking. Though a wide range of manual, mechanical, chemical and biological methods are available for control of these weeds, the manual method is commonly advocated for weed clearance. Mahua oil cake is one of the most commonly used piscicides, which is applied @ 2.0-2.5 t ha^{-1} $^{m-1}$ three weeks before stocking. Besides acting as piscicide the oil cake also serves as organic manure after its decomposition. The application of commercial bleaching powder (30 % Cl) @350 kg ha^{-1} m^{-1} of water is another effective method for killing predatory/ weed fishes. Alternate application of urea @ 100 kg ha^{-1} m^{-1} followed by bleaching powder @ 175 kg ha^{-1} m^{-1} after 18-24 h also effectively controls these fishes. Improving soil and water quality through the use of lime and fertilizers is another important aspect connected with this phase of management. The ponds are usually manured with raw cowdung @ 3-4 t ha^{-1} 15 days before stocking as the basal dose. When mahua oil cake is used for controlling the predatory and weed fishes, the basal cow dung application is avoided. Liming is done in the ponds three to four days before to stocking in order to adjust the pH of the water and soil, aid in the mineralization of organic materials, and disinfect the pond. For liming, calcium oxide (shell lime), dolomite, or calcium carbonate (limestone) are utilized; the dosages are determined by the pH of the surrounding soil and water. For a pond with pH ranges of slightly acidic to neutral, 200–300 kg of lime can often be added. For grow-out operations, it is always crucial to maintain the ideal water quality (Table 8.1).

8.2.1.1. Stocking the ponds

The carp culture ponds are stocked with a proportionate number of surface, column and bottom feeders to minimize the competition among them for food available in the pond. It is often advised to use a ratio of 30–40% bottom feeders (common carp and mrigal), 30–35% column feeders (rohu), and 30–40% surface feeders (silver carp and catla). If aquatic weeds from outside sources are possible, grass carp can be added at a rate of 5–10% in place of other species. After proper acclimation, fingerlings weighing 8 to 10 grams and measuring 80 to 100 mm are frequently transferred to the grow-out ponds. For carp polyculture, a stocking density of 5,000 fingerlings per hectare is advised, with a production aim of 3-5 t ha^{-1} yr^{-1}. In order to achieve production levels of 5-8 t ha^{-1} yr^{-1} and 10-15 t ha^{-1} yr^{-1}, respectively, the densities can be increased to 8,000–10,000 fingerlings per ha and 15,000–25,000 juveniles per ha. The stocking density, however, depends largely on the input use and levels of management.

Fig. 8.2. Indian major carps and exotic carps suitable for polyculture.

Table 8.1. Optimum aquaculture water quality parameters for Indian major carps

Water parameter	Optimum level
Water depth	2.0 m
Temperature	26-32 ^{0}C
Dissolved oxygen	>5.0 ppm
pH	7.2-8.5
Total NH_4-N	<1.0 ppm
Total NO_3-N	<5.0 ppm
NO_2-N	<0.01 ppm
Sulphide	<0.03 ppm
Biological oxygen demand (BOD)	< 5.0 ppm
Chemical oxygen demand (COD)	<70 ppm
Sacchi disc visibility (transparency)	25-50 cm
Ratio of phyto and zooplankton	10:1
Watercolour	Light green/light brown

8.2.2. Post-stocking management

8.2.2.1. Fertilization schedule

The post-stocking pond management primarily involves the aspects of intermittent liming and fertilization, supplementary feeding, water management and health care. In an average productive pond, the post-stocking fertilization includes fortnightly application of cowdung @ 0.5 t ha^{-1}, urea @10 kg ha^{-1}, and SSP @15 kg ha^{-1}. However, the application time and dosage can be deferred depending on the water quality and plankton content of the pond. Poultry litter, duck droppings, pig dung etc., depending on their availability, can be used as an alternative source of organic manure. In intensive carp culture ponds, the application of *Azolla* as biofertilizer @ 40 t ha^{-1} yr^{1}, supplied at weekly split doses, has proven to supply the full supplement of nutrients. i,e. 100 kg N, 25 kg P, 90 kg K and 1,500 kg organic matter. The use of biogas slurry @ 30-45 t ha^{-1} yr^{1} or 80-100 l d^{-1} ha^{-1}) is advantageous due to its lower oxygen consumption, and faster rate of nutrient release.

8.2.2.2. Supplemental feeding

A mixture of mustard/groundnut oil cake (MOC/GOC) with rice bran at 1:1 ratio by weight is the most commonly used supplementary feed in carp polyculture. These are provided at 5% of the stocked biomass in the 1st month of rearing followed by a sliding rate of 3-2% of the estimated biomass in subsequent months. The biomass of fish is estimated through monthly sampling and the daily feed ration is calculated as follows

Feed requirement/day = Estimated fish biomass in the pond x feeding rate%,

where, biomass = Average weight of fish x total number of fish stocked x survival%.

A survival percentage of 75-80% is usually considered for estimation of biomass. The daily ration is divided into two splits and provided in dough form preferably in feed trays or gunny bags hung at uniform distances inside the pond (Fig. 8.3). Aquatic vegetation, *viz.*, hydrilla, *Ceratophylum*, *Najas*, duckweeds, etc. are additionally provided in ponds stocked with grass carps in selected corners of the pond. The nutrient-rich supplementary feed in the form of pellets or floating feed is also available for carps. Feed comprises about 60-65% of production cost. In addition to extra expense, water quality can deteriorate unnecessarily due to the addition of excess feed. Extruded, floating feeds offer the advantage of watching the feeding response as opposed to a sinking, steam-pelleted feed (Fig. 8.4). Nowadays, floating feeds are used widely due to minimal wastage of feed, minimal deterioration of water quality, low FCR, easy to digest, faster growth, higher yield and higher water productivity.

Fig. 8.3. Feeding fish using gunny bags hung at uniform distance inside the pond

Fig. 8.4. Different types of fish feed used for carp polyculture.

8.2.2.3. Aeration and water exchange

Normally dissolved oxygen is not a critical factor in ponds stocked at a density of 5,000-10,000 fingerlings per ha with 3-5 t ha^{-1} yr^{-1} production targets. However, ponds with higher stocking density require additional aeration

(Fig. 8.5) to increase the DO content in water. Paddle wheels, aspirators and submersible types of aerators are commonly used in carp culture ponds. Usually, a pond of a one-ha water area requires 4 aerators (1 HP each). In order to prevent metabolite buildup and decomposed leftover feed from causing water quality to deteriorate and cause delayed growth and disease outbreaks, water exchange is necessary in high-density culture ponds. On the other hand, low-density culture merely needs compensation for seepage and evaporation loss throughout the culture period—it does not require any water exchange. Maintaining water quality in intense culture procedures necessitates replacing a specific volume of water at regular intervals, particularly in the latter stages of the culture period.

Fig. 8.5. Pond aeration using paddle wheel aerators.

8.2.2.4. Plankton management

In addition to reducing water quality oscillations, phytoplankton are important for maintaining the equilibrium of the entire pond ecosystem. A healthy population of phytoplankton reduces the concentrations of CO_2, NH_3, NO_2, and H_2S while enriching the system with oxygen during the day through photosynthesis. Because phytoplankton may bind up heavy metals and absorb NH_4, a good phytoplankton bloom can lower hazardous levels. It can stop filamentous algae from growing because phytoplankton can obstruct light from penetrating the bottom. In addition to providing appropriate turbidity, a healthy bloom stabilizes the water's temperature and reduces temperature loss in the winter. An ideal ratio of phyto and zooplankton (Fig. 8.6) in a aquaculture pond should be 10:1. Ponds typically contain an richness of phytoplankton. These organisms play an important role in pond ecology and influence water quality.

Aquaculture ponds often have 50 or more species of phytoplankton and 3-5 species of zooplankton at any given time, but only a few species usually less than 10, make up most of the plankton community. Phytoplankton communities in ponds typically undergo rapid succession, and the species composition of phytoplankton communities frequently changes over periods of a few weeks. The abundance of phyto and zooplankton usually is controlled by concentrations of nutrients especially those of inorganic nitrogen and phosphate. These organisms impart a colour to the water this situation is called a plankton bloom. Phytoplankton is an important source of dissolved oxygen. In the daytime, these plants produce oxygen by photosynthesis at a much faster rate than oxygen can diffuse from the atmosphere into the pond water. Phytoplankton rapidly removes NH_4-N from the water lessening the concentration of this potentially toxic substance. Finally, turbidity created by phytoplankton helps in restricting light penetration to the pond bottom, and a phytoplankton bloom is a good control measure for aquatic macrophytes that grow below the water surface.

An abundance of certain species of phytoplankton in ponds can have other adverse effects notably the production of odorous and toxic compounds. Some phytoplankton mainly blue-green algae produce the odorous compounds *geosmin* and *methyl-isoborneol* (MIB) that can be absorbed by fish and prawns resulting in a bad odour and taste commonly referred to as off-flavor. *Geosmin* has an earthy odor reminiscent of freshly ploughed moist soil or a damp earthen cellar. This compound is produced most commonly by species of *Anabaena* and *Aphanizomenon* that often are abundant in freshwater aquaculture ponds. The odour of MIB is musty and medicinal somewhat like that of camphor. The common source of MIB appears to be the species of *Planktothrix* (formerly called *Oscillatoria*). Some producers have reported success in preventing off-flavour by stocking species such as Chinese carp and tilapia along with the primary culture species; these planktivorous fish eat blue-green algae and minimize the off-flavour risk. Nevertheless, it sometimes is necessary to treat ponds with copper sulphate ($CuSO_4$) to kill off-flavor-producing algae. The usual treatment rate for $CuSO_4$ in mg per liter is 0.01 times the total alkalinity concentration.

Toxic algae are less common in freshwater water than in coastal water. Nevertheless, there are confirmed instances of toxic algae in freshwater aquaculture ponds that resulted from species of certain genera of blue-green algae (*Microcystis*, *Anabaena*, and *Planktothrix*), Eugleanoids (*Euglena*), and Haptophytes (*Chrysochromulina*). Managers have sometimes attempted to kill toxic algae by $CuSO_4$ or other algicide applications with variable success. There also have been attempts to detoxify algal toxins by application of potassium permanganate ($KMnO_4$). Based on the use of $KMnO_4$ for other purposes, around 2 mg L^{-1} would be needed in relatively clear water and up to 6 mg L^{-1} may be required in water with a high concentration of organic matter to oxidize these compounds.

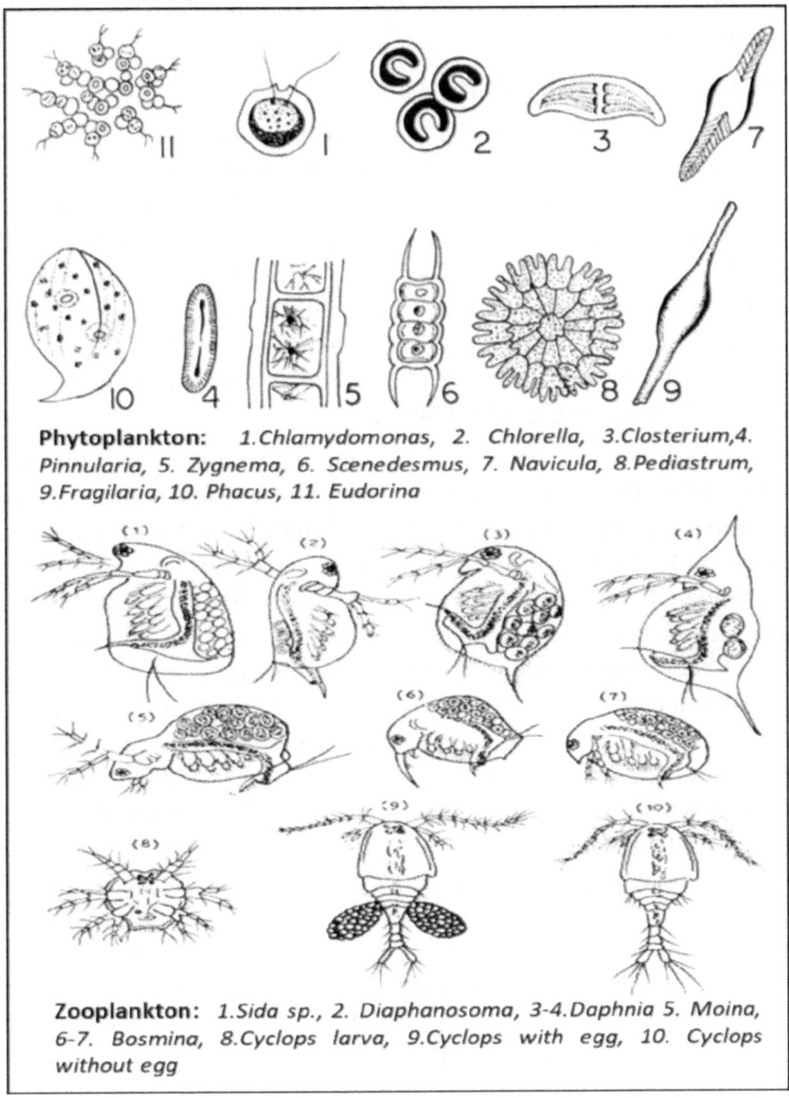

Phytoplankton: *1.Chlamydomonas, 2. Chlorella, 3.Closterium,4. Pinnularia, 5. Zygnema, 6. Scenedesmus, 7. Navicula, 8.Pediastrum, 9.Fragilaria, 10. Phacus, 11. Eudorina*

Zooplankton: *1.Sida sp., 2. Diaphanosoma, 3-4.Daphnia 5. Moina, 6-7. Bosmina, 8.Cyclops larva, 9.Cyclops with egg, 10. Cyclops without egg*

Fig. 8.6. Common phyto and zooplankton encountered in grow-out pond culture.

In pond management, efforts should be made to avoid excessive phytoplankton blooms by adding no more fertilizer or feed to ponds than necessary.

Nevertheless, in feed-based aquaculture, ponds typically develop dense phytoplankton blooms. Phytoplankton blooms often become excessive at feeding rates above 30 kg ha^{-1} per day in ponds without aeration. At greater feeding rates it is important to apply sufficient mechanical aeration to avoid low dissolved oxygen concentration at night. Low night time dissolved oxygen concentration is no doubt the most serious water quality issue in feed-based pond aquaculture.

8.3. Air-breathing catfish culture

Catfishes belong to four families: Ictaluridae, Claridae, Pangasidae and Siluridae, and are widely distributed around the world. The traditional culture of catfishes has been practiced in Southeast and South Asia. Catfishes are considered high-value species in certain areas whereas those are considered lower and medium-value market species in other areas. Catfishes are hardy; they can live longer time without water and have an accessory respiratory organ. There is a specialized trade of live catfish in some places of Eastern India. The most important aquaculture species of catfish is the Asian catfish, *Clarias batrachus*. It belongs to Claridae family.

Because catfish are so popular, they are quite important commercially. Both the air-breathing catfishes (Fig. 8.7), *Clarias batrachus* (magur) and *Heteropneustes fossilis* (singhi), are acclimated to harsh environmental conditions and can be cultured in both mono- and polyculture systems. Although the methods for producing these two significant catfish species' seeds and through induced breeding have been refined, their large-scale production has not yet been implemented. Production levels of 2.5-3.5 t ha^{-1} are attained in the grow-out culture of magur, which gain 100-200 g weight in 6-8 months, with pond management techniques that are essentially identical to those used for carp culture methods and stocking with 20,000-50,000 fingerlings per haThe large-scale cultivation of these species is still lacking the attention it deserves, despite the abundance of potential resources in the form of wetlands and abandoned waters that could be exploited for commercial farming and the enormous market demand for these species. Due to its carnivorous eating habits, the production of balanced supplemental feed and the availability of the required number of seed in the appropriate size are the two crucial elements that must be addressed. Research about the development and standardization of induced breeding and grow-out technologies of several other non-air breathing catfishes like *Mystus seenghala, M. aor*, *Pungasius pungasius*, *Wallago attu*, *Ompak pabda* are also being envisaged given the high consumer preference for these in different parts of the country.

Fig. 8.7. Top: *Clarias batrachus* (magur) and bottom: *Heteropneustes fossilis* (singhi)

During larval rearing, at least 50% water exchange is given every alternate day for maintenance of good water quality, which otherwise leads to poor larval growth and sometimes leads to heavy mortality. The water depth of the larval lank is an important factor that needs to be given attention. Initially, a water column of 8-10 cm is kept in the larval tank to avoid heavy loss of energy spent by the larvae for making vertical trips. The larvae are reared in rearing tanks for 12-14 days during which they reach 10-12 mm size (30-40 mg). Then the fry is reared for 30 days usually at a rearing density of 200-3 00 nos per m^2 in outdoor cement tanks provided with 6-8 cm soil base or well-prepared earthen ponds. The tanks are prepared with a schedule alike carp-rearing tanks. Floating weeds like water hyacinth or duckweeds are provided to cover a portion of the water area. These plants not only give shelter but also harbour insects that form good food for the fry. The fishes are fed with pallet feed (about 35% protein) of finely minced trash fish/molluscan meat and rice bran at a 1:1 ratio @ 6-10% of stocked biomass/day during the evening hours. Fingerlings grow up to 3-4 cm (0.8-1.0 g) in size during this period, which are harvested by draining the tank and are stocked in the grow-out ponds with a water depth of 75-100 cm.

A pond size of 1,000–2,000 m^2 is ideal for the grow-out culture of magur. The pond bund should have a suitable slope of 1:2-1:3. The small and shallow (0.75-1.0 m water depth) earthen ponds/stone-pitched ponds/cemented tanks are suitable for the grow-out culture of magur. During rainy season magur used to crawl out of the pond. So that during the construction of the pond, the dyke should be made in such a way that they can't escape. Commercially, most of the farmers stock about 200 fingerlings per m^2, although the recommended stocking density rate is less than half of it (50 – 70 per m^2). Stocking of bigger-sized seeds (3-5 g) shows good survival and growth during the grow-out culture. The pond bottom should be flat to ensure complete netting. Magur is an air breather, they normally come up to the water surface for atmospheric oxygen. This behaviour attracts birds for predation. Therefore, it is required to cover the ponds with nets to protect the fish from bird predation. To minimize the loss due to cannibalism, measures such as i) stocking equal-sized fingerlings, ii) feeding the fish 2-3 times a day in at least 4 feeding areas of the pond, and iii) separating the fastest growing individuals can be adopted.

8.4. Freshwater prawn culture

The culture of the giant freshwater prawn, *Macrobrachium rosenbergii*, has piqued the interest of aquaculturists and fish farmers all over the world due to its rapid growth, adaptability to polyculture environments and artificial feed, greater disease resistance than its marine counterparts, and strong market demand. The giant freshwater prawn, *Macrobrachium rosenbergii,* popularly known as scampi (Fig. 8.8), is the largest and fastest-growing species among freshwater prawns. The technology of hatchery of the species has been developed and standardized. Monoculture of freshwater prawns at stocking densities of 40,000-50,000 per ha has shown production levels of 1.5-2.0 t ha^{-1} in a culture period of 5-6 months. Further, the polyculture of freshwater prawns along with carp has also been demonstrated to be a technologically sound cultural practice and economically viable option for enhancing the farm income of the farmers.

The grow-out mono-culture of the species is carried out in earthen ponds of 0.2-0.5 ha in size with a depth of 1.2-1.5 m, preferably with a sloping bottom and a sump at the lower end of the pond for easy harvesting. The grow-out ponds are often stocked with post-larvae, which usually leads to poor survival and hence low yield. Therefore, a nursery phase of 45-60 days is advocated before the grow-out phase for successful farming of scampi. The pre-stoking pond preparation for nursery and grow-out production involves several steps including the eradication of competitors and predators, liming and pond fertilization, which are similar to the grow-out carp culture.

Fig. 8.8. The giant freshwater prawn (*Macrobrachium rosenbergii*)

8.4.1. Nursery pond management

Smaller earthen ponds or cemented tanks with an average water depth of one meter are preferably used for nursery rearing. Floating weeds with dense root systems like *Eichhornia, Pistia* etc. may be provided as shelter for the post-larvae, covering approximately 10% of the water surface. A frame of bamboo or PVC pipe may be used to prevent the spreading of weeds. The nursery ponds are stocked with post-larvae (8-10 mg) at usual stocking density of 20 per m^2. Crumbled feed pellets may be provided @ 60-80% of the biomass per day for the first fortnight followed by 30-40 % per day in the subsequent period as supplementary feed. In case of non-availability of formulated pellets, a mixture of groundnut oil cake and rice bran may be given. The feed may be provided through broadcasting in two rations, i.e., during morning and evening hours. Aeration may be provided for 6-8 h d^{-1} to keep the optimum dissolved oxygen level of 5-7 ppm. While water exchange is not a necessity, evaporation seepage loss may be compensated at least at fortnight intervals. During 45-60 days of the rearing period, the post-larvae grow to juveniles of 1-2 g with survival levels of 65-75%.

8.4.2. Grow-out pond management

Grow-out culture of *M. rosenbergii* is normally carried out for 4-6 months. Prawn juveniles obtained from nursery ponds are stocked at a density of 40,000-50,000 per ha (4-5 per m^2) under monoculture. Supplementary feed in the form of pellets (2-3 mm diameter. 25-30% protein) with water stability of 2 hours is broadcasted around the peripheral area @ 8-10% of the biomass during the initial month, followed by 3% of biomass subsequent rearing period. The daily ratio is provided in two to three splits with the major share provided at evening hours. The amount of daily feed is monitored through the provision of 4 cheek trays placed at regular intervals in the pond. Since feed consumption reduces during the moulting period as well as on rainy/ overcast days, the daily ration may be reduced during these periods by 10-20%.

Among the various intrinsic and extrinsic factors associated with the culture of scampi, heterogeneous individual growth (HIG) represents a serious threat to farmers, resulting in heavy economic losses. *Macrobrachium rosenbergii* is known to exhibit a complex social organizational hierarchy consisting of morphologically distinct dominant, subdominant and subordinate groups. The predominance of a definite social hierarchy among the male morphotype increases the differential growth pattern within these prawns. The males show a high degree of HIG, while the size distribution is rather homogeneous of the female population. Three male morphotypes [stunted male (SM), orange-clawed male (OCM) and blue-clawed male (BCM)] usually represent the three developmental stages of the male maturation process, transforming SM>> OC >> BC. The SMs are subordinate and not territorial, and they are the initial stage of the developmental pathway; OCMs are subdominant, representing highly somatic growth; BCMs represent the terminal stage in the morphotypic transformation pathway. Once a set of prawns reaches the terminal morphotype stage, it inhibits the transformation of other morphotypes to successive stages. This leads to a wide range of variations in the growth pattern. To overcome this problem, a few steps can be adopted such as i) periodic removal of berried females that helps in facilitating morphotypic transformation and the male maturation process, ii) periodic removal of the BCMs (dominant) to minimize the 'runt' population, and iii) phased harvesting of fast-growing individuals to improve the prospects of other smaller individuals to achieve their growth potential.

8.4.3. Water quality management

Constant water depth of 1.2-1.5 m is maintained in the pond through periodic compensation of the seepage and evaporation loss. Water quality in the grow-out ponds should be monitored at least every 15-day interval and appropriate intermittent liming and fertilization measures should be taken (Table 8.2). In the case of intensive farming with high stocking density, intermittent water exchange and aeration are the other additional management measures to increase the yield. Harvesting of bigger size prawns mainly females from the fourth month onwards can increase the yield and improve the growth of smaller prawns. The major problems encountered during grow-out culture include low dissolved oxygen content which often occurs due to heavy plankton bloom. This problem gets precipitated during continuous overcast, which can be overcome with water exchange or aeration. Usually, the prawns reach marketable size of about 50-60 g within 5-6 months. Harvesting needs to be done only during the inter-moult period, as prawns of the pre-moulting stage and just after moulting fetch very low market prices.

Table 8.2. Optimum aquaculture water quality parameters for *M. rosenbergii*

Water parameter	**Optimum level**
Water depth	1.2-1.5 m
Temperature	28-32 ^{0}C
Dissolved oxygen	>5-7 ppm
pH	7.0-8.5
Alkalinity ($CaCO_3$)	20-60 ppm
Total hardness ($CaCO_3$)	30-150ppm
Non-ionized ammonia (NH_3-N)	<0.3 ppm

Nitrite nitrogen (NO_2-N)	<0.2 ppm
Nitrate nitrogen (NO_3-N)	<10 ppm
Iron (Fe)	<1.0 ppm
Hydrogen sulphide (H_2S)	<0.01 ppm
Biological oxygen demand (BOD)	< 3.0 ppm
Chemical oxygen demand (COD)	<70 ppm
Transparency	25-40 cm
Ratio of phyto and zooplankton	10:1
Watercolour	Light green/light brown

8.5. Freshwater pearl culture

Lamellidens marginalis, L. corrianus and *Parreysia corrugata* are the three generally accessible freshwater mussel species that are used in the standardization of freshwater pearl culture. To produce various types of pearl products, three distinct surgical techniques - mantle cavity insertion, mantle tissue implantation, and gonadal implantation - have also been standardized. Depending on the shape of the implanted nuclei, the outcomes of the mantle cavity insertion method are round or designed pearls attached to a shell. But during the mantle tissue implantation process, the products - which range in size from tiny round nucleated pearls to irregular oval graft pearls - are detached. The pearls produced are unattached and slightly larger round pearls in gonadal implantation.

8.5.1. Pond culture

The implanted mussels are stocked in the ponds following post-operative care. The mussels are housed at a rate of two per bag in nylon bags (1 cm mesh-12 cm^2), which are suspended from bamboo or PVC pipes and submerged in ponds at a depth of one meter (Fig. 8.9). During the sweltering summer months, the mussels can be positioned in deeper zones - up to 2.0 m - to prevent surface heating. The stocking density of the mussels is 20,000-30,000 ha^{-1} during their culture. A rectangular pond with the right inlets and outlets is the best shape for raising implanted pearl mussels. Pearl culture operations can be conducted in ponds that do not have aquatic macrophytes or algal blooms, such as those caused by *Microcystis* sp. and *Euglena* sp. Platforms (16 x 8 m) made of PVC tubing with a diameter of 2 inches are installed in the ponds to serve as rafts for pearl mussel culture units. Placed in nylon bags (1.0 cm mesh, 12 x 14 cm) with two mussels per bag, the implanted mussels have a density of 50,000 to 75,000 ha^{-1}. The size of the mesh and the area of the pockets in these basic culture units allow the mussels to open and close their shell valves for feeding and functioning. The P.V.C. floating platform units or the bamboo rafts kept in the cultural environment are then secured to these bags. It has been established that for optimal mussel survival and growth, the ideal hanging depth of the pearl mussel culture units is 1.5 to 2.0 m deep in the ponds. Alternatively, the mussels can also be placed in plastic crates (0.5 x 0.35 x 0.25 m) @ 20 to 25 mussels per crate.

Fig. 8.9. A small-scale freshwater pearl culture in the pond

8.5.2. Water management

Pearl culture operations are best suited for ponds that are 2.0–2.5 meters deep, have a clay soil base, and have slightly alkaline waters. During the mussels' culture stage, the ponds' water level and physicochemical conditions need to be observed. The ideal temperature range is between 25 and 30 °C. The pond must be kept at ideal pH levels (7.5–8.5), total alkalinity levels (75–150 ppm), total hardness levels (40–75 ppm), and dissolved calcium levels (25–50 ppm). The habitat used for carp aquaculture is often the same as the freshwater pearl culture pond. As the benthic invertebrates that make up the ecosystem of the pond, their free movement on the pond bottom should be optimal for the growth and survival of freshwater mussels. Research has shown that mussels kept in bottom-set culture units had a lower survival rate (27%) compared to those kept in surface and column-set units (66%). The lower level of primary output, siltation and the units' restricted space could be the cause of the low survival in bottom units. The degree of macrophyte infestation and water flow in the culture environment are additional crucial factors. However, there are certain issues with sample and pearl harvesting because to the operated mussels' pond bottom distribution. To maintain the productivity of the plankton, the ponds are periodically fed with both organic and inorganic fertilizers. A good way to maintain the pearl-bearing mussel standing crop is to supply green water (*Chlorococcum* sp. and *Scenedesmus* sp.) to the pearl culture ponds on a regular basis as direct mussel feed. The "open culture method" is used to produce the green feed in a series of 200 l tanks that are positioned all over the pond bundh. The following is how the water in the tanks is fertilized:

- Cowdung: 1000 kg $ha^{-1} yr^{-1}$ in 12 equal instalments
- Urea: 100 kg $ha^{-1} yr^{-1}$ in 12 equal instalments
- Single super phosphate (SSP): 100 kg $ha^{-1} yr^{-1}$ in 12 equal instalments

The green water appears once the fertilizer has broken down (7 to 10 days). Water enhanced with algae is fed into the ponds used for pearl cultivation. The mussels may take in a wide range of specific organic material feed since they are mucoid filter feeders. A meal consisting of groundnut oil cake and powdered rice bran (1:1) at 1% of the mussels' weight given every other day can be used to sustain pearl mussels in captivity. During the 12 to 18 month cultivation phase, mussels must be periodically checked, with dead ones removed, and the bags must be cleaned.

8.6. Ornamental fish culture

Ornamental fish forms an important commercial module of the fisheries sector. The brilliant colour and exotic appearance of the fishes attract one and all, greatly enhancing the aesthetic beauty of the houses. The relatively minimum requirement of space or attention compared to other species is the reason for the growing interest in keeping aquaria at household levels. The domestic ornamental fish trade mostly confines to a few exotic live-bearers, goldfish and angel fish despite having more than 100 indigenous species of ornamental value and about equal numbers of exotic species in the country. Considering the increasing domestic demand and ready export market, ornamental breeding and culture offer enormous scope for its adoption both at small scale and as a commercial enterprise.

8.6.1. Infrastructure facilities

The infrastructure facilities mainly include FRP/ferro-cement breeding and rearing tanks of different dimensions, smaller earthen ponds, clean water supply systems, glass aquaria of assorted sizes and proper housing for breeding and larval rearing. The ornamental fish culture unit should have provision for a continuous supply of water and electricity as shown in Fig. 8.10. In a large-scale production unit, there must be provision for water pumps, standby generators, air blowers, live food culture units, etc. An oxygen cylinder may be kept for packing fish during the sale. The dimensions of the tanks may depend on the purpose of their use, species to be bred/ reared and scale of operation. Tanks of 2 x 2 x 2' to 4 x 3 x 3' or circular tanks of 4' diameter and 3' height are commonly used as breeding tanks, while the size of culture tanks normally varies from 5 x 6 x 3' to 10 x 7 x 3'.

8.6.2. Species for culture

Ornamental fishes are broadly classified as livebearers and egg layer. Among the livebearers, *Guppies* are the simplest group to start with as the species can tolerate a wide range of water quality and accept a great variety of foods. There are many varieties of guppy *viz.*, round tail, sphere tail, Fantail, Veil tail, Cobra guppy, and Singapore guppy. etc. The next popular livebearer group is the Molly, which includes sail-fin molly, balloon molly, Chocolate molly, White molly, etc. Another magnificent group of live-bearers is the Swordtail with varieties like red swordtail, green swordtail, Sunset swordtail, tuxedo tail, Double-swordtail etc. Platy is yet another group of small livebearers with many varieties like round tail, Veil tail, Fringe tail, Lyre tail, Pin tail, etc.

Fig. 8.10. A small-scale ornamental fish culture unit

The majority of the ornamental fishes are egg-layers, which include carp, barbs, rasboras, danios, tetras, catfishes, etc. Goldfish have been the most common and popular aquarium fish throughout the world for decades. Many varieties of goldfish like Lionhead, Veil-tail, Fringe-tail, Telescope-eye, Calicogold, Peacock-tail, Red-carp, Pearl-scale etc. are common in the market. Nowadays, Koi carp with different colour combinations have become popular in aquaria and garden pools. The easy seed production and bright body colouration have increased the popularity of barbs for aquarium fish keeping. Many indigenous and exotic varieties of barbs are available including well-known ones like Rosy barbs, Tiger barbs, etc. Danios like Zebra danio, tetras like Black Widow Tetra, Neon tetra, Lemon tetra etc. are some other popular varieties of ornamental fishes. The anabantid groups of fish are air breathers and lay eggs in bubble nests. The group includes Siamese fighting fish, and several varieties of gouramies like dwarf gourami, moonlight gourami, pearl gourami, kissing gourami etc. Angelfish belonging to the chilid group is another major group for aquarium breeding. Discuss fish and its blue and brown forms are highly-priced. Some indigenous fishes like Indian glassfish and loaches have also good demand in the market.

8.6.3. Breeding of livebearers

The livebearers like guppies, platies and swordtails take six to eight weeks to mature while mollies take 12-16 weeks. Matured females and males of these species are selected from the community tanks and stocked inside breeding traps made of perforated nylon bags placed in separate breeding pools meant for each species. Aquatic plants like Cobomba are placed in the breeding pool to facilitate breeding. Fertilization is internal in these livebearers where the male inserts its gonopodium to transfer milt into the female and the eggs are fertilized inside the female. The embryos grow within four weeks to tiny young ones and become ready for free swimming. Soon after their birth, the young ones are collected to avoid parental predation. The larvae are fed with infusoria, zooplankton, tubifex etc. till they attain 25-30 mm in size, after which they are transferred to rearing tanks.

8.6.4. Breeding of egg layers

Matured males and females of each species are selected from the community tanks and stocked in breeding tanks at the ratio 1:2-3. Usually, females are introduced in the breeding tank one day in advance than the males and external fertilization takes place. In general, parents eat away their egg and larvae. Therefore, round pebbles at the bottom and aquatic weeds and nylon threads are provided to prevent parental predation. Immediately after hatching, parents should be removed from the breeding pool. The larvae are fed with infusoria, zooplankton, tubifex etc. till they attain 25-30 mm in size, after which they are transferred to rearing tanks. Generally, dispense the feed once daily, preferably in the morning. The rate of feeding depends on species, size and season. Overfeeding is more harmful than underfeeding as the excess feed deteriorates the water quality.

8.6.5. Water quality management

Ornamental fish production unit required a higher level of expertise for better water quality management as ornamental fishes are more sensitive to poor water quality. In the municipal areas, the farmers use normal taps for farming. Before use, it is aerated for a few days for dichlorination. Tube well water is also used directly in rural areas. The average temperature of the rearing water in the area is 20- 28 ^{0}C and the pH is slightly alkaline. Other parameters are not so crucial. Most of the species cultured prefer soft to medium-hard water. Regular water exchange along with proper aeration is a must to maintain critical water quality parameters like pH (7.2-7.7), DO (6-7 ppm), nitrite (<0.75 ppm), nitrate (<30 ppm), ammonia (almost zero) etc.

8.7. Water use in high-value commercial aquaculture: case studies from India

ICAR-Indian Institute of Water Management (Bhubaneswar, India), a premier central government research organization having PAN India research mandate on water management in agriculture, aquaculture and allied sectors, had conducted field experiments on water requirement, WUE, water productivity and water footprint of high-value aquaculture and aquaculture-based IFS during 2011-2022.

In 'Pacific white shrimp' (*Litopenaeus vannamei*) culture, with best management practice (BMP) at the optimum stocking density of 5,00,000 shrimp seed ha^{-1}, total water use could be minimized to 32,600 m^3 and water exchange minimized to 6,300 m^3 ha^{-1}, without hampering the normal growth and production. With this stocking density, shrimp productivity enhanced to a level of 10.58 t ha^{-1} in 120 days (Table 8.3). Further, with this technique, a total water productivity of □46.6 m^{-3}, and consumptive water productivity of □ 83.3 m^{-3} was generated while only 1.72 m^3 of water was required to produce 1 kg of *L. vannamei* shrimp (Mohanty *et al.*, 2018a; 2018b). In this system, the total water footprint (WF_t) amounted to 1229 m^3 t^{-1}. Various factors that affect aquacultural water management are depicted in Fig. 8.11.

Table 8.3. Technologically validated field trial result of water uses in high-value aquaculture.

Water Management parameter	**Carp poly culture**	**Monoculture of *P. monodon***	**Monoculture of *L. vannamei***
Optimum density, ha^{-1}	8000 fingerlings	200000 PL	500000 PL
Culture duration	180 days	125 days	120 days
Water depth (m) maintained	1.2 up to 90 DOC 1.75 up to 180 DOC	1.0 up to 30 DOC 1.2 up to 125 DOC	1.2 up to 30 DOC 1.5 up to 120 DOC
Evaporation losses, ($\times 10^4$, m^3)	0.58	0.63	0.62
Seepage losses, ($\times 10^4$, m^3)	0.49	0.55	0.52
Regulated outflow, ($\times 10^4$, m^3)	-	0.70	0.63
Other losses, ($\times 10^4$, m^3)	0.06	0.08	0.05
Total loss (CWU), ($\times 10^4$, m^3)	1.13	1.96	1.82
Initial water level, ($\times 10^4$, m^3)	1.20	1.00	1.20
Precipitation, ($\times 10^4$, m^3)	0.6	0.46	0.51
Regulated inflow, ($\times 10^4$, m^3)	1.02	1.40	1.55
Total water use, ($\times 10^4$, m^3)	2.82	2.86	3.26
Consumptive Water Use Index, m^3kg^{-1}	2.88	4.28	1.72
Productivity, t ha^{-1}	3.92	4.58	10.58
Feed conversion ratio, FCR	1.74	1.40	1.63
Sediment load, m^3 t^{-1} biomass	53.0	42.1	46.3
OV-CC ratio	1.81	2.47	2.11
Total water productivity, □ m^{-3}	5.9	26.6	46.6
Consumptive water productivity, □ m^{-3}	10.3	37.9	83.3
Total water footprint (WF_t, m^3 t^{-1})	1633	3079	1229

In carp polyculture (Indian major carps, IMCs), with BMP at the optimum stocking density of 8,000 fingerlings ha^{-1}, total water use could be minimized to 28,200 m^3, without hampering the normal growth and production. With this stocking density, fish productivity enhanced to a level of 3.92 t ha^{-1} in 180 days (Table 8.3) with economic benefit (OV: CC) of 1.81. Further, a total water productivity of □5.9 m^{-3}, and consumptive water productivity of □ 10.3 m^{-3} was generated while only 2.9 m^3 of water was required to produce 1 kg of fish biomass (Mohanty *et al.*, 2016 & 2017a) In this system, total water footprint (WF_t) amounted to 1633 m^3 t^{-1}.

Similarly, in 'black tiger prawn (*Penaeus monodon*)' monoculture, stocking density of 2,00,000 post-larvae ha^{-1} gave significantly higher yield (4.58 t ha^{-1} 125 d^{-1}), economic benefit (OV: CC, 2.47), total water productivity (□26.6 m^{-3}) and consumptive water productivity (□ 37,9 m^{-3}). At this optimum stocking density, total water use could be minimized to 28,600 m^3 and water exchange minimized to 7,000 m^3 ha^{-1}, without hampering the normal growth and production while, only 4.28 m^3 of water was required to produce 1 kg of *Penaeus monodon* biomass (Mohanty *et al.*, 2014, 2015 & 2017b). In this system, the total water footprint (WF_t) amounted to 3079 m^3 t^{-1} (Table 8.3).

Water requirement of different freshwater carp polyculture systems was also carried out to optimize water requirement. Out of four different production systems such as:

i) IMC grow-out culture: *Single Stock-Single Harvest system,*

ii) IMC grow-out culture: *Single Stock-Multi Harvest system,*

iii) IMC grow-out culture: *Multi Stock-Multi Harvest system,*

iv) Intercropping of IMC-Minor carp.

Multi stock-multi harvest system (MSMH) was found to be more efficient, productive and profitable (Table 8.4) in terms of water use efficiency (0.57 kg m^{-3}), productivity (4.6 t ha^{-1}), FCR (1.51), total water footprint (998 m^3 t^{-1}) and net consumptive water productivity (INR 9.2 m^{-3}). Next to multi multi-stock-multi-harvest system, an intercropping system of IMC-Minor carp was considered efficient, productive and profitable.

Table 8.4. Water use in different freshwater carp polyculture systems.

Water use parameters	Single stock-single harvest	Single stock-multi harvest	Multi stock-multi harvest	Intercropping of IMC-Minor carp
Stocking density, ha^{-1}	10,000 IMC fingerlings	10,000 IMC fingerlings	4,000 IMC fingerlings**	8,000 fingerlings Including 50% IMC
Culture duration	360 days	360 days	360 days	360 days
Water depth (m)	1.2 up to 180 DOC 1.5 up to 360 DOC	1.2 up to 180 DOC 1.5 up to 360 DOC	1.2 up to 180 DOC 1.5 up to 360 DOC	1.2 up to 180 DOC 1.5 up to 360 DOC
TWU, ($\times 10^4$, m^3)	2.40	2.53	2.5	2.43
CWU, ($\times 10^4$, m^3)	0.73	0.84	0.82	0.76
CWUI, m^3kg^{-1}	1.97	1.95	1.78	1.77
Productivity, t ha^{-1}	3.7	4.3	4.6	4.3
WUE_c (kg m^{-3})	0.50	0.51	0.57	0.56
FCR	1.76	1.63	1.51	1.72
NCWP, □ m^{-3}	7.7	8.1	9.2	8.3
WF_t, m^3 t^{-1}	1122	1093	998	1026

TWU-total water use; CWU-consumptive water use; CWUI-consumptive water use index, WUE-water use efficiency; NCWP-net consumptive water productivity; WF_t-total water footprint; WUE_c-Consumptive water use efficiency; FCR-feed conversion ratio; DOC-days of culture, **partial harvesting was carried out at 6th and 9th month followed by restocking @1.25 times of the harvested numbers.

Fig. 8.11. Factors affecting aquacultural water management

Water budgeting in the 'Aquaculture-based Integrated Farming System' was carried out by integrating various components such as aquaculture (carp polyculture), agriculture (*kharif* rice followed by green gram and vegetables with sprinkler and drip irrigation), on-dyke horticulture (banana and papaya with drip irrigation) and poultry '*Vanaraja*'. Out of the system's total crop water use (3.14×10^4 m^3 240 d^{-1}), the estimated TWU, consumptive water use index (CWUI) and productivity in carp polyculture alone was 2.3×10^4 m^3 240 d^{-1}, 2.6 m^3 kg^{-1} fish production and 2.86 t ha^{-1}, respectively. The estimated evaporation and seepage losses were 2.8 and 2.1 m^3 water kg^{-1} fish production respectively and contributed significantly to CWU. The system as a whole, resulted in a net profit of Rs.1,77,117 ha^{-1} with an output value - cost of cultivation ratio of 3.1 and net consumptive water productivity of Rs.19.2 m^{-3}.

References

Abakari, G., Wu, X., He, X., Fan, L. and Luo, G. 2022. Bacteria in biofloc technology aquaculture systems: roles and mediating factors. *Reviews in Aquaculture,* 00, 1-25. https://doi. org/10.1111/raq.12649.

Avnimelech, Y. 2007. Feeding with microbial flocs by tilapia in minimal discharge bioflocs technology ponds. *Aquaculture*, 264:140-147.

Avnimelech, Y. 2012. Biofloc Technology. A Practical Guide Book. The World Aquaculture Society, Baton Rouge, p. 272.

Collazos-Lasso, L.F. and Arias-Castellanos, J.A. 2015. Fundamentals of bioflocs technology (BFT). An alternative for fish farming in Colombia: a review. *Orinoquia*, 19 (1):77-86.

Deng, M., Chen, J., Gou, J., Hou, J., Li, D. and He, X. 2018. The effect of different carbon sources on water quality, microbial community and structure of biofloc systems. *Aquaculture*, 482: 103-110.

Emerenciano, M.G.C., Gaxiola, G. and Cuzon, G. 2013. Biofloc technology (BFT): a review for aquaculture application and animal food industry. *In*: Matovic, M.D. (Ed.), Biomass Now: Cultivation and Utilization. InTech, Belfast, Canada, pp. 301-308.

Khanjani, M.H., Eslami, J., Ghaedi, G. and Sourinejad, I. 2022b. The effects of different stocking densities on nursery performance of Banana shrimp (*Fenneropenaeus merguiensis*) reared under biofloc condition. *Annals of Animal Science,* https://doi.org/ 10.2478/aoas-2022-0027.

Khanjani, M.H. and Sharifinia, M. 2020. Biofloc technology as a promising tool to improve aquaculture production. *Reviews in Aquaculture*, 12 (3):1836-1850.

Khanjani, M.H. and Sharifinia, M. 2021. Production of Nile tilapia Oreochromis niloticus reared in a limited water exchange system: the effect of different light levels. *Aquaculture,* 542: 736912.

Khanjani, M.H., Sharifinia, M. and Hajirezaee, S. 2022a. Recent progress towards the application of biofloc technology for tilapia farming. *Aquaculture*, 552:738021.

Khanjani, M.H., Torfi Mozanzade, M. and Foes, G.K. 2022c. Aquamimicry system: a suitable strategy for shrimp aquaculture. *Annals of Animal Science,* https://doi.org/10.2478/ aoas2022-0044.

Komives, T. and Junge, R. 2015. Editorial: on the "aquaponic corner" section of the journal. *Ecocycles*, 1(2):1-2.

Krummenauer, D., Reis, W.G., Abreu, P.C., Bezerra, A., Wasielesky, W. and Advent, B. 2021. Effect of aeration on the development of the microbial community in shrimp biofloc technology systems. *Global Seafood Alliance,* ‹https://www.globalseafood.org/advocate /effect-of-aeration-on-the-development-of-the-microbial-community-in-shrimp-bio floc-technology-systems/›

Kumar, V., Roy, S., Behera, B.K., Swain, H.S. and Das, B.K. 2021. Biofloc microbiome with bioremediation and health benefits. *Frontiers in Microbiology*, 12: 741164.

Martínez-Cordova, ′ L.R., Emerenciano, M., Miranda-Baeza, A. and Martínez-Porchas, M. 2015. Microbial-based systems for aquaculture of fish and shrimp: an updated review. *Reviews in Aquaculture*, 7:131-148.

Meenakshisundaram, M., Sugantham, F., Muthukumar, C. and Chandrasekar, M.S. 2021. Metagenomic characterization of biofloc in the grow-out culture of Genetically Improved Farmed Tilapia (GIFT). *Aquaculture Research,* 52: 4249-4262.

Mohanty, R.K., Ambast, S.K., Panigrahi, P. and Mandal, K.G. 2018a. Water quality suitability and water use indices: Useful management tools in coastal aquaculture of *Litopenaeus vannamei*. *Aquaculture,* 485: 210-219.

Mohanty, R.K., Ambast, S.K., Panigrahi, P., Thakur, A.K. and Mandal, K.G. 2018b. Enhancing water use efficiency in monoculture of *Litopenaeus vannamei*: Impacts on pond water quality, waste production, water footprint and production performance. *Aquacultural Engineering*, 82: 46-55.

Mohanty, R.K., Mishra, A., Ambast, S.K., Mandal, K.G., Panda, D.K. and Panigrahi, P. 2017b. Effects of various shrimp (*Penaeus monodon*) densities on their growth, water and sediment quality, and water budget *Aquaculture International,* 25:2161-2176.

Mohanty, R.K., Mishra, A., Panda, D.K. and Patil, D.U. 2015. Effects of water exchange protocols on water quality, sedimentation rate and production performance of *Penaeus monodon* in earthen ponds. *Aquaculture Research,* 46: 2457-2468.

Mohanty, R.K., Mishra, A. and Patil, D.U. 2014. Water Budgeting in Black Tiger Shrimp *Penaeus monodon* Culture Using Different Water and Feed Management Systems. *Turkish Journal of Fisheries and Aquatic Sciences,* 14:487-496.

Monroy-Dosta, M.D.C., Lara-Andrade, D., Castro-Mejía, J., Castro-Mejía, G. and Emerenciano, M.G. 2013. Microbiology community composition and abundance associated to biofloc in tilapia aquaculture. *Revista de Biologia Marina y Oceanografia,*, 48 (3): 511-520.

Panigrahi, A., Esakkiraj, P., Jayashree, S., Saranya, C., Das, R.R. and Sundaram, M. 2019a. Colonization of enzymatic bacterial flora in biofloc grown shrimp Penaeus vannamei and evaluation of their beneficial effect. *Aquaculture International,* 27:1835-1846.

Panigrahi, A., Saranya, C., Sundaram, M., Vinoth Kannan, S.R., Das, R.R., Satish Kumar, R., Rajesh, P. and Otta, S.K. 2018. Carbon: Nitrogen (C: N) ratio level variation influences microbial community of the system and growth as well as immunity of shrimp (*Litopenaeus vannamei*) in biofloc based culture system. *Fish and Shellfish Immunology,* 81:329-337.

Panigrahi, A., Sundaram, M., Saranya, C., Swain, S., Dash, R. and Dayal, J.S. 2019b. Carbohydrate sources deferentially influence growth performances, microbial dynamics and immunomodulation in Pacific white shrimp (*Litopenaeus vannamei*) under biofloc system. *Fish and Shellfish Immunology,* 86:1207-1216.

Suita, S.M., Ballester, E.L.C., Abreu, P.C.O.V. and Wasielesky, W. 2015. Dextrose as carbon source in the culture of Litopenaeus vannamei (Boone, 1931) in a zero exchange system. *Latin American Journal of Aquatic Research,* 45 (3):526-533.

Wang, C., Pan, L., Zhang, K., Xu, W., Zhao, D. and Mei, L. 2016. Effects of different carbon sources addition on nutrition composition and extracellular enzymes activity of bioflocs, and digestive enzymes activity and growth performance of *Litopenaeus vannamei* in zero-exchange culture tanks. *Aquaculture Research,* 47 (10): 3307-3318.

9

People's Awareness of Ecosystem Services

9.1. Introduction

Human dependence on natural processes for the goods and services provided by nature, especially food, fuel and fibre, has been a known fact. In recent times, the value of less tangible services, such as climate control, water filtration, and soil fertility, as well as recreational and cultural services has been emphasized also. Ecosystem services underpin human existence, health, and prosperity. It has been beneficial to humans through the transformations of resources (or environmental assets, including land, water, vegetation, and the atmosphere) into a flow of essential goods and services like clean air, water, and food (Costanza *et al.*, 1997). Ecosystem services are those functions of ecosystems- including agroecosystems that are useful to humans or supportive of human well-being (Kremen, 2005). Services provided by ecosystems receive little notice for their support to human civilization often leading to poor management of the ecosystems. Land use systems influence ecosystem functions, which determine the supply of ecosystem services. Land and water resources are under considerable pressure from the expansion of cultivated and residential lands. Diverse land use in agrarian landscapes not only fosters biodiversity, but also provides potential ecosystem services such as air cleaning by vegetation, natural water filtration by plant roots, natural pest suppression, soil conservation, nutrient retention, and crop pollination by insects, etc. Lowland floodplains or waterlogged land areas are generally more conscious of the benefits provided by ecosystem services. It is important to create awareness among the people on provisioning services provided by ecosystems, such as crops, biofuel, fresh water, natural medicine, wildlife, and several cultural services. Managing ecosystem services, which link human well-being to conservation outcomes, is essential for achieving the sustainable development goals (SDGs) of the United Nations (Zhang, 2017).

According to the Millennium Ecosystem Assessment (MEA), ecosystem services are "the benefits people obtain from ecosystems" (Fig. 9.1). The MEA also delineated the four categories of ecosystem services- provisioning, regulating, cultural and supporting. The Millennium Ecosystem Assessment Report 2005 defined ecosystem services as benefits people obtain from ecosystems and distinguishes four categories of ecosystem services, where the so-called supporting services are regarded as the basis for the services of the other three categories. *Provisioning services* consist of all "the products obtained from ecosystems", such as food, fuel, fibre, raw materials, genetic resources, ornamental resources, energy, by-products, and other harvestable goods. *Regulating services* are the "benefits obtained from the regulation of ecosystem processes" like biological control of pest and disease control, pollination, soil retention, carbon sequestration and climate regulation, waste decomposition and water quality regulation, and weather disturbance regulation i.e., flood protection. *Cultural services* relate to the non-material world, as they have recreational and cultural benefits including aesthetic, cognitive and spiritual activities, which are not easily quantifiable in monetary terms. *Supporting services* are the services that allow for the other ecosystem services to be present, such as soil formation, nutrient cycling, habitat provision, hydrological flow, and soil fertility that maintain favourable conditions for life on the earth. They have indirect impacts on humans that last over a long time. Supporting services make it possible for the ecosystems to continue providing services such as food supply, flood regulation, and water purification (MEA, 2003).

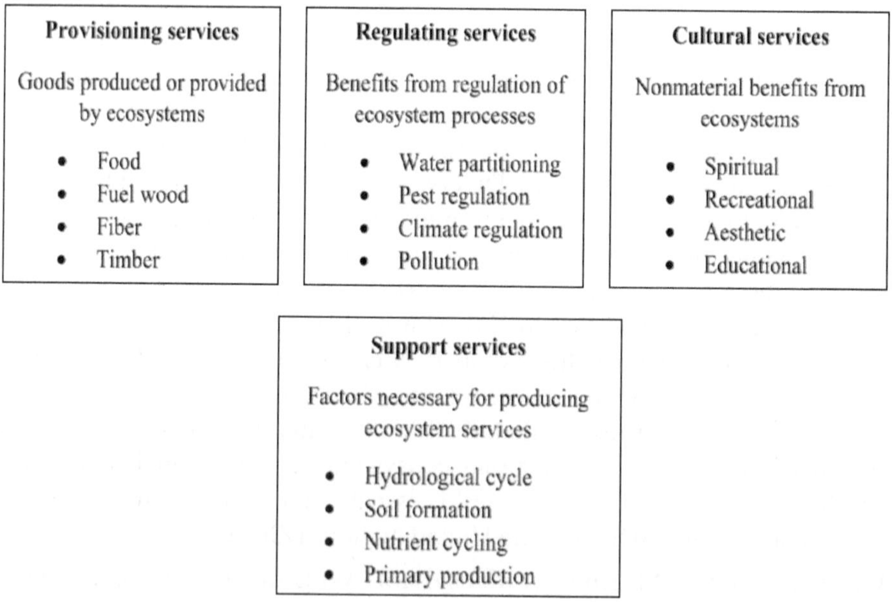

Provisioning services

Goods produced or provided by ecosystems

- Food
- Fuel wood
- Fiber
- Timber

Regulating services

Benefits from regulation of ecosystem processes

- Water partitioning
- Pest regulation
- Climate regulation
- Pollution

Cultural services

Nonmaterial benefits from ecosystems

- Spiritual
- Recreational
- Aesthetic
- Educational

Support services

Factors necessary for producing ecosystem services

- Hydrological cycle
- Soil formation
- Nutrient cycling
- Primary production

Fig. 9.1. Ecosystem services (*Source*: MEA, 2003)

9.2. Agro-ecosystem

An agro-ecosystem (or agricultural-socio-economic-ecological system) is an ecological system partially modified by human to meet their present and future demands for agricultural produce, for example, a rice field. Humans intentionally design agricultural ecosystems to generate essential sustenance for survival and growth, primarily in the form of food. In addition to nourishment, these ecosystems yield other valuable resources like fibre and energy, as well as various unquantifiable environmental benefits (Costanza *et al.*, 1997; Millennium Ecosystem Assessment (MEA) 2005; de Groot *et al.*, 2012). The provision of ecosystem services (ES) in agricultural settings is heavily influenced by the farming inputs and practices adopted by growers (Wossink and Swinton, 2007; Ma *et al.*, 2012).

9.2.1. Features of an agro-ecosystem

i) Well-defined boundaries
   - Bund made up of soil
   - External linkage limited for irrigation

ii) Dominating to human-introduced component
   - Rice, dominating in number and area
   - Reduction in natural and biological components

iii) Sustenance of ecological processes
   - Competition between rice crops and weeds
   - Herbivorous of rice by pests
   - Predation of pests by natural enemy

iv) Human intervention alters ecological forces
   - Cultivation and harvesting of rice crop
   - Subsidies in the form of inputs
   - Control of water and pests
   - Co-operation and competition between people in management

9.2.2. Properties of an agro-ecosystem

Four interconnected properties distinguish one agro-ecosystem from another:

Productivity: It refers to the output per unit of input. It is measured as yield or net income per unit area. Input refers to land, labour, capital, technology etc.

Productivity can be measured for land, labour, capital, and technology.

- Land: output/ha or profit per ha
- Labour: output/man hour

- Capital: return/unit of capital invested
- Technology: output/unit of technological intervention

Levels of measurement can be plant, crop, field, farm, village, region etc. Productivity may remain constant, increase, or decrease, with similarly measured productivity of different agro-ecosystems. Productivity varies with space, time, flow, and decision patterns changes.

Stability: Stability refers to the degree to which production remains constant despite normal small-scale environmental fluctuations. The environment includes physical, biological, social and economic factors. Fluctuation may be due to natural (rainfall, temperature, etc.) or artificial (market fluctuations, subsidies) factors. The coefficient of variation of productivity helps to measure stability. Time series of productivity measurements and trend lines are also used to measure stability. Stability may remain constant, increase or decrease with similarly measured stability of different agroecosystems. Stability will also show variation with space, time, flow, and decision pattern changes.

Sustainability: Sustainability is the ability of a system to maintain productivity when subject to stress or perturbations in the surrounding environment. Stress refers to times-continuous but relatively small and predictable disturbances. It has a cumulative effect on productivity, for example, waterlogging, salinity, indebtedness etc. Perturbations refer to relatively large and unpredictable but less frequent disturbances that have an immediate impact such as drought, flood or pest attack, for example, flood, drought, new pest, price fall etc. Qualitative measurement is normally used. Quantitative measurements are difficult. With shock or stress, the productivity of an agro-ecosystem may remain the same or go down but return soon to its original level or go down and settle at a lower level or become very low

Equitability: It is the measure of how evenly the production of the agroecosystem is distributed among people or beneficiaries. The production is measured in terms of total yield or net income for the concerned agro-ecosystem. Production may be uniformly distributed with everyone obtaining an equal share (high), or with some receiving more than others (low). Equitability may show variation due to space, time, flow and decision pattern changes.

Agro-ecosystems provide and rely on ecosystem services (Fig. 9.2). Services that help to support the production of harvestable goods can be considered services to agriculture (Zhang *et al.*, 2007). Management of agricultural ecosystems by humans to optimize the provision of food, fibre, and fuel, which are classified as provisioning services by the MEA. These provisions from agriculture depend on several supporting and regulating services as inputs to production (*e.g.*, fertilizers and pollination).

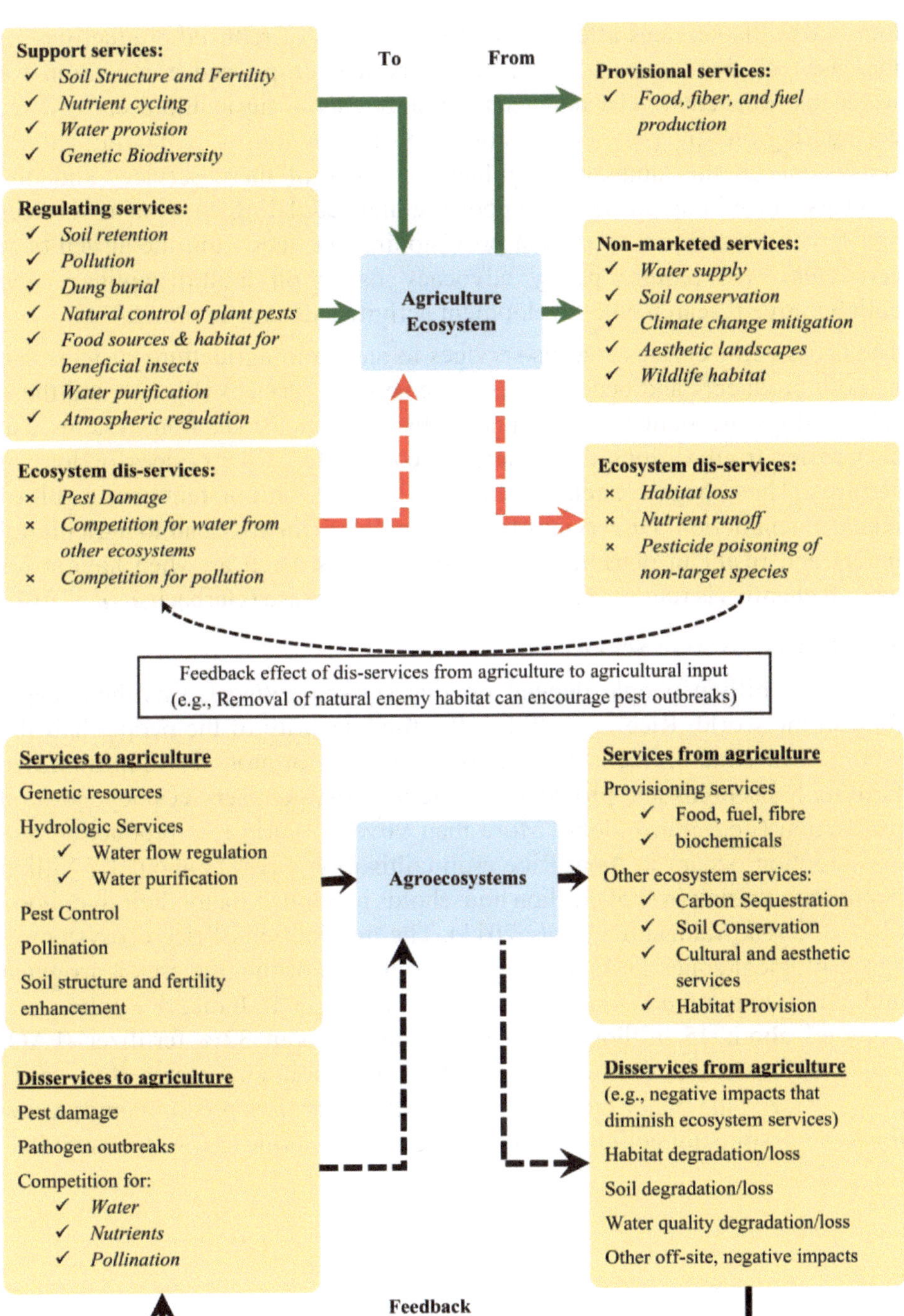

Fig. 9.2. Ecosystem services and dis-services from agriculture (*Source:* Zhang *et al.*, 2007; Garbach *et al.*, 2014)

Ecosystem dis-services affect agriculture in terms of reduced productivity or increased production costs (*e.g.*, herbivory and crop-weed competition for water and nutrients). The management strategies of agricultural ecosystems considering the diversity, composition, and functioning of remaining natural ecosystems in the landscape regulate the flows of these services and dis-services. Providing sufficient supporting and regulating ecosystem services and fewer dis-services to sustain provisioning services from the agricultural ecosystem will require policy advocacy based on multidisciplinary and collaborative research and development activities (Zhang *et al.*, 2007).

The ecosystem services and dis-services to and from agriculture systems can be both producers and beneficiaries of ecosystem services (Fig. 2), and these relationships are deliberately managed by farmers. Production systems can also suffer from various disservices or contribute to disservices or loss of services. These negative relationships are usually the unintentional result of management action. Disservices from agriculture can also lead to agricultural inputs and result in detrimental on-farm impacts, such as when habitat for natural enemies is removed and pest outbreaks increase (Garbach *et al.*, 2014).

9.3. Rice ecosystem services

About 2.5 billion people depend on rice (*Oryza sativa* L.) as their staple food in the world. Rice is central to the food security of the people half the world. Furthermore, rice is a source of livelihood for around 140 million rice farming households and provides a range of ecosystem services beyond food production (i.e., grain) alone. More than 90% of world rice production and consumption occurs in Asia. Rice crop cultivation engages about 1.7 billion people comprising over 50 million households in South Asia for their livelihood (TEEB, 2016; Manzanilla *et al.*, 2011). The distribution of rice ecosystem in agroecological zones in Asia indicates about 57% of the rice area is irrigated and 26.7% in rainfed lowland ecosystem (Table 9.1). In India, rice cultivation occupies about 45 million ha area consuming about 32% fertilizer (FAO, 2005), 22% pesticide (Krishna *et al.*, 2003), and about 60% water (Raju *et al.*, 2005). Rice is the major crop in eastern India as farmers predominantly grow this crop despite not getting adequate returns from it due to low productivity.

Table 9.1. Rice ecosystem distribution in agro-ecological zones in Asia

Agro-ecological zones	Rice crop area (million ha)	Rice area in ecosystems (%)			
		Irrigated	Rainfed lowland	Rainfed upland	Flood-prone
Warm tropics:					
· semi-arid	9.68	75.0	12.4	10.8	1.8
· subhumid	28.94	23.3	53.9	10.6	12.1
· humid	44.52	42.2	32.0	10.3	15.5
Warm subtropics:					
· semi-arid	7.47	99.7	0.0	0.3	0.0
· subhumid	23.91	76.6	13,8	5.2	4.4
· humid	18.35	92.1	6.4	1.5	0.0
Cool subtropics	0.4	100.0	0.0	0.0	0.0
Total	133.27	56.9	26.7	7.7	8.7

Source: Garrity *et al.* (1996).

Rice-based agricultural systems have a significant impact on both the aspects of ecosystem services related to supply and demand (Zhang *et al.*, 2007). Within rice ecosystems, three main categories of ecosystem services are evident: provisioning, regulating, and cultural services. Nevertheless, maintaining the productivity of these systems also necessitates the presence of supporting services. Provisioning services directly offer tangible benefits from farming, such as food, fibre, or fuel (Zhang *et al.*, 2007). Farmers rely on supporting and regulating services to enable the provisioning of services, underscoring the importance of appropriate management strategies to ensure the continuous provision of these benefits.

Ecosystem services within rice cultivation encompass various categories: regulating services (such as natural pest control, carbon management, prevention of soil erosion, and nitrogen fixation), provisioning services (including food and associated by-products), and supporting services (like soil fertility maintenance, hydrological regulation, nutrient cycling, and soil formation). The significance of these ecosystem services extends beyond their immediate contributions to provisioning. For instance, their direct impacts are evident in the valuation of provisioning services, including aspects like nitrogen fixation, soil fertility, and nutrient cycling. Additionally, their indirect effects, such as ensuring long-term soil fertility for future generations and reducing greenhouse gas emissions throughout the production-to-application cycle of chemical fertilizers, also carry value and should be taken into consideration (Spangenberg and Settele, 2010).

Rice ecosystems can also result in adverse effects, including soil erosion, nutrient exhaustion, the harbouring of pests and diseases, and the release of greenhouse gases (GHGs) (Zhang *et al.*, 2007; MEA, 2005; Stallman, 2011). Besides providing a range of ecosystem services, rice production has been linked to a range of different environmental impacts such as high GHG emissions, air and water pollution as well as an increase in water consumption. Policymakers need to make decisions on how to manage and mitigate these impacts while providing affordable, nutritious, equitably accessible and safe food for a growing global population with changing patterns of consumption (TEEB, 2016).

The prevailing agricultural crisis, as indicated by Devi *et al.* (2017), is primarily attributed to low farm income. To address this issue and enhance farmers' income, one potential approach is to provide compensation or allowances based on the ecosystem services they generate through their farming practices. The valuation of these ecosystem services is considered a crucial step in raising awareness and advocating for policies related to such services (Devi *et al.*, 2017). Therefore, it becomes essential to evaluate all services to ensure sustainable rice farming across various agroclimatic regions and with the adoption of improved production techniques. Additionally, it's important to investigate the disparities between the total value of ecosystem services provided by rice farming and the economic returns to farmers, particularly in terms of provisioning services.

Rice cultivation and usage has immense cultural importance in countries such as India, particularly in eastern Indian states like Odisha, where it is seen as a symbol of affluence and plays a vital role in various ceremonies marking important life transitions and personal growth (Nayak *et al.*, 2019). While previous research has explored the connections between cultural aspects and rice cultivation, there is a scarcity of literature addressing the methods for quantifying the monetary worth of these cultural services. The evaluation of ecosystem services serves a dual purpose: not only does it aim to preserve and enhance these services but it also establishes a firm foundation for compensating land managers who safeguard, nurture, and uphold ecosystem services. In the case of rice fields in eastern India, the total economic value of ecosystem services per unit area ranged from US$ 1238 to 1688 per ha per year. Of this, marketed ecosystem services (primarily production-related) accounted for 66–89%, while non-marketed services constituted 11–34% of the total value.

If a comprehensive assessment were conducted to include ecosystem services related to the aesthetic, artistic, educational, and spiritual aspects (cultural), gas regulation (CO_2/O_2 balance), and the preservation of biodiversity (biodiversity) within rice-based systems, the value of non-marketed ecosystem services would likely be significantly higher. The variability in ecosystem services among different agro-climatic zones in eastern India can be attributed to different parameter values. Grouping locations based on this variability in ecosystem services assists in pinpointing areas for intervention to sustain and enhance these services while allowing for sustainable agroecological intensification, as discussed by Nayak *et al.* (2019).

The depletion of global natural resources caused by excessive exploitation has led to the erosion of ecosystem services, as noted by Reid *et al.* (2005). Consequently, it is imperative to ensure that ecosystem services are duly acknowledged and valued through appropriate assessments of goods and services derived from ecosystems. Furthermore, there is a need to actively advocate for the concept of ecosystem services in governmental policies, as emphasized by Loc *et al.* (2016).

9.4. Assessment of rice ecosystem services

The ecosystem services and their associated values and methodologies are important for assessing the economic and environmental impacts of various agricultural practices and strategies. According to Sandhu *et al.* (2008), the elements used to gauge ecosystem services encompass food, secondary products, natural pest control, the creation of soil, the conversion of plant nutrients, carbon circulation, nitrogen fixation, soil quality, hydrological processes, and soil erosion (Table 9.2). The measurement procedure for each component is discussed below.

Table 9.2. Components of rice ecosystem services and their measurements

Services	Methodology of service	Conversion factor	Economic value	Reference
Food	1st crop-rice, 2nd crop-rice, green gram, black gram, groundnut, turmeric	Whole farm produce in quintals	MSP of crops	http://eands.dacnet.nic.in/MSP.htm
By-products	Rice straw	1.5 times of grain	US$ 0.0155 kg^{-1}	Farm gate price
Biocontrol of pest	Economic threshold level (ETL)	The market cost of recommended dose of pesticide at ETL level	1 Spider = US$0.038, 1 Miridbug = US$0.008, 1 Ladybird beetle = US$0.06, 1 Ground beetle = US$ 0.0043	Mala *et al.* (2009); Shepard *et al.* (1987); Gurr *et al.* (2004); Sandhu *et al.* (2005)
Soil formation	Earthworms and Dung beetle	1 tonne of earthworms forms 1 tonne of soil $ha^{-1}yr^{-1}$ 2g soil accumulated per gram of dung beetle (dry wt basis)	Top-soil value is US$ 2093 ha^{-1}	Herrick and Lal (1996); Sandhu *et al.* (2008), Kathuria (2015)
Mineralization of plant nutrients	Assessment of soil N mineralization	Total N = 77.78% of NH4	Equivalent price of N = US$ 0.082 kg^{-1}	Shahid *et al.* (2017)
Carbon flow	Difference of only change of C and C emission	Carbon accumulation is 40% of total biomass incorporated in the field, Percentage leftover C of the applied amount from organic amendments and crop residues is 28.8% in rice field	CER (Carbon Emission Reduction) is about US$ 21.71	Sandhu *et al.* (2008); Johnson *et al.* (2006); Bhola and Malhotra (2014); Mandal *et al.* (2008)
Nitrogen fixation	Rice-19 kg N ha^{-1} $crop^{-1}$ Green gram- 140 kg^{-1} ha^{-1} $year^{-1}$		Equivalent price of N = US$ 0.082 kg^{-1}	Franco and Balieiro (2000); Lodger and Ladha (1992)

Soil fertility	Nutrient (NPK) uptake from soil (Net soil contribution)	NUE = 34%; PUE = 25%; KUE = 60% for rice	The market price of fertilizers	Roy *et al.* (2006); Ravichandran and Sriramachandrasekharan (2011); Lakshmi *et al.* (2012)
Hydrological flow	Rain and irrigation water recharge	About 45% of total rainfall and irrigation as recharge to groundwater	US$ 1.5 per 1000 m^3	Water charging in irrigated agriculture (FAO, 2004); Tripathi (2016)
Erosion	District-level data on soil erosion	Sediment delivery ratio (SDR) = 0.3	Topsoil value as US$ 2093 ha^{-1}	http://www.dowrorissa.gov.in/DIP/DIPIndex.htm, Sreenivasalu (2012)

Source: Nayak *et al.* (2019)

Food and its by-products

To assess the ecosystem services provided by the food generated through rice farming, the value is determined by multiplying the minimum support prices (MSP) set by the Indian government for the given year with the total farm produce. MSP is established to facilitate the government's procurement of food grains from farmers. Similarly, the service from by-products is derived based on the quantity of by-products, specifically rice straw, derived from rice crops based on a factor of 1.5 times the crop grain yield, as recommended by Johnson *et al.* (2006) and Sandhu *et al.* (2008). The economic worth was then determined using the market prices of straw bales at local markets, which amounted to US$ 15 per ton of straw (Table 9.3).

Biological pest control

The management of rice insect pests can be evaluated through the activities of predators and parasitoids, employing both actual pests and surrogate "prey" as indicators to establish a "predation rate". In order to calculate the predation rate within rice ecosystems, the reduction in "prey" populations in specific fields is gauged by counting the presence of bio-control agents, drawing from the research of Nayak *et al.* (2019), Mala *et al.* (2009), Sandhu *et al.* (2005), Gurr *et al.* (2004), and Shepard *et al.* (1987). The economic threshold level (ETL), which advises farmers to apply pesticides only when pest populations exceed a certain threshold, is employed to compute insect-pest infestation per ha at the ETL level. The recommended pesticide doses for these pests are multiplied by market prices, inclusive of application costs, to ascertain the total expense of pest control per ha. This cost figure is then utilized to estimate the expenditure involved in managing each specific insect pest. Predation rates are determined based on the amount of prey consumed during the life cycle of each predator, as described by Shepard *et al.* (1987). These predation rates are subsequently used to compute the number of pests eliminated. The overall value of biological pest control is established by multiplying the number of pests eliminated by the cost associated with managing each pest (Table 9.3).

Formation of soil

The ecosystem's contribution to soil formation is assessed by examining the populations of earthworms and dung beetles in rice fields. The numbers of earthworms and dung beetles is determined by using the Tullgren funnel method (Southwood, 1978). On average, an earthworm weighs about 0.2 grams (Fraser *et al.*, 1996). It is established that 1 ton of earthworms contributes to the formation of 1 ton of soil per ha per year (Pimentel *et al.*, 1995; Sandhu *et al.*, 2008). It has been reported that 2 grams of soil are formed for each gram of dung beetle on a dry weight basis, as mentioned by Herrick and Lal (1996). To determine the economic value associated with the creation of soil, a calculation involves multiplying the value of topsoil by the annual soil production. It's worth noting that the economic value of one foot (approximately 30 cm) of topsoil in India has been established at US$ 2093 per ha, as reported by Kathuria (2015). This figure can be employed in the assessment of ecosystem services.

Mineralization of plant nutrients

Economically assessing the mineralization of plant nutrients involves utilizing data gathered from soil samples regarding the soil's nitrogen mineralization. The process to determine nitrogen mineralization (referred to as N_{min}) involves a one-month incubation of the soil. Afterward, the mineral nitrogen (NH_4-N) is extracted from the incubated soil using a 2 (M) KCl solution (following the method by Keeney and Nelson, 1982) and quantified through colourimetry using the indophenol blue method, based on the Berthelot reaction.

The net mineralized nitrogen is then calculated as per the method outlined by Shahid *et al.* (2017), which involves determining cumulative N_{min}. This cumulative value is subsequently converted into an annual nitrogen mineralization rate. The total quantity of nitrogen mineralized is assigned a value based on the equivalent cost of nitrogen fertilizer, which stands at US$ 0.0824 per kg of nitrogen.

Carbon flows and greenhouse gas emissions

To estimate carbon flow, one must consider the total carbon input into the soil, which includes remnants like stubble, roots, and rhizodeposition, alongside emissions of carbon dioxide (CO_2), methane (CH_4), and nitrous oxide (N_2O) measured in carbon equivalents. In rice fields, the proportions of stubble, roots, and rhizodeposition are typically set at 2.5%, 19%, and 15%, respectively, based on total biomass production (Mandal *et al.*, 2008). Residues added to the soil are assumed to contain 40% carbon content (Johnson *et al.*, 2006), with the remaining carbon estimated to be 28.8% of the total carbon added to the soil (Mandal *et al.*, 2008).

Greenhouse gas emissions from rice fields primarily comprise CO_2, CH_4, and N_2O, and they can be quantified using a partial life cycle assessment (LCA) method (Hillier *et al.*, 2009; Dubey and Lal, 2009; Yan *et al.*, 2015). This approach accounts for both direct and indirect greenhouse gas emissions within the farm gate, spanning from sowing to harvest. Direct emissions are calculated from nitrogen (N) fertilizer application (resulting in N_2O emissions), CH4 emissions (Yan *et al.*, 2003; IPCC 2006), and manure application (Zou *et al.*, 2007). Indirect emissions are determined based on reported values from agrochemical manufacturing processes (Soni *et al.*, 2013; Yan *et al.*, 2015).

The economic value of carbon flow in rice fields, calculated as carbon sequestration minus carbon emissions, is contingent on the value of one carbon emission reduction (CER) or carbon credit. This value is approximately US$ 21.7 per ton of carbon (Bhola and Malhotra, 2014).

Nitrogen fixation

One can apply the techniques outlined by Roger and Ladha (1992) to assess the nitrogen fixation carried out by beneficial microorganisms in rice. Additionally, when legume crops are cultivated preceding a rice crop (as indicated in Table 3), the nitrogen fixed by these legumes (as documented by Franco and Balieiro, 2000) can also be factored into the calculation. The economic value of nitrogen fixed within such land is determined by quantifying the nitrogen fixation per ha and then assigning it a value based on the unit price of urea, which stands at US$ 0.0824 per kilogram in rice-based ecosystems.

Soil fertility

The evaluation of the soil fertility service offered by rice fields involves an examination of the soil's role in providing accessible nitrogen, phosphorus, potassium, sulphur, and various micronutrients in representative soils. This assessment also takes into account the nutrient absorption from the soil per unit of economic yield, as described by Roy *et al.* (2006). The contribution of nutrients from fertilizer is calculated using standard nutrient use efficiency for rice, which is N (34%), P_2O_5 (25%), and K_2O (60%) of the region (Ravichandran and Sriramachandrasekharan, 2011; Lakshmi *et al.*, 2012). Plant uptake of nutrients from the soil is computed by subtracting the nutrient contribution from fertilizer from the total uptake. The economic value of these soil nutrients is based on the unit price of respective local fertilizers.

Hydrological flow

Rice cultivation is typically conducted in fields with bunds, a practice that reduces runoff and enhances the replenishment of groundwater. To quantify this ecosystem service in bunded rice fields, it is estimated that 45% of the total precipitation and irrigation contribute to groundwater recharge, as described by Tripathi (2016). This recharged groundwater can subsequently be extracted and utilized for irrigation, thereby obviating the need for purchasing irrigation water. This service holds significant value in rice farming. Consequently, the economic worth of the recharged groundwater is determined by referencing the purchase price of irrigation water, which amounts to US$ 1.5 per 1000 m^3 of water, as per the FAO (2004).

Soil erosion

Soil erosion has a detrimental impact on the ecosystem and needs to be accounted for by subtracting it from the overall ecosystem service values. The assessment of soil erosion relies on data provided by the Department of Water Resources and the state's soil erosion map. However, it's worth noting that in this region, rice is cultivated within a bunded system, with bund heights ranging from 0.50 to 0.75 meters. Given the relatively small size of the fields (averaging less than 0.1 ha), it is assumed that the sediment delivery ratio (SDR) is 0.3, as indicated by Sreenivasalu (2012). This SDR figure represents the portion of soil lost from the rice fields. It is applied to the reference data on soil erosion for the region to compute the actual soil loss, which is then multiplied by the value of topsoil, which is US$ 2093 per ha of soil, as determined by Kathuria (2015). This calculation yields the economic value associated with soil loss.

Farm income and land holding identification/classification

Household farm income is determined by considering the total value of the primary and related crop commodities produced on the farm, all of which are assessed based on local market prices within a given year. When sampled farmers are surveyed, they may be asked to provide detailed information regarding the farm's outputs and the prices at which these commodities were sold in the local market during the preceding 12-month period. This data collection method aligns with the established approach used in the National Sample Survey Office (NSSO) survey, as outlined in NSSO 2013. Household farm income is then calculated by multiplying the quantity of total outputs produced by the average price of the corresponding commodities over the preceding 12 months.

The total value of ecosystem services (ES) is determined by summing the values of both the marketed (tangible) and non-marketed (intangible) ES (Nayak *et al.*, 2019; Sandhu *et al.*, 2008). The marketed value of ES includes the economic worth of products (such as grains) and by-products (like rice straw) produced by farmers and directly traded in the market. The remaining services fall under the category of non-marketed ES values.

9.5. Biodiversity and ecosystem services

Biodiversity is a significant determinant of ecosystem function and provision of ecosystem services. Despite the complex and varied relationships between biodiversity and ecosystem services in different types of ecosystems, broadly, an increased native biodiversity results in higher levels of ecosystem services (Balvanera *et al.*, 2006; Cardinale *et al.*, 2012). Biodiversity is central to the production of ecosystem services. Biodiversity plays a crucial role in delivering essential ecosystem services like nutrient cycling, carbon storage, pest control, and pollination, all of which support the sustainability of agricultural productivity. By fostering the well-being of ecosystems, we ensure the robustness of agriculture, especially as it intensifies to meet the increasing demands for food production.

Table 9.3. Relationships between the six agroecological intensification systems in terms of ecosystem services provided by them

Ecosystem services	Conservation agriculture	Integrated Farming System	Organic agriculture	Holistic heritage agriculture	System of rice intensification (SRI)	Integrated pest management (IPM)
Diet diversity	●	●	●	●		
Carbon sequestration	●	●	●	●		
Cultural services		●		●		
Energy provision		●				
Genetic diversity			●	●		
Mitigation of GHGs	●	●	●	●		●
Pest control		●	●	●	●	●
Soil structure, fertility, erosion control	●	●	●	●	●	●
Resilience to climate disturbance				●	●	
Water quality	●	●		●	●	
Water quantity	●			●	●	●
Weed control		●	●	●	●	
Wild biodiversity & habitat provisioning			●	●		

Source: FAO (2014)

Biodiversity plays a critical role in regulating various functions within agroecosystems to meet the diverse needs of both farmers and society. Farmers effectively manage these vital ecosystem services provided by biodiversity through the adoption of sound farming practices that align with ecosystem-based approaches, ensuring sustainable agricultural production. The conservation and improvement of biodiversity within crop systems are integral components of sustainable farming practices. These measures also result in enhanced biodiversity in water bodies and the broader agricultural landscape, which may be adjacent to but not directly part of the cultivated land. Ecosystem-based production practices encompass a range of approaches, including conservation agriculture (CA), integrated rice management systems like the system of rice intensification (SRI), integrated pest management (IPM) for both plant and animal diseases, integrated production and pest management (IPPM), integrated plant nutrition systems (IPNS), and integrated weed management (IWM). The ecosystem services provided by these six systems are detailed in Table 9.3.

Preserving biodiversity is essential to secure the ongoing provision of benefits and services in agroecosystems related to food and agriculture, often referred to as agricultural biodiversity. This includes vital functions such as the evolution of crops and improvements through breeding efforts. Biodiversity also plays a crucial role in enabling adaptation to changing environmental conditions. It offers biological support to agricultural production through the various organisms present within the agroecosystem, including soil fauna, microorganisms, plant roots, and trees. These elements collectively contribute to nutrient cycling and help control pests and diseases through natural predators and disease control agents. Moreover, they provide genetic resistance within crop plants themselves. Additionally, insect pollinators aid in the cross-fertilization of outcrossing crop plants. Beyond these agricultural-specific roles, biodiversity serves broader ecological functions. This includes fostering interactions between species and between species and their environment, maintaining soil fertility, ensuring water quality, and contributing to climate regulation. (http://www.fao.org/agriculture/crops/core-themes/theme/biodiversity0/en/#bio2).

FAO's "Save and Grow" highlights the value of ecological approaches to food production – using land, water, biodiversity, and nutrients efficiently and in regenerative ways, minimizing negative impacts. It calls for agroecological approaches that advocate a shift of agricultural development focusing on singular focal areas – e.g., improved seed, pest control, and water management – to solutions that integrate all components of the farming system (Table 9.4). Core principles include maintaining soil health, recycling biomass and nutrients, increasing biological diversity and beneficial interactions among species, and optimizing the use of water, energy, nutrients, and genetic resources. Agroecological systems should be farmer-based, aiding farmers to consider interactions between the different parts of the farming system (FAO, 2014).

Table 9.4. Ecosystem service descriptions and related on-farm and off-farm benefits

Ecosystem service	Description	On-farm benefits	Off-farm benefits
Provision of food, fuel, fibre, and biochemicals	Harvestable goods from agroecosystems	Food and other goods for on-farm consumption or sale	Goods for agricultural markets
Soil structure and fertility enhancement	Soil structure and processes of nutrient cycling and delivery of nutrients to plants; processing organic matter and transforming detritus and wastes	Support for crop growth; can limit the need for chemical fertilizers	This may limit the need to mine or manufacture chemical fertilizer
Erosion protection	Soil retention; limiting soil loss through wind and water erosion	Maintain soil, and the nutrients it contains, to support production	Potential reduction of sediment transfer to downstream systems & users
Hydrologic services: Water flow regulation	Buffering and moderation of the hydrologic cycle, including water infiltration into soils and aquifers, moderation of runoff, and plant transpiration	Water in soils, aquifers, and surface bodies available to support plant growth	Stabilize stream base flow and mitigate flooding to downstream areas; recharge into aquifers and bodies of water; plant transpiration may support precipitation patterns
Hydrologic services: Water purification	Filtration and absorption of particles and contaminants by soil and living organisms in the water and soil	Clean water available for human consumption, irrigation, and another on-farm uses	Clean water available to downstream users
Pollination	Transfer of pollen grains to fertilize flowers	Necessary for seed set and fruit production in flowering plants and crops	Necessary for outcrossing in non-cultivated flowering plants

Pest control	Control of animal and insect pests by their natural enemies – predators, parasites, and pathogens	Minimize crop damage and limit competition with crops	May limit the need for pesticides that threaten environmental and human health
Weed control	Botanical component of pest control; suppressing weeds, fungi, and other potential competitors through physical and chemical properties of cover crops, intercrops, and other planted elements	Minimize weed competition with crops	May limit the need for herbicides that threaten environmental and human health
Carbon sequestration	Atmospheric carbon dioxide is taken up by trees, grasses, and other plants through photosynthesis and stored as carbon in biomass and soils	Few demonstrable on-farm benefits. Example: On-farm carbon sequestration can be associated with on-farm benefits in closely related ecosystem services, such as increased soil organic matter and microclimate regulation (e.g., shade provided by trees).	Regulation of the carbon cycle; mitigation of greenhouse gas contributions to atmospheric change
Genetic resources	A pool of genetic diversity needed to support both natural and artificial selection	Distinct genotypes (cultivars) allow fruit set in orchards and hybrid seed production; trait diversity (from landraces and wild relatives) supports disease resistance, new hybrids, and climate adaptations	Prevention against large-scale crop failure
Cultural and esthetical services	Maintaining landscapes that support: aesthetics and inspiration; spiritual and religious values; sense of place; cultural heritage; recreation and ecotourism	Aesthetics and inspiration; spiritual and religious values; sense of place; cultural heritage; recreation and ecotourism	Aesthetics and inspiration; spiritual and religious values; sense of place; cultural heritage; recreation and ecotourism

Source: Garbach *et al.* (2014)

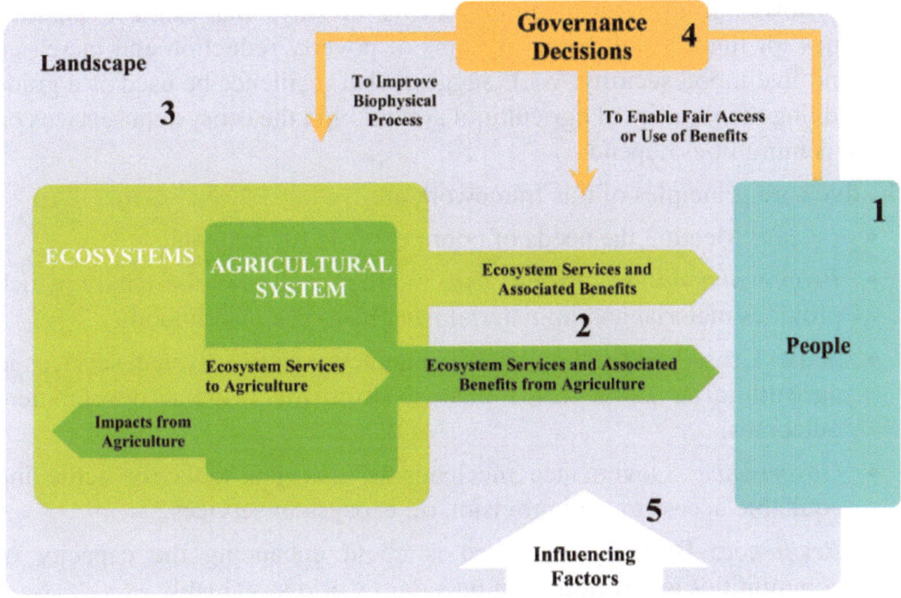

Fig. 9.3. Ecosystem services and resilience (ESR) framework
(*Source*: CGIAR Research Program on Water, Land and Ecosystems (WLE), 2014)

The CGIAR Research Program (CRP) on Water, Land and Ecosystems (WLE) combines the resources of 11 CGIAR centres, FAO, and numerous national, regional, and international partners to provide an integrated approach to natural resource management research (Fig. 9.3). WLE supports an approach to sustainable intensification in which healthy functioning ecosystems are seen as a prerequisite to agricultural development, food security and human well-being. Ecosystem Services and Resilience (ESR) is a crosscutting core theme within WLE that focuses on the role of ecosystem service-based approaches in building community resilience and helping WLE achieve its development outcomes. This program is led by the International Water Management Institute (IWMI), a member of the CGIAR Consortium and is supported by CGIAR, a global research partnership for a food-secure future. It focuses on increasing food and livelihood security through the creation of resilient socio-ecological systems that secure the sustainable provision and equitable distribution of ecosystem services (CGIAR Research Program on Water, Land and Ecosystems (WLE, 2014).

The ESR Framework, under CGIAR Research Program on Water, Land and Ecosystems (WLE), is centred on the notion that people can manage ecosystem service flows through agricultural systems in ways that achieve positive outcomes for human well-being in terms of poverty reduction and increased food and livelihood security. WLE suggests that resilience be used as a guide for studying the stability of agricultural systems and the ecosystem services on which communities depend.

The five core principles of this framework are:

- *People:* Meeting the needs of poor people is fundamental.
- *People and nature:* People use, modify, and care for nature which provides material and immaterial benefits to their livelihoods.
- *Scale:* Cross-scale and cross-level interactions of ecosystem services in agricultural landscapes can be managed to positively impact development outcomes.
- *Governance:* Governance mechanisms are vital tools for achieving equitable access to, and provision of, ecosystem services.
- *Resilience:* Building resilience is about enhancing the capacity of communities to develop in an uncertain world sustainably.

9.6. Strengthening resilience of rural communities: ecosystem-based adaptation

Ecosystem-based adaptation is a strategy for community development and environmental management that seeks to use an ecosystem services framework to help communities adapt to the effects of climate change. The Convention on Biological Diversity defines it as "the use of biodiversity and ecosystem services to help people adapt to the adverse effects of climate change", which includes the use of "sustainable management, conservation and restoration of ecosystems, as part of an overall adaptation strategy that takes into account the multiple social, economic and cultural co-benefits for local communities". Ecosystem-based adaptation sought to use the restoration of ecosystems as a stepping-stone to improve the quality of life in communities experiencing the impacts of climate change. Specifically, it involved the restoration of such ecosystems that provide food and water and protection from climate change events. The interventions combine elements of both climate change mitigation and adaptation to global warming to help address the community's current and future needs. Collaborative planning between scientists, policymakers, and community members is an essential element of ecosystem-based adaptation. By drawing on the expertise of outside experts and residents alike, it seeks to develop unique solutions to unique problems, rather than simply replicating past projects (https://en.wikipedia.org/wiki/Ecosystem_service).

9.7. Role of farm women in the management of rice-based ecosystems

The majority of the world's agricultural producers are women. They produce more than 50% of the food that is grown worldwide. Women are usually responsible for food processing and make a major contribution to food storage. Women farmers play a distinct and well-accepted role in all activities of rice cultivation and 93% of the farm women are actively involved in storing rice grains (Sumathi and Budhar, 2003).

Women in Asia play a major role in rice cultivation. New rice technologies have had and will continue to have a substantial impact on women's income-earning opportunities both within and outside the household. Women-headed households benefit from new rice technologies that increase household labour or land productivity. Women in landless households, however, can only benefit from technologies that generate employment opportunities for hired labour. Women's participation is much greater in rice-based cropping systems than in dryland farming. A comparison of women's participation in northern and southern India also indicates that rice systems have traditionally used more female labour, and within rice farming, the more intensive irrigated systems use more female labour. Within Asian rice farming, women contribute most of the labour in transplanting, harvesting, and weeding. One exception to this pattern is Bangladesh, where women do not participate in fieldwork but do almost all of the post-harvest processing. Their contribution is highest at one-half in Nepal and India and is roughly a third in Southeast Asian countries (Unnevehr and Stanford, 1983). Poor rural women play important roles in rice-based farming systems as unpaid family workers, hired laborers, income earners, savers of expenditures and major caretakers of family health and nutrition (Paris, 2007).

On the occasion of the United Nations General Assembly (UNGA) Declaration of the International Year of Rice (IYR) in 2004, FAO organised a Regional Consultation on Gender Dimensions in Asian Rice Livelihood Systems in the Changing Milieu of Technologies and Economy (9-12 March 2004), where country as well as region-specific recommendations were made. Such recommendations to strengthen roles of farm women in management of rice-based ecosystems in India were:

- Give women cultivators better access to resources by providing executable property rights to land.
- Train women in rice farming systems with skills in water conservation and water management to enhance productivity.
- Enforce equal wages for women in the rice growing areas; this is of utmost importance to bringing about gender equity.

- Diversify women labourers' livelihoods by enhancing their employment opportunities in rice-based biomass industries during the off-season.

Female participation increases with poverty and in unfavourable rice environments. Based on the micro-level studies of the Women in Rice Farming System (WIRFS) Network based at the International Rice Research Institute (IRRI), women's contributions in rice production ranged from 27-84%. Female labour participation is higher than men in Thailand, Nepal, and India. Women's labour contributions were the highest in rainfed rice ecosystems in eastern India, which used to suffer from abiotic stresses (drought, submergence/ floods and salinity) and where the majority of the women were illiterate. Women from the upper castes do not work in the fields while women from the lower castes work in their fields as well as exchange and hire laborers in other fields. Except for land preparation, irrigating the fields and spraying chemicals, women dominate rice operations. Women are also repositories of indigenous knowledge of crop and natural resource management. Post-harvest activities such as seed cleaning, selection, storing seeds for the next cropping season, dehusking the grains, cooking rice or preparing rice into products for home consumption or for sale are women's domain. They add value to rice by preparing them into products for sale and consumption, and use rice by-products and biomass for livestock and other livelihood uses to save on expenditures. The emerging changes in rainfed rice ecosystems in Asia have been affecting rice production and the roles of women in rice-based agriculture viz., abiotic stresses, which are further aggravated by natural disasters, climate change and increasing out-migration of labour from agriculture (Paris, 2007).

9.8. Way forward

To support the Sustainable Development Goals (SDGs), sustainable natural resource management and land use policies increasingly look at people's perceptions of ecosystem services at the local level. Diverse land use in agricultural ecosystems not only fosters biodiversity but also provides potential ecosystem services. People are mostly aware of provisioning services; however, generally unaware of regulating and supporting services. Although modern agricultural practices have increased food production worldwide, it has also raised concerns about unsustainable resource use, environmental damage and land degradation. The first step towards protecting vital natural resources is to create awareness of specific ecosystem services in rural communities. People's awareness of a greater number of ecosystem services would inspire them to consider a wider range of benefits that nature provides, enabling them to habitat management that better balances trade-offs between different services. Ecosystem services are the benefits that nature provides to people and people differ in their interests on specific services; therefore, people's management decisions depend on how they prioritize the relevant ecosystem services. Thus, details of ecosystem services rather than describing the number of ecosystem services would help in people's effective decision-making.

9.9. Conclusion

Ecosystem services represent the intricate and invaluable web of benefits that nature bestows upon humanity. These services, ranging from clean air and water to pollination and climate regulation, underpin our well-being and economic prosperity. Recognizing and valuing these services is essential for sustainable development and conservation efforts. As we face the challenges of a rapidly changing world, understanding and safeguarding ecosystem services has never been more critical. By acknowledging their significance, we can forge a more harmonious relationship with the natural world, fostering a future where both human and ecological needs are met in balance, ensuring a healthier planet for generations to come.

The rice ecosystem provides an impressive array of vital services that extend far beyond the cultivation of this staple crop. From food security and livelihoods for millions of people to water purification, carbon sequestration, and habitat provision, the multifaceted benefits of rice ecosystems underscore their importance in sustaining both human well-being and environmental health. As we navigate the complexities of a changing climate and growing global population, it is imperative that we recognize and appreciate the diverse services rice ecosystems offer. By employing sustainable practices and innovative approaches, we can harness the full potential of these systems while preserving their delicate balance, ensuring a prosperous and harmonious coexistence between humans and nature.

Rice ecosystems offer a wide array of services, categorized into regulating, provisioning, and supporting services. These services have both direct and indirect values, encompassing aspects like pest control, carbon flow, soil fertility, and greenhouse gas reduction. Nonetheless, rice cultivation also presents environmental issues, such as greenhouse gas emissions, pollution, and high-water consumption, necessitating a careful balance between productivity and sustainability. Low farm incomes have driven discussions about compensating farmers based on the ecosystem services they provide through rice farming, necessitating the valuation of these services to bridge the gap between their total value and the economic return to farmers.

Understanding the complex interplay between rice ecosystems, their multifaceted services, and the need for sustainable management is critical to strike a balance between agricultural productivity, environmental preservation, and cultural significance. This holistic perspective emphasizes the need for comprehensive policies and practices that account for the true value of rice ecosystems in our global landscape.

References

Balvanera, P., Pfisterer, A.B. and Buchmann, N. 2006. Quantifying the evidence for biodiversity effects on ecosystem functioning and services. *Ecology Letters*, 9: 1146-1156.

Bhola, K. and Malhotra, R. 2014. Carbon credit capital: Indian commodity market 2020. *Middle-East Journal of Scientific Research*, 22:1622-1629. https://doi.org/10.5829/idosi.mejsr.2014.22.11.21626

Cardinale, B.J., Duffy, J.E. and Gonzalez, A. 2012. Biodiversity loss and its impact on humanity. *Nature*, 486: 59-67.

CGIAR Research Program on Water, Land and Ecosystems (WLE) 2014. *Ecosystem services and resilience framework*. Colombo, Sri Lanka: International Water Management Institute (IWMI). CGIAR Research Program on Water, Land and Ecosystems (WLE). 46p. doi: 10.5337/2014.229

Costanza, R., d'Arge, R., de Groot, R., Farber, S., Grasso, M., Hannon, B., Limburg, K., Naeem, S., O'Neill, R.V., Paruelo, J., Raskin, R.G., Sutton, P. and van den Belt, M. 1997, The value of the world's ecosystem services and natural capital. *Nature*, 387: 253-260.

de Groot, R., Brander, L., van der Ploeg, S., Costanza, R., Bernard, F. and Braat, L. 2012. Global estimates of the value of ecosystems and their services in monetary units. *Ecosystem Service*, 1: 50–61.

Devi, P.I., Kumar, L., Kumar, D.S., Manjula, M., Mukhopadhyay, P., Raghu, P., Sharma, D., Sridhar, R. and Venkatachalam, L. 2017. Payment for ecosystem services guaranteed farm income and sustainable agriculture. *Economic Political Weekly,* 52 (17): 12-14.

Dubey, A. and Lal, R. 2009. Carbon footprint and sustainability of agricultural production systems in Punjab, India, and Ohio, USA. *Journal of Crop Improvement,* 23: 332-350.

FAO. 2004. *Water charging in irrigated agriculture*, FAO water reports 28, accessed at http://www.fao.org/docrep/008/y5690e/y5690e0b.htmFAO (2005) Fertilizer use by crop in India. First version, published by FAO, Rome. http://www.fao.org/docrep/009/a0257e/A0257E05.htm

FAO. 2014. *The Multiple Goods and Services of Asian Rice Production Systems*. Plant Production and Protection Division, Food and Agriculture Organization of the United Nations (FAO), Rome.

Franco, A.A. and Balieiro, F.D.C. 2000. The role of biological nitrogen fixation in land reclamation, agroecology and sustainability of tropical agriculture. *Transition to global sustainability: the contribution of Brazilian science*, 209-233.

Fraser, P.M., Williams, P.H. and Haynes, R.J. 1996. Earthworm species, population size and biomass under different cropping systems across the Canterbury Plains, New Zealand. *Applied Soil Ecology*, 3:49-57.

Garbach, K., Milder, J.C., Montenegro, M., Karp, D.S. and De Clerck, F.A.J. 2014. Biodiversity and Ecosystem Services in Agroecosystems. *Encyclopaedia of Agriculture and Food Systems*, Volume 2: 21-40. doi:10.1016/B978-0-444-52512-3.00013-9

Garrity, D.P., Singh, V.P. and Hossain, M. 1996. *Rice ecosystems analysis for research prioritisation.* In: Evenson, R.E. (Ed.) Rice research in Asia. Progress and priorities, p. 35-58. 418 pp.

Gurr, G.M., Wratten, S.D. and Altieri, M.A. (Eds.) 2004. *Ecological engineering for pest management: advances in habitat manipulation for arthropods*. CSIRO, Victoria.

Herrick, J.E. and Lal, R. 1996. Dung decomposition and pedoturbation in a seasonally dry tropical pasture. *Biology and Fertility of Soils*, 23:177-181

Hillier, J., Hawe, C., Squire, G., Hilton, A., Wale, S. and Smith, P. 2009. The carbon footprints of food crop production. *International Journal of Agricultural Sustainability*, 7:107-118.

IPCC, 2006. National Greenhouse gas inventories: Land use, land use change and forestry. Institute of Global Environmental Strategies, Hayama, Japan

Johnson, J.M.F., Allmaras, R.R. and Reicosky, D.C. 2006. Estimating source carbon from crop residues, roots and rhizodeposits using the national grain-yield database. *Agronomy Journal*, 98: 622–636. https://doi.org/10.2134/agronj2005.0179

Kathuria, V. 2015. Environmental cost of using top-soil for brick making-a case study from India problem/issue? *Review Market Integration*, 5:171–201. https://doi.org/10.1177/0974929214521892

Keeney, D.R. and Nelson, D.W. 1982. *Nitrogen-inorganic forms*. In: Page, A.L., Miller, R. H., Keeney, D.R. (Eds.), Methods of soil analysis. Part 2. Chemical and microbiological properties-agronomy monograph, No. 9. ASA and SSSA, Madison, Wisconsin U.S.A. p. 643–698.

Kremen, C. 2005. Managing ecosystem services: what do we need to know about their ecology? *Ecology Letters*, 8: 468-479.

Krishna, V.V., Byju, N.G. and Tamizheniyan, S. 2003. *Integrated pest management in Indian agriculture: a developing economic perspective*. In: Radcliff, E.B. and Hutchson, W.D. (eds) IPM World Text book, St Paul, Minnesota.

Lakshmi, C.S.R., Rao, P.C., Sreelatha, T., Madahvi, M., Padmaja, G., Rao, P.V. and Sireesha, A. 2012. Nitrogen use efficiency and production efficiency of rice under rice-pulse cropping system with integrated nutrient management. *Journal of Rice Research*, 5:42-51.

Loc, H.H., Diep, N.T.H., Can, N.T., Irvine, K.N. and Shimizu, Y. 2016. Integrated evaluation of ecosystem services in prawn-rice rotational crops, Vietnam. *Ecosystem Services*, 26: 377-387.

Ma, S., Swinton, S.M. and Lupi, F. 2012. Farmer's willingness to participate in payment for-environmental-services programmes. *Journal of Agricultural Economics*, 63: 604-626.

Mandal, B., Majumder, B., Adhya, T.K., Bandyopadhyay, P.K., Gangopadhyay, A., Sarkar, D., Kundu, M.C., Choudhury, S.G., Hazra, G.C., Kundu. S., Samantaray, S.N. and Mishra, A.K. 2008. Potential of double-cropped rice ecology to conserve organic carbon under subtropical climate. *Global Change Biology*, 14: 2139-2151. https://doi. org/10.1111/j.1365-2486.2008.01627.x

Mala, M., Jahan. M. and Islam, K.S. 2009. Comparative prey consumption of a predatory spider, *Pardosa pseudoannulata* (bosenberg and strand) on three different diets. *Progressive Agriculture*, 20: 57–62.

Manzanilla, D.O., Paris, T.R., Vergara, G.V., Ismail, A.M., Pandey, S., Labios, R.V., Tatlonghari, G.T., Acd, R.D., Chi, T.T.N., Duoangsila, K., Siliphouthone, I., Manikmas, M.O.A. and Mackill, D.J. 2011. Submergence risks and farmers' preferences: Implications for breeding Sub1 rice in Southeast Asia. *Agricultural Systems*, 104: 335-347.

MEA (Millennium Ecosystem Assessment). 2003. *Ecosystems and Human wellbeing: a Framework for Assessment*. World Resources Institute, Washington, D.C.

MEA (Millennium Ecosystem Assessment). 2005. *Ecosystems and human wellbeing: biodiversity synthesis*. World Resources Institute, Washington, D.C.

Nayak, A.K., Shahid, Md., Nayak, A.D., Dhal, B., Moharana, K.C., Mondal, B., Tripathi, R., Mohapatra, S.D., Bhattacharyya, P., Jambhulkar, N.N., Shukla, A.K., Fitton, N., Smith,

P. and Pathak, H. 2019 Assessment of ecosystem services of rice farms in eastern India. *Ecological Processes*, 8, 35. https://doi.org/10.1186/s13717-019-0189-1

NSSO. 2013. *Key indicators of land and livestock holdings in India*. 70th National Sample Survey Office, Ministry of Statistics and Programme Implementation. Government of India.

Paris, T.R. 2007. *Women's roles and needs in changing rural Asia with emphasis on rice-based agriculture*. Food and Fertilizer Technology Center (FFTC). Extension Bulletin 593. November 2007. pp. 1-15.

Pimentel, D., Harvey. C., Resosudarm. P., Sinclair, K., Kurz, D., McNair, M., Crist, S., Shpritz, L., Fitton, L., Saffouri, R. and Blair, R. 1995. Environmental and economic costs of soil erosion and conservation benefits. *Science*, 267:1117-1125.

Raju, D.S.R., Reddy, V.R., Reddy, P.P. and Reddy, M.S. 2005. Water use efficiency: a study of system of rice intensification (SRI) adoption in Andhra Pradesh. *Indian Journal of Agricultural Economics*, 60: 458-472.

Ravichandran, M. and Sriramachandrasekharan, M.V. 2011. Optimizing timing of potassium application in productivity enhancement of crops of Karnataka. *Journal of Agricultural Sciences,* 24:75-80.

Reid, W.V., Mooney, H.A., Cropper, A., Capistrano, D., Carpenter, S.R., Chopra, K., Dasgupta, P., Dietz, T., Duraiappah, A.K., Hassan, R., Kasperson, R., Leemans, R., May, R.M., McMichael, A.J., Pingali, P., Samper, C., Scholes, R., Watson, R.T., Zakri, A.H., Shidong, Z., Ash, N.J., Bennett, E., Kumar, P., Lee, M.J., Raudsepp-Hearne, C., Simons, H., Thonell, J. and Zurek, N.B. 2005. *Millennium Ecosystem Assessment Synthesis Report*. Island Press, Washington.

Roger, P.A. and Ladha, J.K. 1992. Biological N2 fixation in wetland rice fields: estimation and contribution to nitrogen balance. *Plant and Soil*, 141: 41-55.

Roy, R.N., Finck, A., Blair, G.J. and Tandon, H.S. 2006. Plant nutrition for food security: a guide to integrated nutrient management. FAO Fertilizer & Plant Nutrition Bulletin, No. 16, Food and Agriculture Organization of the United Nations, Rome, Italy.

Sandhu H.S., Wratten S.D., Cullen R. 2005. Evaluating ecosystem services on farm land: a novel, experimental, bottom-up approach, shaping sustainable systems. Proceedings of the 15th IFOAM Organic World Congress. 20-23 September 2005, Adelaide Australia, (September 2005). http://orgprints.org/view/projects/int-isofar.html.

Sandhu, H.S., Wratten, D., Cullen, R., Case, B. and Case, B. 2008. The future of farming: the value of ecosystem services in conventional and organic arable land, An experimental approach. *Ecological Economics*, 64: 835-848.

Shahid, M., Nayak, A.K., Puree, C., Tripathi, R., Lal, B., Gautam, P., Bhattacharyya, P., Mohanty, S., Kumar, A., Panda, B.B., Kumar, U. and Sukla, A.K. 2017. Carbon and nitrogen fractions and stocks under 41 years of chemical and organic fertilization in a sub-humid tropical rice soil. *Soil and Tillage Research,* 170:136-146.

Shepard, B.M., Barrion, A.T. and Litsinger, J.A. 1987. *Helpful insects, spiders, and pathogens*. International Rice Research Institute, Manila, Philippines, 127p.

Soni, P., Taewichit, C. and Salokhe, V.M. 2013. Energy consumption and CO2 emissions in rainfed agricultural production systems of Northeast Thailand. *Agricultural Systems*, 116:25-36.

Southwood, T.R.E. 1978. *Ecological Methods*. Chapman and Hall, London, England

Spangenberg, J.H. and Settele, J. 2010. Precisely incorrect? Monetising the value of ecosystem services. *Ecological Complexity*, 7: 327-337.

Sreenivasalu, V.U.B. 2012., Sediment yield estimation and prioritization of watershed using RS and GIS. Inter. Arch. Photogrammetry, *Remote Sensing Spatial Infor. Sci.* XXXIX-B8, 529–533. https://doi.org/10.5194/isprsarchives- XXXIX-B8-529-2012

Stallman, H.R. 2011. Ecosystem services in agriculture: determining suitability for provision by collective management. *Ecological Economics*, 71: 131-139.

Sumathi, P. and Budhar, M.N. 2003. Extent of women's involvement in Indian rice cultivation. *International Rice Research Notes*, 28(1):74-76.

The Economics of Ecosystems and Biodiversity (TEEB) 2016. *Valuation of rice agro-ecosystems: Phase One, Pilot Study for TEEB.* Final Report. The Economics of Ecosystems and Biodiversity (TEEB) secretariat of UNEP.

Tripathi, K.P. 2016. *Field bunding: foremost rain water management measure.* In: Singh, N., Kumar, A. *et al.* (Eds.), Water Conservation and Management in India, The institute of Engineers (India), Annual Technical, Vol II, pp.34-40.

Unnevehr, L.J. and Stanford, M.L. 1983. *Technology and the demand for women's labor in Asian rice farming.* Paper presented at the conference on 'Women in Rice Farming Systems', IRRI, September 26-30, 1983.

Wossink, A. and Swinton, S.M. 2007. Jointness in production and farmers' willingness to supply non-marketed ecosystem services. *Ecological Economics*, 64: 297-304.

Yan, X.Y., Cai, Z.C., Ohara, T. and Akimoto, H. 2003. Methane emission from rice fields in mainland China: amount and seasonal and spatial distribution. *Journal of Geophysical Research,* 108(D16): 4505

Yan, M., Luo, T., Bian, R., Cheng, K., Pan, G., Rees, R. 2015. A comparative Study on carbon footprint of rice production between household and aggregated farms from Jiangxi, China Environ Monit Assess, 187:332. https://doi.org/10.1007/s10661-015-4572-9

Zhang, W. 2017. *What makes people aware of ecosystem services? Lessons from a survey of Nigerian rural communities.* IFPRI Blog: Issue Post, CGIAR Research Program on Water, Land and Ecosystems. https://wle.cgiar.org/thrive/2017/02/12/what-makes-people-aware-ecosystem-services

Zhang, W., Ricketts, T.H., Kremen, C., Carney, K. and Swinton, S.M. 2007. Ecosystem services and dis-services to agriculture. *Ecological Economics*, 64(2): 253-260.

Zou J.W., Huang, Y., Zheng, X.H., Wang, Y.S. 2007. Quantifying direct N_2O emission in paddy fields during rice growing season in mainland China: dependence on water regime. Atmos Environ, 41: 8030-8042.

[illegible]

[illegible] management. *Agricultural [illegible]*, 71, 135–149. [illegible]

[illegible] *Ecological Economics*, [illegible]

[illegible] Economics of Ecosystems and Biodiversity (TEEB). 2010. [illegible] Synthesis Report. The Economics of [illegible] UNEP.

[illegible] Singh, N. [illegible] Management [illegible] The Institute of [illegible] pp. 1–40.

[illegible] paper presented at the conference on Women in Rice Farming Systems. [illegible]

[illegible]

[illegible]

10

Water Governance & Participatory Irrigation Management for Rice-based Agro-Ecosystems

10.1. Introduction

Irrigation facilities are created by the Government at huge cost for supplying water to the farmers for raising crops to increase agricultural production. The major and medium irrigation projects are planned, constructed, managed, operated and maintained by the Irrigation Departments, wherein certain procedures and rules are evolved and followed to the delivery schedule of canals, frequency of supply and distribution of water to the farmers, and closing of canals. The creation of such facilities resulted in ushering the green revolution in the sixties and seventies in the production of food grains and also employment generation in the country, thereby improving the socio-economic condition of Indian farmers. However, optimized management of irrigation water is the key to the sustainable development of agriculture and livestock. Since management of water for irrigation is highly location-specific, depending upon agro-ecological and socio-economic conditions, type of irrigation project and land use pattern, it differs from project to project and region to region in the country.

Irrigation is critical to agricultural growth and development, bringing self-sufficiency in food production in India. It has always been emphasized due to its yield-augmenting impact and is treated as one of the priority areas of India's agricultural development strategies. Irrigation development in India has been quite remarkable with an increase in irrigation potential from 22.6 million hectares in 1950–51 to about 123 million hectares by 2007 including 42 million ha under major and medium irrigation projects, 14 and 67 million ha under minor surface and minor groundwater lift projects, respectively (MoSPI, Govt. of India, 2010), making it a world leader in irrigation. Irrigated agriculture in about 51% net sown area (net sown and net irrigated area 139.3 and 71.6 million ha, respectively) contributes to 60% of India's food grain production (DAC & FW, Govt. of India, 2023).

Phases in irrigation development and management in India reveal several paradigm shifts: 1950-1970- the era of capital-intensive expansion of irrigation, 1970-1980- the era of irrigation improvement, and from the late 1980s until present- the era of reform. The central challenge facing irrigated agriculture today and in the foreseeable future is 'per drop more crop' (MoWR, RD & GR, Govt. of India, 2018; DAC & FW, Govt. of India, 2018). The problems in the irrigation sector in India include low irrigation efficiency (30-35%), deteriorating physical structures, inadequate maintenance, low-cost recovery, under-utilization of created potential (74%), uncontrolled water delivery, tail-end water deprivation, seepage loss, siltation, waterlogging, and soil salinity. Inequitable and unpredictable water supply among the farmers over space and time led to the injudicious use of water in the irrigation commands and an increase in inequity within the same unit of command area.

In search of the solutions to aforesaid problems, farmers' participation in irrigation management has taken centre stage and the irrigators who were considered beneficiaries are now considered partners in the planning, development, operation and maintenance of irrigation systems. The National Water Policies of India (1987, 2002) have emphasized farmers' participation in irrigation management based on the concept of people's management of developmental infrastructures that require local solutions to local problems affecting them. Most of the states in India have been implementing the participatory irrigation management (PIM) reform and transferring the irrigation management to water user associations (WUAs) to provide equitable, timely and assured irrigation.

10.2. Rationale of PIM

In India, the irrigation sector faces several problems. A vicious cycle has existed in the Indian irrigation sector that needs to be transformed into a virtuous cycle to enhance farmers' income in irrigated areas through effective implementation of PIM (Fig. 10.1). Inequitable and unpredictable water supply among farmers over space and time is also leading to injudicious use of water in the irrigation commands and increasing inequity within the same unit of command area (Ghosh *et al.*, 2005).

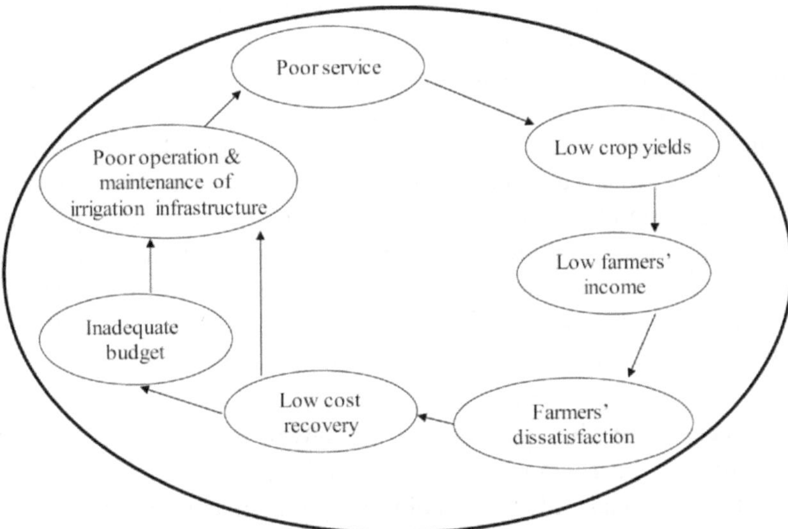

Fig. 10.1. Vicious cycle in the Indian irrigation sector

The optimum benefits from irrigation water are not obtained as it was expected. There is considerable wastage of water and over-irrigation in the upper reaches of the command area with consequent non-availability of water to the tail-end farmers which is a common phenomenon. The canal is designed for 24-hour irrigation water supply; however, the farmers are accustomed to daytime irrigation water supply. Consequently, one-third of water delivered at the outlet is wasted during night time. Injudicious use of irrigation water has resulted in huge losses in the command and consequently, many irrigation schemes are operating sub-optimally. The average yield of food grains is about 2.5 t ha^{-1} under irrigation where whereas it should be at least 4.5 t ha^{-1}.

Another problem is drainage in canal commands. In the planning of most of the existing irrigation projects in our country, provision was not made for adequate drainage in the command areas. The water table has built up steadily due to high seepage and percolation losses, consistent over-irrigation, spillage from the canal system and not using groundwater. Waterlogging and salinization have thrown considerable acreage in the command area out of production. About 5.76 million ha of land has degraded due to waterlogging, soil salinity and alkalinity.

There are deficiencies in the design, operation and maintenance of the irrigation system. This is mainly due to assuming unrealistic duties for crops, assuming low seepage losses in the distribution system, inadequate maintenance of the system etc. By and large, the indiscipline attribute of the users is the main reason for tail-end fields in localized ayacut not receiving water. Further, to irrigate their lands in the shortest possible time, the farmers do not hesitate to damage the sluices or cause breaches in canals. This calls for a holistic approach to efficient irrigation management so that more and more area can be added under irrigation command which is possible only when the performance of the irrigation system and management of irrigation is improved with proper crop planning, increased cropping intensity and share of food grains in the gross cropped area. To achieve this, farmers' active participation in irrigation water management is essential as the farmers are the ultimate users of irrigation water and their participation will ensure successful operation, maintenance of the irrigation system and collection of water rates.

The performance of government-managed irrigation systems has been sub-optimal because of the deterioration of physical infrastructure due to a rigid, top-down, bureaucratic approach, deferred maintenance, poor water service delivery, lack of accountability, poor incentives (financial), and weak institutional arrangements for infrastructure management. PIM has emerged as an important approach for improving the performance of the irrigated agricultural sector, including productivity and financial & physical sustainability.

As water delivery function influences the profitability of agricultural operations due to increases in irrigated area, cropping intensities and/ or crop diversity, yields and economic returns, farmers are interested in participation in irrigation management. Thus, a consensus has emerged regarding WUAs to be the best strategy for the long-term sustainability of irrigated agriculture. Other reasons for promoting PIM include reduction of the burden of costs, staff requirements and technical or management problems faced by governments, farmers' access and control on operation, maintenance, water delivery and fixation of water rate, improvements in the agricultural productivity and economic profitability of irrigation systems. PIM forms a strong basis for collective action in related areas, such as the adoption of modern agricultural practices and input management. The establishment of WUAs builds social capital through improved leadership and capacity building (FAO, 2007).

10.3. What is PIM?

PIM refers to the participation of water users, the farmers, in the management of the irrigation systems. It ensures the involvement of irrigation users in all aspects of irrigation management including planning, design, construction, operation, maintenance, financing, governance and monitoring and evaluation of the irrigation system, at the primary, secondary and tertiary levels. A paradigm shift in the policies for irrigation development and management has been happening during the past two decades through PIM. The centralized control and management responsibility of the irrigation resources are being transferred to the local farmer groups or WUAs for better management. Participation of beneficiaries facilitates the optimal upkeep of irrigation system and effective utilization of irrigation water. With greater participation of the farmers, the investments in the irrigation infrastructure and other irrigated agriculture-related services will be need-based and hence more effective. In many places, PIM reforms lead to irrigation management transfer (IMT). IMT is the full or partial transfer of responsibility and authority for the governance, management and financing of irrigation systems from the Government to WUAs. In India, IMT is mostly followed in the case of minor irrigation projects.

Irrigation is a state subject in India and its management is being done by the State Irrigation Departments but the involvement of individual farmers or a group of farmers directly or indirectly may not be denied. The private sector management or contractual management approach is not popular, particularly in major or medium irrigation projects but in the case of tube-well irrigation, it may not be denied. In India, farmers were managing irrigation projects but after independence, the management of irrigation was taken over by the Irrigation Departments. It is now realized that the PIM is the only answer through the involvement of farmers; however, the introduction of PIM is very slow.

Any farmers' managed system with full operations & maintenance (O&M) and regulatory control is rarely found in the country. The appropriate approach may be a Joint Management System, wherein the O&M and regulation with ownership of the system may continue to remain with Irrigation Department till farmers are fully trained to take over the system. The formal group of farmers called WUA, may carry out all other functions relating to management of irrigation in their operational area. The participation of farmers in department-managed activities and the department's participation in farmers managed activities should be ensured till the satisfactory performance of WUAs.

The participation of farmers in department managed activities and the department's participation in farmers managed activities should be ensured till the performance of WUAs becomes satisfactory.

Involvement of irrigation users in all aspects of irrigation management *viz.*,

- Planning and design of the project
- Construction, operation and maintenance
- Irrigation water distribution
- Fixation and recovery of water rates
- Agricultural development

Involvement of irrigation users at various levels of irrigation management *viz.*,

- Tertiary- Watercourse/field channel
- Secondary- Branch canal/distributary/minor
- Main system- Main canal

Involvement of irrigation users through various approaches *viz.*

- Irrigation Department Management
- Private sector management
- Users'/farmers' organization management
- Joint management

Various approaches may be adopted suiting the local condition or type of project. But it is important to understand the functions for management of irrigation, which are summarized as under:

- Operation and maintenance of all irrigation systems and structures *viz.*, head works, main, secondary, minor, tertiary and drainage systems
- Regulation/rostering of main and secondary system
- Equitable distribution of irrigation water amongst users
- Fixation and recovery of water rates
- Improvement in water use efficiency
- Maintenance of drainage system and prevention of water pollution
- Reduction in conveyance, distribution and field application losses
- Economic use of water
- Conjunctive use of surface and groundwater
- Payment of land and water rates to the Department as applicable through the approach adopted in the transfer agreement
- To develop a suitable cropping pattern, improve cropping intensity and agricultural production
- Conflict resolution.
- Administrative and financial functions.

10.3.1 Structure of WUAs

Structural development, size and organization of WUA are used to be at the convenience of farmers and the irrigation department. The jurisdiction of a WUA is between 500 to 750 ha preferably under a minor irrigation command covering one or more villages. A minor irrigation project (command area up to 2000 ha) is managed by one WUA- a single-tier system. The medium irrigation project (command area of more than 2000 to 10,000 ha) is managed by a two-tier system having WUAs at the minor level and apex body at the project level. In a major irrigation project (command area of more than 10000 ha), there is a three-tier system, the WUA at the minor level, a federation or a distributary committee at the branch canal or the distributary level and the apex body at the project level. The WUAs at the minor level are federated at the distributary level to coordinate their functions.

The structural arrangements of WUA/ *Pani Panchayat* indicate the involvement of both the irrigation and agriculture departments to facilitate proper planning of irrigation and agriculture in the irrigation command (Fig. 10.2).

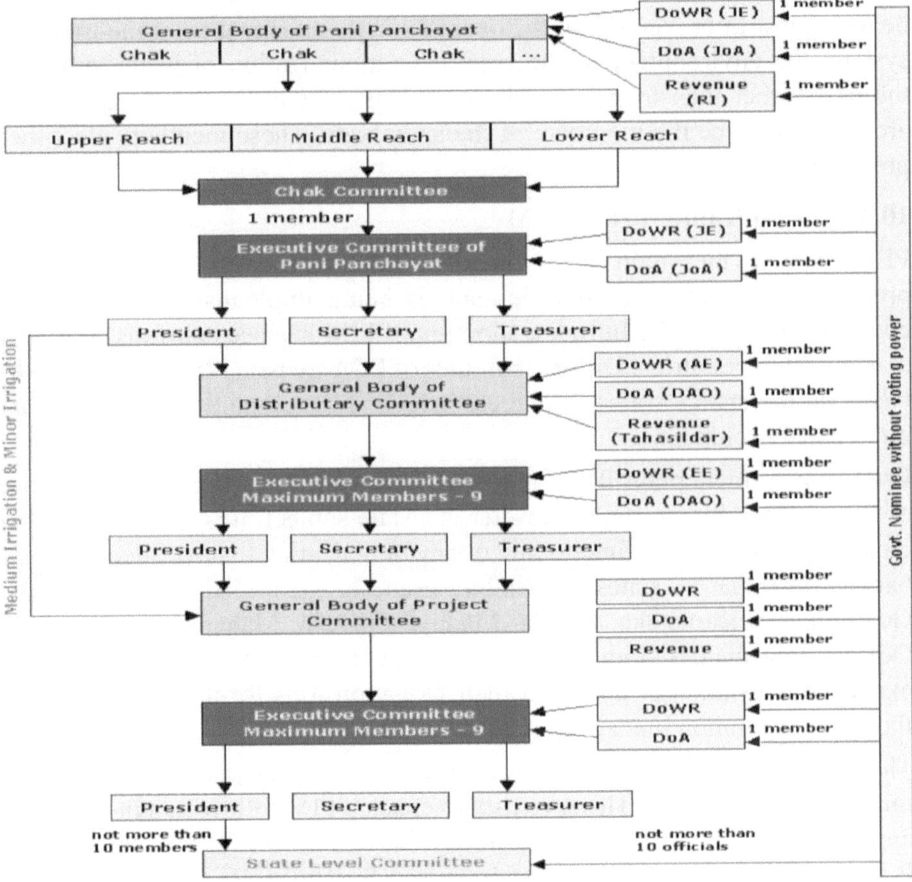

Chak (Outlet), Pani Panchayat (WUA), DoWR: Department of Water Resources DoA: Department of Agriculture, RI: Revenue Inspector, JE: Junior Engineer, AE: Assistant Engineer, EE: Executive Engineer, JoA: Junior Agriculture Officer, DAO: District Agriculture Officer

Fig. 10.2. Structural arrangement of WUA (following the Pani Panchayat in Odisha)

WUA is the registered body and all farmers in its jurisdiction command area become the members. The managing committee of WUA consists of a president, vice president, treasurer and secretary in addition to at least five members. State Irrigation, Agriculture, Command Area Development and Revenue Departments also nominate one officer from each department for inclusion in the managing committee to provide help in its functioning.

The presidents of all WUAs are the members of the distributary committee and these members may elect their president. The state departments may also nominate their officers to the distributary committee. The distributary committees of the project may be further federated into the project-level committee called the apex body. The main function of the apex body would be to suggest improvement in the operation and maintenance of the irrigation system, the environment of the area and rationalization of water supply to the users other than irrigation. The presidents of the distributary committee are required to be the members of the apex body. These members elect their president.

10.3.2. Legal framework of PIM

PIM remains the main institutional solution for irrigation management problems in developing countries and is being implemented in nearly 60 countries (FAO, 2007). In India, adoption of PIM has been slow as compared to some other countries. Different models of PIM are being tried in the country based on the state's water resources, irrigation development, and social and political environment.

Union government plays the role of a facilitator but the actual implementation of PIM is done by the states as water is a state subject. It is important for the creation and successful functioning of any organization to have adequate legal backup. In India, 16 states *viz.* Andhra Pradesh, Assam, Bihar, Chhattisgarh, Goa, Gujarat, Karnataka, Kerala, Madhya Pradesh, Maharashtra, Nagaland, Odisha, Rajasthan, Sikkim, Tamil Nadu and Uttar Pradesh have either enacted exclusive legislation or amended their Irrigation Acts for the involvement of farmers in irrigation management (Table 10.1).

The states, which are yet to do the enactment, follow various approaches to the implementation of PIM. However, states enacted PIM follow an almost similar approach. Andhra Pradesh (undivided?) has done the enactment (Andhra Pradesh Farmers' Management of Irrigation System Act) in 1997 as the first state of India. These enabling laws and/ or bylaws ensure formation of WUAs for undertaking the management of irrigation, participation of farmers in irrigation management to manage the irrigation systems within the operational area of WUA, legal rights to WUA to receive irrigation water and distribute the same among the members in the operational area, empowering WUA in developing a suitable crop pattern, fixation of water rates for different crops on season wise area basis, collection of water charges from the farmers for utilization of irrigation water, generation of resources from donor agencies other than the water charges or maintenance grant, resolving conflicts among stakeholders, etc. (Ministry of Water Resources, Govt. of India, 2007).

Table 10.1. Position of enactment for PIM in different states of India

Sl. No.	Name of State	Position of issue/amendment of Irrigation Act
1.	Andhra Pradesh	Enacted "Andhra Pradesh Farmers Management of Irrigation Systems Act, March, 1997"
2.	Assam	The Assam Irrigation Water Users Act 2004
3.	Bihar	"The Bihar Irrigation, Flood Management and Drainage Rules, 2003" under the Bihar Irrigation Act, 1997
4.	Chhattis-garh	Enacted "*Chhattisgarh Sinchai Prabandhan Me Krishkon Ki Bhagidari Adhiniyam, 2006*".
5.	Goa	Enacted "Goa Command Area Development Act 1997 (Goa Act 27 of 1997)"
6.	Gujarat	Gujarat Water Users Participation Management Act, 2007
7.	Karnataka	Promulgated an Ordinance on 7th June 2000 for amendment of the existing Karnataka Irrigation Act 1957.
8.	Kerala	Enacted "The Kerala Irrigation and Water Conservation Act 2003".
9.	Madhya Pradesh	Enacted "*Madhya Pradesh Sinchai Prabandhan Me Krishkon Ki Bhagidari Adhiniyam, 1999*" during September 1999.
10.	Maharash-tra	"The Maharashtra Management of Irrigation Systems by Farmers Act, 2005"
11.	Nagaland	Nagaland Farmer's Participation in Management of Irrigation Systems Act, 2013
12.	Odisha	Enacted "The Orissa *Pani Panchayat* Act, 2002".
13.	Rajasthan	Passed the "*Rajasthan Sinchai Pranali Ke Prabandh Me Krishkon Ki Sahabhagita Adhiniyam, 2000*".
14.	Sikkim	"Sikkim Irrigation Water Tax 2002" and "Sikkim Irrigation Water Tax (Amendment) Act 2008"
15.	Tamil Nadu	Enacted the "Tamil Nadu Farmers" Management of Irrigation Systems Act, 2000".
16.	Uttar Pradesh	Enacted the "Uttar Pradesh Irrigation Management Act, 2009"

Source: MoWR, RD & GR, Govt. of India (http://mowr.gov.in/sites/default/files/CADWM_Status_of_PIM_0.pdf)

The by-laws of WUA are defined as regulations of the society, which may be summarized as follows:

- Objective of water users' association
- Definition of the member
- Size of the water users' association
- Functions to be carried out by the WUA
- Managing committee size, function and legal powers
- Responsibilities and rights of the WUA
- Responsibilities and rights of the members
- Details of the general body
- Election procedure of the managing committee.
- The structure of the WUA and its internal management
- Conducting meetings of the general body and managing committee.
- Amendment of by-laws
- Liquidation of WUA.

In addition to the by-laws, another important legal document is the Transfer Agreement or the Memorandum of Understanding between the WUA and the Irrigation Department, wherever necessary. This agreement contains the terms and conditions of the transfer of the system and will spell out the duties and responsibilities of the Irrigation Department and WUA. Handing over the system, method of its operation and supply of irrigation water in bulk at agreed times and quantities in different crop seasons would be included in the agreement. The transfer agreement/MOU will clearly define the area of operation of WUA, the details of the irrigation system its present condition and the existing structures with their technical details. The plan map is to be attached delineating the area of operation, irrigation distribution network and location of structures with the transfer agreement. Technical details like longitudinal and cross section of the system are also to be attached to the document. It will also specify the procedure to adopt for the management of irrigation in the operation area. There will be a legal provision of termination of agreement but the procedure would be defined in detail in the transfer agreement. The agreement would be valid for a specified period but having been satisfied with the performance; the agreement may be renewed for a further mutually agreed period. The grounds for termination of the agreement should be mentioned in the agreement.

10.4. Progress of PIM

PIM in India has followed two approaches - legislative and motivational. Andhra Pradesh and Madhya Pradesh first enacted legislation and opted for a fast and extensive introduction of PIM. Maharashtra and Gujarat adopted motivational strategies followed by legislation. So far, 84,779 WUAs have been formed in various states covering an area of 17.84 million ha under various commands of irrigation schemes (Table 10.2). One-time functional grants @Rs.1200 per ha shared by the Centre, State and Farmers in the ratio of 45:45:10, respectively, are being paid to the WUAs as an incentive, the interest accrued is used for maintenance charges. Apart from this, an amount of Rs. 3.00 lakh (central share 60%, state 40%) is being provided to each WUA as a one-time infrastructure grant for the creation of suitable facilities (MoWR, RD & GR, Govt. of India, 2023).

Table 10.2. Progress of PIM in different states of India

Sl. No.	Name of State	Number of WUAs Formed	Area Covered ('000 ha)
1	Andhra Pradesh	10884	4179.25
2	Arunachal Pradesh	43	10.97
3	Assam	847	95.02
4	Bihar	80	209.47
5	Chhattisgarh	1324	1244.56
6	Goa	84	9.54
7	Gujarat	8278	662.99
8	Haryana	8490	1616.27
9	Himachal Pradesh	1173	140.56
10	J & K	383	32.794
11	Jharkhand	0	0
12	Karnataka	2787	1418.66
13	Kerala	4398	191.22
14	Madhya Pradesh	2062	1999.64
15	Maharashtra	2959	1156.22
16	Manipur	69	29.4
17	Meghalaya	159	20.17
18	Mizoram	390	18.23
19	Nagaland	24	3.44
20	Odisha	20794	1757.71

21	Punjab	4845	610.29
22	Rajasthan	1994	1144.45
23	Sikkim	0	0
24	Tamil Nadu	1910	935.664
25	Telangana	0	0
26	Tripura	0	0
27	Uttar Pradesh	802	318.69
28	Uttarakhand	0	0
29	West Bengal	10000	37
Total		84779	17842.208

Source: MoWR, RD & GR, Govt. of India (http://mowr.gov.in/sites/default/files/CADWM_Status_of_PIM_0.pdf)

According to the Annual Report 2022-23 of MoWR, RD & GR, Govt. of India, presently 18 States have either enacted a new Act or modified their existing Act to fulfil the objective of the PIM. As per information made available by the State Governments, about 93,000 WUAs, covering an area of 17.84 million hectares, have been formed in India.

Command Area Development and Water Management (CADWM) work is being implemented in 99 prioritized Accelerated Irrigation Benefit Projects (AIBP) under *Pradhan Mantri Krishi Sinchai Yojana* (PMKSY), *Har Khet Ko Pani* (HKKP). Under the restructured CADWM Programme, more emphasis is being given to the participatory approach; thus, payment of central assistance to the state is linked with the formation of WUAs. As of April, 2016 a total of 4223 WUAs had been created. About 1250 WUAs are formed in the states of Assam, Gujarat, Jammu & Kashmir, Karnataka, Madhya Pradesh, Maharashtra, Odisha and Punjab during 2016-17. Again during 2017-18, another 1372 WUAs had been formed in the states of Assam, Bihar, Chhattisgarh, Gujarat, Jharkhand, Karnataka, Maharashtra, Manipur, Odisha, Punjab and Rajasthan. Further in 2018-19 (up to 2nd Quarter), a total of 890 WUAs were formed in the states of Andhra Pradesh, Bihar, Chhattisgarh, Gujarat, Karnataka, Maharashtra, Odisha and Telangana. Thus, a total of 7735 WUAs have been formed (MoWR, RD & GR, Govt. of India, 2019). Under CAD&WM for the 99 prioritized projects during 2016-22, 14,685 WUAs were targeted to be created under the ongoing 88 projects, out of which 9,272 WUAs have been formed, and about 2,900 WUAs the CAD assets have also been transferred to respective WUAs (MoWR, RD & GR, Govt. of India, 2023).

PIM has increased in irrigation intensity, cropping intensity and yield with spatial and temporal variations. However, it is debated that the shift from government-managed to farmers' managed irrigation systems may be one of the contributing but not the only factors responsible for better agriculture performance. Irrigation systems having non-rice cropping can be more productive than the system having rice crops in the command. The effect is varied between sources of irrigation as well as across the command areas of different irrigation systems.

10.5. Performance of WUAs

The gap between the prescribed and performed functions of WUAs is attributed to certain factors, which need to be addressed. The effect is found to be varied between sources of irrigation as well as across the command areas of different irrigation systems (Ghosh and Kumar, 2012).

10.5.1. Water user group dynamics

An approach was evolved by Ghosh *et al.* (2010) to measure the group dynamics effectiveness of different WUAs under different irrigation systems identifying different dimensions and their relative importance in it. They found that many of the parameters in the group dynamics effectiveness index (participation, decision-making procedure, group atmosphere, empathy, interpersonal trust and social support) were quite low in the case of major and medium irrigation WUAs whereas these were high in the case of minor irrigation (both flow and lift) WUAs. It also means that the smaller the irrigation system, the better the group dynamics.

The reasons for this kind of differential group performance may be attributed to the fact that in the case of minor irrigation, irrigation management transfer to the WUA has inculcated a sense of ownership with full access and control of the system. In contrast, in the case of major and medium irrigation systems, the efforts are undertaken based on the assumption that things could be set right by the organization of the irrigators at the local level despite the availability of insight that without changes in management at higher levels it is very unlikely that these local organizations (WUAs) will be successful over time.

10.5.2. Unpredictability and unreliability of water supply

The unpredictability and unreliability of water supply from the main system to the local units is a major cause of the problems at the local level. The small and marginal farmers are unaware of the WUAs in many cases, especially in larger irrigation systems, where they face more problems in accessing water, as their lands are concentrated in the tail-ends unlike those of large farmers. It has resulted in the dissatisfaction of the small and marginal farmers towards the decision-making process, thus the reluctance to participate in WUA activities.

10.5.3. Institutional challenges in transferring powers and responsibilities

Transfer of irrigation management responsibility from the government irrigation authority to local management demands both allocative and investment decisions by the farmers' group. The problem at the main system level is mainly the problem of allocation of rights and entitlements and therefore, of governance. Several socio-economic and political factors outside the water sector influence irrigation management. The success of irrigation management transfer depends on the whole set of institutional arrangements and the willingness to comply and enforce and/ or change the rules in light of changing circumstances. The issues of rights are beginning to enter into the debate on Indian irrigation which needs a lot of attention. Inefficient water use, social power capture by rural elites in the name of participation, inadequate support from government agencies and reluctance to fully transfer powers and functions to the WUA are some of the bottlenecks. Currently, there is inadequate understanding of the linkage between socio-cultural, institutional and ecological factors affecting the outcome of the PIM reforms in India (Saravanan, 2010). There are also concerns about the effectiveness of the PIM approach and the sustainability of WUAs (Reddy and Reddy, 2005; Kulkarni *et al.*, 2009).

10.5.4. Other uses of water

The concept of WUA does not consider other uses of water (domestic, industrial use, etc.) and also the needs of landless people in irrigation command areas that hampers social support and accentuates rural inequity. Without proper education and interface with all categories of farmers, there will not be widespread acceptance of the idea of farmers assuming management and maintenance responsibilities beyond the on-farm level. Poor people's relative benefit typically depends on the allocation of water rights, which is in proportion to land size rather than, for example, based on an equal quantity of water to every farm household.

10.6. Impacts of PIM in rice-based ecosystems

Globally, there have been tangible impacts of PIM in many countries. The effect of PIM on irrigation system performance is realized in terms of bettering water resource management. The performance in managing water scarcity and use efficiency, promoting empowerment and equity, decreasing negative environmental impact, and boosting financial viability are only a few of the anticipated effects. Many studies have looked at this. Gandhi and Namboodiri (2011) looked at six broad indicators to determine how effective the WUA is at managing the irrigation system: timely and adequate water availability, increase in irrigated area, change in crop pattern, better maintenance of the irrigation structure, and finally decrease in maintenance cost. Besides hydraulic impact evaluation, the impact of PIM in improving agricultural productivity, production, and incomes through more effective use of water resources is one of the most studied aspects. The influence of PIM and the WUAs on agriculture includes changes to the amount of land that is irrigated, the way that crops are grown, how inputs are used, the use of enhanced and high-yielding varieties, and productivity changes. This has been investigated in various irrigation schemes. Farmers may choose water-intensive but more lucrative crops if there is water available for irrigation, and the availability of irrigation may also affect how many different inputs are used. Rice is the major crop in many countries; thus, impact evaluation studies have largely focused on the perspective of rice-based cropping systems.

10.6.1. Global experience

The impact of PIM is varied across different regions in the world. In an evaluation of 29 irrigation schemes around the world, the most often reported positive impacts were reduction in the cost of irrigation to farmers and government, enhanced financial self-reliance of irrigation schemes, expansion of service areas, reduction in the amount of water delivered per hectare, and increases in cropping intensity and yields (Vermillion, 1997).

The study by NIACONSULT (National Irrigation Administration Consultancy, a subsidiary corporation of the National Irrigation Administration, The Philippines, 1993) reported that participatory systems had higher dry season rice yields and increased income after farmer participation was introduced in three irrigation systems in the Philippines, with clearest gains for tail-end farmers, due to improvements in the equity of water deliveries. A study of management transfer of a Dawhenya Smallholder Rice Irrigation Project in Ghana reported sustained high rice yields (Sam-Amoah and Gowing, 2001). Rice productivity increased by more than 30% and the overall profits from agriculture have been doubled due to IMT in Nepal. Significant increases in rice production along with profits from agriculture may also be due to increased use of inputs, improved varieties and better management practices besides changes in irrigation managers from the government to farmers' associations (Bhatta *et al.*, 2005). Results from the field study carried out in the Niger Irrigation Scheme (rice is the major crop) in Mali (Vandersypen *et al.*, 2009) revealed that collective action has improved irrigation efficiency by 14%. Increased total rice output was due to the expansion of irrigated area after the operation, maintenance and management responsibility of the irrigation system taken up by the farmers. Increased yields and incomes arising from better water delivery services and maintenance are, ultimately, the most compelling reasons. International Water Management Institute (IWMI) examined the impact of PIM in 11 countries with the help of an indicator called standardized gross value of production (SGVP). The four basic comparative performance indicators considered are output per cropped area, output per unit command, output per unit irrigation supply and output per unit water consumed. These indicators were applied to 18 irrigation systems located in 11 countries. Measurements of SGVP per unit of cropped land indicated that non-rice-producing irrigation systems can be more productive than rice-producing irrigation systems by 100–200%. SGVP per unit irrigation supply indicates the lowest value for purely rice-based systems. Irrigation systems which grow rice during rainy seasons and other crops during the dry season obtain middle-range values. The highest value is obtained by systems which grow orchards, industrial crops and vegetables (Restrepo, 2004).

10.6.2. Indian experience

Despite the widespread and continued application of PIM as a remedy for 'poor' irrigation system performance, the actual effectiveness of transfer programs in terms of improving irrigation, agricultural and management efficiency is still a subject of much debate. Irrigation has dictated largely the pace and the process of agricultural development in India. The irrigation sector has always been emphasized due to its yield-augmenting impact and treated as one of the priority areas of India's agricultural development strategy with substantial financial support in successive Five-Year Plans.

Andhra Pradesh is the first Indian state to witness farmers' participation in irrigation management. The PIM has resulted in a significant expansion of irrigated areas, reduced flooding losses, an early cropping calendar and higher yields of rice crops (Svendsen and Huppert, 2003). Analyses from a sample of 222 WUAs covering 22 districts in Andhra Pradesh indicate that there are differences in the number of irrigations required and actual number of irrigations given in rice (Reddy and Reddy, 2005). The productivity of rice has been increased; however, the rate of change is higher in the head and middle reaches than in the tail reaches. By implementing a modernization programme in minor irrigation tanks in Tamil Nadu, an increase in rice yield was reported due to improvement in conveyance, distribution, application and irrigation efficiencies, with a concomitant increase in water productivity and gross income (Anbumozhi *et al.*, 2001). Conjunctive use of groundwater and tank water is also one of the important factors that influence rice yield in tank command areas. In rainfed tanks, dependency on well water is high towards higher rice production. From a detailed survey covering 566 tanks, it was found that a higher proportion of farms in the tail reaches are found to have more supplemental irrigation from wells than the head reach and rice yield was progressively higher with supplemental irrigation, rather than tank water alone (Palanisamy *et al.*, 1994). The impact was found positive to the cultivated area, cropping intensity, irrigated area, irrigation intensity, crop diversification and crop productivity in three minor irrigation systems in Odisha rehabilitated with the aid from the European Commission and turned over to WUAs in 2005 (Mishra *et al.*, 2011). Rice is the predominant crop in the command areas of all three selected systems; however, there is a major shift in the cropping pattern from paddy to non-paddy crops due to assured water availability. During the dry season, pulses, oilseeds and vegetables are grown by the farmers with a reduction in fallow area by 50%. On average, yield increased by 45% in rice, 57% in pulses, 80% in oilseeds, 40% in sugarcane and 43% in vegetables. Thus, it is worth concluding that irrigation systems having non-rice cropping can be more productive than the rice-producing irrigation systems (Ghosh and Kumar, 2012).

In a study in the states of Andhra Pradesh, Gujarat, and Maharashtra, a sample of 29 WUAs (canal: 13, tank: 4, tube well: 4, river lift: 3, pani panchayat: 2, and check dams: 3) and a sample of 435 beneficiary farmer households were considered to evaluate the impact of PIM (Gandhi and Namboodiri, 2011). The average rate of participation (Active to Very Active) by members in the WUAs was found to be high at approximately 80% when all forms of WUAs were taken into consideration. This demonstrates that members' average engagement in WUAs is extremely high. The effect of the WUAs on the agricultural economy was investigated in terms of increasing the amount of land irrigated, the amount of inputs used, and the improvement of yields. The cropped area grows by 8.28%, and the irrigated area grows by 31.43%, according to the average figures. This suggests that PIM action has had a significant influence. The findings show that the irrigated area increased significantly in the summer season, followed by the kharif season, and very little in the rabi season. PIM also significantly increases the use of inputs that increase production, such as HYV, better seeds, and fertilizers. Asian Development Bank (2008) reports that crop yields in Andhra Pradesh rise by 15–25% following WUAs, even in the tail end. After the implementation of PIM, the cultivation of food grains under irrigated conditions became more common among the sample households. It is reported that the utilization of local seed varieties and bullock labour has decreased, regardless of the irrigation scheme. However, there have been considerable advances in the use of farm machinery, fertilizers, and other contemporary inputs including high-yielding seed varieties. Under the selected major canal command area in Andhra Pradesh under WUA's jurisdiction, the yields of major crops such as paddy, banana, and pulses have increased. On average, 78% of respondents report a positive to highly positive impact of PIM on the efficiency of water use, equity, and user empowerment. They also report a positive to highly positive impact on timely and adequate water availability, better maintenance, more equitable distribution, and user empowerment. According to the Asian Development Bank (2008), water flows to the tail end in 50% of the cases following WUA activities.

One of the metrics to assess the success of PIM is to assess its impact on food grain production and or cropping intensity. If PIM has contributed to increased water use efficiency through better management of available irrigation water at the local level, then it should have reflected in overall food grain productivity at the local level. Spatial variations in approaches to PIM implementation might result in spatial variations in its overall impact. An overall understanding of irrigation development in India helps understand the reasons for spatial variations in the implementation and impact of PIM. This also helps to answer questions such as whether PIM is the best strategy for improving irrigation management in all the states in India or whether there are any spatial variations in its impact. Additionally, empirical evidence is lacking in the existing literature on the effects of PIM on crop productivity and irrigation efficiency at the aggregate level. Therefore, after a decade of implementation of PIM policy in India, the impact of PIM on agricultural and irrigation scenarios is studied in six states viz. undivided state of Andhra Pradesh (presently Andhra Pradesh state has been bifurcated into two states *viz.* Telangana and Andhra Pradesh), Karnataka, Tamil Nadu, Madhya Pradesh, Odisha, and Rajasthan) implementing the PIM reform. In the case of each state, district-level panel data set of 10 years each before and after the PIM enactment is analysed for the variables depicting agricultural and irrigation scenarios. Rice-based cropping system is found in all the selected states except Rajasthan. Feasible Generalized Least Squares (FGLS) regression models are worked out to understand the impact of PIM reform on food grain productivity in each state (Ghosh *et al.*, 2019).

The share of area under food grains has decreased in four out of six selected states while it has remained almost stationary in Karnataka and Odisha. The growth rate is found negative in the case of all the states (Table 3). In fact, despite no remarkable changes in the area under food grains in states like Karnataka and Odisha, the growth rate has become negative during the post-PIM period which otherwise was positive during the pre-PIM period. This may be attributed to the fact of a shift in cropping patterns from food grains to more remunerative non-food grains crop with better irrigation scenarios after the implementation of the PIM. The area under non-food grains crops like cotton in Andhra Pradesh, vegetables in Karnataka, sugarcane in Tamil Nadu, and chickpeas in Madhya Pradesh. The cumulative annual exponential growth rate of area under HYVs in Karnataka turned positive after the implementation of PIM while that has remained positive in other states. The mean productivity of food grains has increased in all selected states with the maximum increase in Andhra Pradesh (1631 to 2119 kg ha^{-1}) and minimum increase in Madhya Pradesh (1111 to 1242 kg ha^{-1}) and Odisha (1024 to 1157 kg ha^{-1}). The growth rate is positive for all states during the post-PIM period; in fact, it has been shifted from negative to positive in the case of Odisha (Table 10.3).

The cropping intensity has increased in all the six selected states except in Tamil Nadu. The extent of the increase is between three per cent in Karnataka and 12 per cent in Odisha. All the states have depicted a positive growth rate during the post-PIM implementation period (Table 10.4). Fertilizer consumption has increased with the maximum increase in Andhra Pradesh (77.13 to 120.14 thousand tonnes) and the minimum in Madhya Pradesh (19.83 to 32.18 thousand tonnes) after the implementation of PIM in the respective states. The growth rate is found positive during both pre- and post-PIM periods. All the selected states have shown an increase in the growth rate of fertiliser consumption during the post-PIM period (Table 10.4).

The decadal (10 years) mean values calculated before and after the implementation of PIM indicate an increase in net irrigated area in five out of six selected states after the implementation of PIM except in Tamil Nadu (decreased from 2794 to 2764 thousand ha). However, the decadal mean of irrigation coverage (% of net irrigated area to net sown area) has increased in all selected six states after the implementation of PIM. The maximum increase was observed in Odisha (12%) and the minimum in Andhra Pradesh (3%). The growth rate has been positive (Table 10.5).

Groundwater irrigated areas registered an increase in all the states during the post-PIM period as compared to the pre-PIM period with a positive growth rate throughout. The decadal (10 years) mean values calculated before and after the enactment of PIM indicate that the share of groundwater irrigation area to net sown area has increased in all selected six states during the PIM period (Table 10.6).

Table 10.3. Area under food grains and the productivity of food grain in six states of India during the pre-and post-PIM enactment period

States enacted PIM	Percentage of food grains area to gross sown area						Productivity of food grains (kg/ha)					
	Pre-PIM Period			Post-PIM Period			Pre-PIM Period			Post-PIM Period		
	Mean	SD	CAEGR	Mean	SD	CAEGR	Mean	SD	CAEGR	Mean	SD	CAEGR
Andhra Pradesh	58.23	3.36	-1.77	54.98	1.76	-0.40	1630.94	148.43	2.63	2119.33	306.85	2.70
Karnataka	59.08	1.50	0.26	58.83	4.18	-0.74	1326.71	154.55	3.50	1562.80	199.47	1.45
Tamil Nadu	52.59	4.66	-0.03	49.15	0.87	-0.26	2275.11	326.20	2.54	2465.94	350.86	2.12
Madhya Pradesh	66.71	2.55	-1.20	63.58	1.63	-0.02	1110.94	101.40	2.70	1242.11	164.51	1.75
Odisha	75.81	2.01	0.62	75.21	0.96	-0.30	1023.80	128.47	-0.84	1156.68	186.51	3.91
Rajasthan	64.83	2.99	-0.95	61.45	2.82	-0.50	931.68	168.12	3.59	1290.80	258.89	4.11

Source: Ghosh *et al.* (2019)

Table 10.4. Cropping intensity and fertilizer consumption in six states of India during the pre-and post PIM enactment period

States enacted PIM	Cropping intensity (%)						Fertilizer consumption (,000 tonnes)					
	Pre-PIM Period			Post-PIM Period			Pre-PIM Period			Post-PIM Period		
	Mean	SD	CAEGR	Mean	SD	CAEGR	Mean	SD	CAEGR	Mean	SD	CAEGR
Andhra Pradesh	123.66	2.69	0.68	127.90	2.93	0.33	77.13	10.72	3.90	120.14	31.30	5.09
Karnataka	115.43	0.77	0.13	118.21	2.47	0.47	50.22	8.76	4.59	81.82	21.94	6.68
Tamil Nadu	118.28	1.87	-0.35	114.35	1.91	-0.22	72.35	7.60	2.08	87.48	15.98	4.39
Madhya Pradesh	126.90	6.41	1.63	135.58	6.80	1.03	19.83	3.78	4.69	32.18	11.19	8.49
Odisha	144.78	7.03	-0.80	156.71	8.58	1.75	20.49	4.56	6.67	33.29	7.08	7.28
Rajasthan	126.07	6.15	1.06	133.83	7.70	1.38	19.92	8.28	12.89	35.83	9.43	7.07

Source: Ghosh *et al.* (2019)

Table 10.5. Net irrigation area in six states of India during the pre- and post-PIM enactment period

States enacted PIM	Net irrigation area (,000 ha)						Percentage of net irrigation area to net sown area					
	Pre-PIM Period			Post-PIM Period			Pre-PIM Period			Post-PIM Period		
	Mean	SD	CAEGR	Mean	SD	CAEGR	Mean	SD	CAEGR	Mean	SD	CAEGR
Andhra Pradesh	4039.86	300.90	0.88	4296.95	447.11	1.22	42.48	2.37	1.23	45.73	2.69	0.69
Karnataka	2336.28	129.99	1.67	2955.72	389.77	3.48	23.28	1.44	1.90	29.31	2.81	2.60
Tamil Nadu	2794.02	209.09	2.21	2763.49	262.04	1.35	47.01	3.49	2.47	51.45	2.92	1.14
Madhya Pradesh	4340.39	927.17	7.56	5972.76	1070.84	4.26	30.19	5.96	6.91	40.65	6.32	3.71
Odisha	3052.47	622.96	5.29	4139.07	303.73	2.42	33.07	8.27	6.97	45.18	2.93	1.80
Rajasthan	4615.45	833.40	4.82	5910.30	796.65	3.26	36.67	5.14	3.90	44.57	4.18	2.16

Source: Ghosh *et al.* (2019)

Table 10.6. Groundwater irrigation scenario in six states of India during the pre- and post-PIM enactment period

States enacted PIM	Groundwater irrigation area (,000 ha)						Percentage of groundwater irrigation area to net sown area					
	Pre-PIM Period			Post-PIM Period			Pre-PIM Period			Post-PIM Period		
	Mean	SD	CAEGR	Mean	SD	CAEGR	Mean	SD	CAEGR	Mean	SD	CAEGR
Andhra Pradesh	1331.28	223.04	5.72	2003.97	252.33	2.58	14.65	2.87	6.85	22.17	2.05	1.86
Karnataka	848.96	98.02	3.65	1380.31	248.73	5.13	7.88	0.89	3.37	13.11	2.14	4.69
Tamil Nadu	1356.57	168.26	4.19	1573.42	137.96	2.02	21.33	2.78	4.41	27.78	2.16	1.99
Madhya Pradesh	2698.62	661.13	8.77	3953.11	761.14	4.64	18.74	4.29	8.13	26.69	4.40	3.96
Odisha	1170.74	281.43	6.19	1692.05	190.88	3.77	13.20	3.81	8.12	19.25	1.90	3.13
Rajasthan	2997.35	668.12	5.98	4268.64	543.44	3.33	23.36	4.43	4.84	31.74	3.23	2.40

Source: Ghosh *et al.* (2019)

Decadal mean canal irrigated area has increased in Karnataka, Madhya Pradesh, and Odisha during the post-PIM period as compared to the pre-PIM period with the growth rate remaining positive throughout. The same has decreased in the states of Andhra Pradesh and Tamil Nadu; however, there has been a shift in growth rate from negative during pre-PIM to positive during the post-PIM period in both these states. To the share of canal irrigated area to net sown area (% of canal irrigated area to net sown area), an increase of 4% is found in Odisha during the post-PIM period. A marginal increase (less than 1%) is found in Madhya Pradesh, Karnataka, and Rajasthan. However, it has decreased in Andhra Pradesh (17.08 to 15.50%) and Tamil Nadu (13.81 to 12.55%) during the post-PIM period. All the states except Andhra Pradesh and Tamil Nadu have positive growth rates during the pre as well as post-PIM period. The growth rate of Andhra Pradesh has changed from negative to positive; however, that remained negative in Tamil Nadu even during the post-PIM period (Table 10.7).

The analyses have revealed variations in the impact of PIM in the states of India. In Andhra Pradesh, Tamil Nadu and Madhya Pradesh, descriptive analysis has indicated varied improvements in agricultural and irrigation scenarios after implementation of PIM. However, according to the Feasible Generalized Least Squares (FGLS), regression models worked out to understand the impact of PIM reform on food grain productivity (Table 10.8), the PIM has made a significant impact on food grain productivity as evidenced by the significant coefficient value for the interaction term between year and PIM dummy in Karnataka, Odisha and Rajasthan. The impact of PIM is not significantly visible in the case of Andhra Pradesh, Tamil Nadu and Madhya Pradesh (Ghosh *et al.*, 2019).

Table 10.7. Canal irrigation scenario in six states of India during the pre- and post-PIM enactment period

States enacted PIM	**Canal irrigation area (,000 ha)**						**Percentage of canal irrigation area to net sown area**					
	Pre-PIM Period			**Post-PIM Period**			**Pre-PIM Period**			**Post-PIM Period**		
	Mean	**SD**	**CAEGR**	**Mean**	**SD**	**CAEGR**	**Mean**	**SD**	**CAEGR**	**Mean**	**SD**	**CAEGR**
Andhra Pradesh	1718.45	132.45	-1.44	1547.39	187.41	0.77	17.08	1.13	-1.75	15.50	1.41	0.32
Karnataka	926.79	34.78	0.94	992.75	136.26	3.12	10.04	0.49	1.51	10.95	1.04	2.15
Tamil Nadu	800.56	40.62	-0.52	672.21	113.24	0.55	13.81	0.42	-0.29	12.55	1.52	-0.44
Madhya Pradesh	930.61	95.08	3.03	1021.71	128.48	2.82	6.76	0.56	2.39	7.53	0.90	2.74
Odisha	1413.73	292.82	5.41	1850.89	80.96	1.36	14.31	3.54	6.91	18.87	0.95	0.75
Rajasthan	1412.87	153.61	2.76	1558.65	187.45	2.79	10.09	0.87	1.82	10.80	1.62	0.05

Source: Ghosh *et al.* (2019)

Table 10.8. FGLS regression model showing the impact of PIM on food grain productivity

Parameters	Selected states implementing PIM						
	Andhra Pradesh	**Tamil Nadu**	**Karnataka**	**Odisha**	**Madhya Pradesh**	**Rajasthan**	**Pooled**
Dependent variable: Food grain productivity							
Explanatory variables:							
Year	30.81**	8.03	6.15	-6.29	-6.40*	-6.21	-2.62
	(6.66)	(14.95)	(8.49)	(5.62)	(3.14)	(3.86)	(2.83)
Fertilizer Consumption	2.78**	3.03*	-0.99	5.53**	6.78**	4.48**	6.19**
	(0.28)	(1.40)	(0.70)	(1.22)	(0.47)	(0.60)	(0.25)
HYV Area	-0.25	-0.88	-1.09**	0.32		-0.05	
	(0.14)	(0.61)	(0.24)	(0.19)		(0.09)	
Share of NIA to NSA	22.34**	26.27**	16.24**	-3.99**	12.53**	21.78**	15.37**
	(0.92)	(2.89)	(2.03)	(1.09)	(0.50)	(0.66)	(0.56)
Interaction PIM Year	-1.22	-10.37	16.14**	11.26**	-1.74	11.40**	0.23
	(0.14)	(10.05)	(6.02)	(3.83)	(2.31)	(2.68)	(1.99)
Constant	339.23**	724.40**	1062.33**	993.78**	568.06**	150.82**	476.45**
	(46.73)	(164.72)	(86.32)	(48.03)	(19.24)	(35.03)	(24.62)

** significance at 1%; * significance at 5%; figures within parentheses are standard errors of estimated parameters

Source: Ghosh *et al.* (2019)

Based on panel data minimum for 10 years each before and after implementation of PIM, findings indicate betterment to areas under non-food grains crops, the productivity of most of the major crops both food grains crops and non-food grains crops, areas under HYVs, cropping intensity and fertilizer consumption during post-PIM period. The growth rate has been varied; however, positive growth rates are mostly observed for most of the parameters of agriculture during the post-PIM period. These improvements may be attributed to the better irrigation regimes in the respective states after the implementation of the PIM. The government has initiated a major shift in its approach through WUAs as the most appropriate institute for efficient irrigation management and agricultural development in irrigated areas.

10.7. PIM and extension advisory services

Views on extension advisory services have changed from technology dissemination with emphasis on increasing agricultural production to helping farmers organize themselves, building their capacities for operation, maintenance and management of productive resources or assets, providing them with all necessary information services, linking them to markets and value chains resulting in bettering income and improved livelihood. Thus, community-based approaches, which involve farmers' groups, have gained increasing importance in recent years for extension services with the belief that small and poor farmers can benefit from these arrangements only when they are organized into groups. Community-based extension is perceived as an important strategy because it promises to overcome both the state failures and the market failures inherent in extension (World Bank, 2005). Farmers associations can play an important role in aggregating farmers' demands for extension and in representing farmers in participatory models of extension management to make extension more demand-driven (Feder *et al.*, 2010). This holds significance for sustainable and efficient management of water resources as a collective effort by all farmers. Extension should therefore ideally work with farmer groups to strengthen their capacities to articulate demands, help them link to new sources of knowledge and enable innovation. However, experiences from the irrigation sector in India reveal that extension has not engaged sufficiently with the existing large number of WUAs in the irrigation command areas.

Enhancing farmers' income would depend on harnessing all possible sources of growth within as well as outside of the agricultural sector. The major sources of growth operating within the agriculture sector are productivity improvement, resource use efficiency or saving in cost of cultivation, increasing in cropping intensity and diversification towards high-value crops (Chand, 2017). All of these sources of growth in the agriculture sector are integral part of PIM that can be a potential option for community-based extension and a vehicle of enhancing farmers' income in irrigated areas.

The presence of WUAs in the irrigated area offers great potential for extension to improve its effectiveness. Extension can play an important role in capacity building of the WUAs not only on the operation, maintenance and management of irrigation systems but also in implementing efficient crop planning, synchronization of farm operations, linking WUAs to other sources of knowledge, support and services and establishing effective forward and backward linkages. Networking of WUAs would also serve the farm households in a better manner by sharing of information and expertise between them. As use of productivity enhancing inputs in agriculture is often influenced by the available irrigation water regime, the PIM under the WUA umbrella would complement the other extension services well with more assured benefits from the extension efforts.

The extension also needs to give attention on the equal benefits sharing among the landholders and landless by working out the modus operandi of involving the landless in WUA's activities like repair and maintenance works in irrigation network, allowing the cultivation on common lands in the command, fishing rights in the minor flow irrigation system's reservoir, etc to ensure equal benefits to all. Present structural arrangements of WUA ensure the involvement of both agriculture and irrigation department functionaries; thus, efficient cropping and water use in the irrigation command. Involvement of other line department functionaries (livestock, fishery) would help in developing the capacities of the rural households in a holistic manner with diversified livelihood options in the irrigated areas.

10.8. Women's participation

Representation of women in the WUAs at all levels has been brought in the guidelines issued by the Ministry of Water Resources. Accordingly, many states have amended the enactment to implementation of PIM. Considering the importance of women in terms of their significant contribution to the agricultural labour force, the participation of women in the management of water resources is ensured through their membership in the WUAs. Some states have already implemented it; for example, in Madhya Pradesh, all farm owners, be they men or women, are rightful members of the outlet committees in WUAs. Where there are no women members, at least one woman from the area must be taken even if she is not a land owner. Further, at least one woman shall be nominated to the Governing Body of the Association.

10.9. Way forward

International Commission on Irrigation and Drainage (ICID) Working Group on Institutional and Organizational Aspects organized an international symposium with the theme 'Global Review of Institutional Reforms in the Irrigation Sector for Sustainable Agricultural Water Management' including WUAs, during the 23rd ICID Congress (8-14 October 2017) in Mexico City. The symposium deliberated upon the legal framework and organizational structure including WUAs for water supply services, PIM and management transfer- approaches and conditions for successful irrigation and drainage, and public-private partnership (PPP) in irrigation and drainage operation and maintenance towards sustainable irrigated agricultural water management. Based on this, a review of institutional reforms in the irrigation sector was done in India among other countries. Despite the diverse geography and institutional frameworks in the irrigation sector, food security remains a major concern in all the countries and regions reviewed. The formation and functioning of WUAs, ageing irrigation and drainage infrastructure and its operation and management remain the major institutional and PPP challenges for sustainable agricultural water management in the immediate future.

WUAs like any other organization may sustain only when

- The earnings are more than the expenditure funds available as and when required
- It has some reserve funds at its disposal
- Ensured supply of irrigation water from the system
- Effective operation plan of the project
- Equitable distribution of water
- Improvement in the recovery of water charges from the farmers
- Recovery of water charges for using other sources of water in the operational area
- Willingness of the farmers and irrigation agency to make it a success

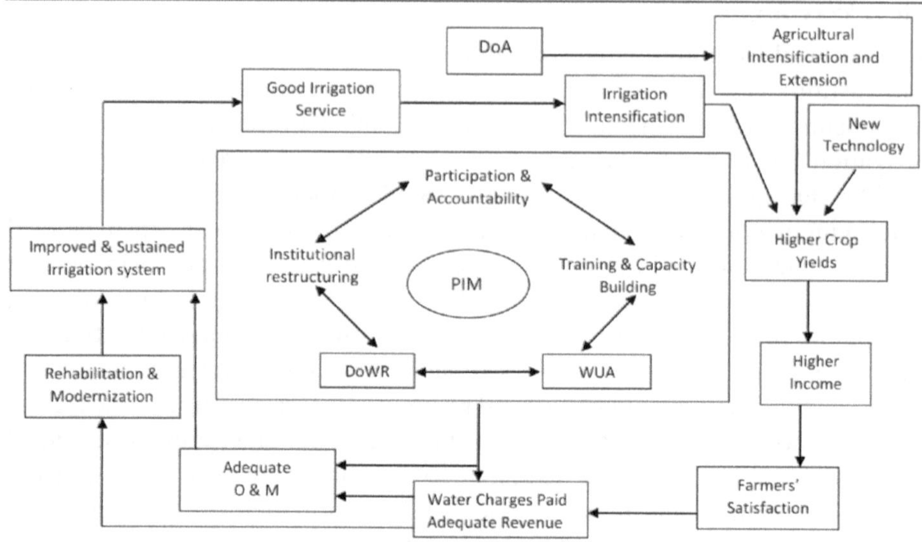

Fig. 10.3. PIM- potential option in transforming vicious cycle to virtuous cycle in the irrigation sector

The issues of rights are beginning to enter the debate on Indian irrigation which needs a lot of attention. Poor people's relative benefit typically depends on the allocation of water rights, which is in proportion to land size rather than, for example, based on an equal quantity of water to every farm household. The concept of WUA does not consider other uses of water (domestic, industrial use, etc.) and also the needs of landless people in irrigation command areas that hampers social support and accentuates rural inequity. Without proper education and interface with all categories of farmers, there will not be widespread acceptance of the idea of farmers assuming management and maintenance responsibilities beyond the on-farm level. Collective efforts by all concerned are required for successful and effective PIM to pave the way for creating a virtuous circle in the irrigation and farm sector (Fig. 10.3).

10.10. Conclusion

Water forms the backbone for all future endeavours to achieve the vision of food security. Water management is by nature beyond the work of individuals and thus collective effort by all farmers concerned is required for successful management. Farmers' participation in irrigation management has taken centre stage and the irrigators who were considered as beneficiaries are now considered partners in the planning, development, operation and maintenance of irrigation systems. Transfer of irrigation management responsibilities from government agencies to farmers is now an important policy that has resulted in variable impact over space and time. The most ideal situation may be when the demand to take over the management of the irrigation system comes strongly from the farmers.

PIM's effects on food grain productivity, overall irrigation development, and agriculture have varied across the different states and regions. In traditionally rainfed states, PIM's effect on food grain productivity is particularly noticeable. Results suggest that efforts to restructure India's irrigation sector should focus on rainfed states with low food grain output. A key policy now is the handover of irrigation management responsibilities from government agencies to farmers, not just in India but also in other emerging nations. The positive effects of PIM reform would be more dependent on ongoing efforts to increase the effectiveness of surface irrigation while maintaining the sustainability of groundwater irrigation.

Participatory management of irrigation sources and systems, and devolution of authority to irrigation management have advanced significantly, but there is still much work to be done to improve performance. It has been discovered that greater participation frequently has considerable positive effects on water resource management performance; however, some WUAs have done better than others. For many of the WUAs to be robust and long-lasting, they need more than mere setup; they also need contributions to institution design, institution building, and training. The financial viability and sustainability of these institutions must be improved by both local resource mobilization and outside development support, and greater accountability must be integrated through effective financial audits, performance reviews, and social audits.

References

Anbumozhi, V., Matsumoto, K. and Yamajil, E. 2001. Towards Improved Performance of Irrigation Tanks in Semi-Arid Regions of India: Modernization Opportunities and Challenges. *Irrigation and Drainage Systems*, 15: 293-309.

Asian Development Bank. 2008. *Irrigation Management Transfer: Strategies and Best Practices*. Sage: New Delhi.

Bhatta, K.P., Ishida, A., Taniguchi, K. and Sharma, R. 2005. Performance of agency-managed and farmer-managed irrigation systems: A comparative case study at Chitwan, Nepal. *Irrigation and Drainage Systems*, 20: 177-191.

Central Water Commission. 2010. *Water and related statistics*. CWC, Ministry of Water Resources, New Delhi, Government of India.

Chand, R. 2017. *Doubling farmers' income: rationale, strategy, prospects and action plan.* National Institute for Transforming India, NITI Aayog, NITI Policy Paper No. 1/2017, P. 33.

DAC & FW. 2023. *Annual Report 2022-23*. Department of Agriculture, Cooperation & Farmers Welfare, Ministry of Agriculture & Farmers Welfare, Government of India Krishi Bhawan, New Delhi.

(https://agricoop.gov.in/Documents/annual_report_english_2022_23.pdf)

DAC & FW. 2018. *Annual Report 2017-18*. Department of Agriculture, Cooperation & Farmers Welfare, Ministry of Agriculture & Farmers Welfare, Government of India Krishi Bhawan, New Delhi.
(http://agricoop.nic.in/sites/default/files/Krishi%20AR%202017-18-1%20for%20web.pdf)

FAO. 2007. Irrigation management transfer: worldwide efforts and results. In: FAO Water Reports 32, Food and Agricultural Organization of the United Nations.
(http://www.fao.org/3/a1520e/a1520e00.pdf)

Feder, G., Anderson, J.R., Birner, R. and Deininger, K. 2010. *Promises and Realities of Community-Based Agricultural Extension*. IFPRI Discussion Paper 00959, Development Strategy and Governance Division, International Food Policy Research Institute, Washington, DC.

Gandhi, V. P., and Namboodiri, N. V. 2011. Participatory Irrigation Management in India: An Evaluation of the Performance in Andhra Pradesh, Gujarat and Maharashtra. *CMA Publication*, (237).

Ghosh, S., Singh, R. and Kundu, D. K. 2005. Evaluation of irrigation-service utility from the perspective of farmers. *Water Resources Management*, 19: 467-482.

Ghosh, S., Kumar, A., Nanda, P. and Anand, P.S.B. 2010. Group dynamics effectiveness of water user associations under different irrigation systems in an eastern India state. *Irrigation and Drainage*, 59: 559-574.

Ghosh, S. and Kumar A. 2012. How participatory irrigation management reform influences agriculture performance. *Current Science*, 103 (4): 359-360.

Ghosh, S., Kolady, D.E., Das, U., Gorain, S., Srivastava, S.K. and Mondal, B. 2019. Spatio-Temporal Variations in Effects of Participatory Irrigation Management (PIM) Reform in India: A Panel Data Analysis. *Agricultural Water Management*, 222: 48-61.

ICID. 2017. *Global Review of Institutional Reforms in the Irrigation Sector for Sustainable Agricultural Water Management, Including Water Users' Associations, Irrigation and Drainage*.(https://doi.org/10.1002/ird.2305) (https://onlinelibrary.wiley.com/doi/abs/10.1002/ird.2305)

Kulkarni, S.A.; Sinha, P.K.; Belsare, S.M. and Tejawat, C.M. 2009. *Participatory irrigation management in India: achievements, threats and opportunities*. International Commission on Irrigation and Drainage, Ministry of Water Resources, New Delhi.

Ministry of Water Resources. 2002. *National Water Policy*. Government of India, New Delhi.

Ministry of Water Resources. 2007. *Annual Report,* Ministry of Water Resources, Govt. of India, New Delhi

Ministry of Water Resources. 2012. *Annual Report 2011-12 of Ministry of Water Resources*, Govt. of India, New Delhi (http://wrmin.nic.in/writereaddata/linkimages/AR2011-122204676989.pdf).

Mishra, A., Ghosh, S., Nanda, P. and Kumar, A. 2011. Assessing the impact of rehabilitation and irrigation management transfer in minor irrigation projects in Orissa, India: A case study. *Irrigation and Drainage*, 60: 42-56.

MoSPI. 2010. *Irrigation Infrastructure*. Ministry of Statistics and Programme Implementation, Government of India.
(http://mospi.nic.in/sites/default/files/reports_and_publication/cso_research_and_publication_unit/Infrastructure_Statistics/infra_stat_2010/11.ch_irrigation.pdf)

MoWR, RD and GR. 2023. *Annual Report 2022-23*. Ministry of Water Resources, River Development and Ganga Rejuvenation, Government of India, New Delhi.
(https://cdnbbsr.s3waas.gov.in/s3a70dc40477bc2adceef4d2c90f47eb82/uploads/2023/04/2023041116.pdf)

MoWR, RD and GR. 2018. *Annual Report 2017-18*. Ministry of Water Resources, River Development and Ganga Rejuvenation, Government of India, New Delhi.
(http://mowr.gov.in/sites/default/files/Annual_Report_MoWR_2017-18.pdf)

MoWR, RD and GR. 2019. *Annual Report 2018-19*. Ministry of Water Resources, River Development and Ganga Rejuvenation, Government of India, New Delhi.
(http://mowr.gov.in/sites/default/files/Annual_Report_MoWR_2018-19.pdf)

NIACONSULT. 1993. *An evaluation of the impact of farmers' participation on the national irrigation systems (NISs) performance*. NIACONSULT, Manila.

Palanisamy, K., Murali, N. and Ali, M. 1994. *Tank System Performance*. TNAU, Coimbatore, pp. 120-143.

Planning Commission. 2013. *Draft Twelfth Five Year Plan 2012-2017 Vol. I, II & III*. Planning Commission, Government of India, New Delhi. (http://planningcommission.nic.in/plans/planrel/12thplan/welcome.html).

Reddy, V.R. and Reddy, P.P. 2005. How participatory is participatory irrigation management? Water users' associations in Andhra Pradesh. *Economic and Political Weekly, December*, 5587-5595.

Restrepo, C.G. 2004. *Effect of participatory irrigation management on agricultural yields and system performance*. International Network on Participatory Irrigation Management; (http://www. inpim.org/leftlinks/FAQ/Newsletters/N9/ n9a5.htm)

Sam-Amoah, L.K. and Gowing, J.W. 2001. The experience of irrigation management transfer in Ghana: A case study of Dawhenya Irrigation Scheme. *Irrigation and Drainage Systems*, 15: 21-38.

Saravanan, V.S. 2010. Negotiating participatory irrigation management in Indian Himalayas. *Agricultural Water Management*, 97:651-658.

Svendsen, M. and Huppert, W. 2003. Maintenance under institutional reform in Andhra Pradesh. *Irrigation and Drainage Systems*, 17: 23-46.

Vandersypen, K., Verbist, B., Keita, A.C.T., Raes, D. and Jamin, J.Y. 2009. Linking Performance and Collective Action- the Case of the Office du Niger Irrigation Scheme in Mali. *Water Resources Management*, 23: 153-168.

Vermillion, D.L. 1997. *Impacts of irrigation management transfer: A review of the evidence*. IWMI Research Report 11, IWMI, Colombo, Sri Lanka.

World Bank. 2005. *Agriculture Investment Sourcebook*. Washington, DC. (http://go.worldbank.org/CCC68JMIZ0).

Index